# Nanoscale Nonlinear PANDA Ring Resonator

**Preecha P. Yupapin**
King Mongkut's Institute of Technology Ladkrabang
Bangkok, Thailand

**Chat Teeka**
Suan Dusit Rajabhat University
Bangkok, Thailand

**Muhammad Arif Jalil**
Ibnu Sina Institute for Fundamental Science Studies
Nanotechnology Research Alliance
Universiti Teknologi Malaysia
Johor Bahru, Malaysia

**Jalil Ali**
Institute of Advanced Photonics Science
Nanotechnology Research Alliance
Universiti Teknologi Malaysia
Johor Bahru, Malaysia

Published by Science Publishers, an imprint of Edenbridge Ltd.

- St. Helier, Jersey, British Channel Islands
- P.O. Box 699, Enfield, NH 03748, USA

E-mail: *info@scipub.net*  Website: *www.scipub.net*

**Marketed and distributed by:**

CRC Press
Taylor & Francis Group
an **informa** business
www.taylorandfrancisgroup.com

6000 Broken Sound Parkway, NW
Suite 300, Boca Raton, FL 33487
711 Third Avenue
New York, NY 10017
2 Park Square, Milton Park
Abingdon, Oxon OX14 4RN, UK

Copyright reserved © 2012

ISBN    978-1-57808-746-4

**Library of Congress Cataloging-in-Publication Data**

```
Nanoscale nonlinear PANDA ring resonator / Preecha P. Yupapin ... [et
   al.].
       p. cm.
   Includes bibliographical references and index.
   ISBN 978-1-57808-746-4 (hardcover)
 1. Integrated optics. 2. Optical resonance. 3. Resonators. 4. Nonlinear
   waves. 5. Nanoelectronics.  I. Yupapin, Preecha P.
 TA1660.N355 2012
 621.36'93--dc23                                           2011052093
```

The views expressed in this book are those of the author(s) and the publisher does not assume responsibility for the authenticity of the findings/conclusions drawn by the author(s). Also no responsibility is assumed by the publishers for any damage to the property or persons as a result of operation or use of this publication and/or the information contained herein.

All rights reserved. No part of this publication may be reproduced, stored in a retrieval system, or transmitted in any form or by any means, electronic, mechanical, photocopying or otherwise, without the prior permission of the publisher, in writing. The exception to this is when a reasonable part of the text is quoted for purpose of book review, abstracting etc.

This book is sold subject to the condition that it shall not, by way of trade or otherwise be lent, re-sold, hired out, or otherwise circulated without the publisher's prior consent in any form of binding or cover other than that in which it is published and without a similar condition including this condition being imposed on the subsequent purchaser.

Printed in the United States of America

# Preface

Microring/nanoring resonator has now become a useful device which has been widely studied and investigated by researchers from a variety of specializations. One of them has shown that the use of ring resonator device in the form a PANDA ring resonator can have interesting applications such as using the trapped/retrieved photon/atom in a microring device system to carry photons/atoms for long distance transmission. Several researchers have shown that dynamic optical trapping tools can be formed by controlling the dark-bright soliton behaviors within a semiconductor add/drop multiplexer (ring resonator). This book begins with the basic background of linear and nonlinear ring resonators. A novel design of nano device known as a PANDA ring resonator is proposed. The use of the device in the form of a PANDA in applications such as nanoelectronics, measurement, communication, sensors, optical and quantum computing, drug delivery, hybrid transistor and a new concept of electron-hole pair is discussed in detail.

**Preecha P. Yupapin**
**Chat Teeka**
**Muhammad Arif Jalil**
**Jalil Ali**

# Contents

**1. Linear and Nonlinear Ring Resonators**     **1–52**

1.1 Introduction   *1*
1.2 A Linear Microring Resonator   *3*
1.3 A Linear Add/Drop Filter   *4*
    1.3.1 Characteristics of Complementary Ring-resonator Add/drop   *4*
    1.3.2 Graphical Representation of Photonic Circuits   *5*
    1.3.3 Photonic Transfer Functions   *6*
    1.3.4 Simulation Results   *9*
1.4 Vernier Effect   *13*
    1.4.1 Transfer Function of a Single-ring Resonator Filter   *15*
    1.4.2 Transfer Function of Double-ring Resonator Vernier Filter   *16*
    1.4.3 Transfer Function of a Triple-ring Resonator Vernier Filter   *17*
    1.4.4 Simulation Results   *19*
1.5 All-Pass and Add/Drop Filter   *24*
1.6 A Nonlinear Microring Resonator   *31*
1.7 A Nonlinear Add/Drop Filter   *35*
    1.7.1 The effect of TPA   *39*
    1.7.2 The Impact of Coupling Coefficient   *40*
    1.7.3 The Impact of Wavelength   *41*
    1.7.4 The impact of TOE   *41*
    1.7.5 The Impact of Photon Lifetime   *42*
1.8 Conclusion   *43*
*References*   44

**2. A PANDA Ring Resonator**     **53–82**

2.1 Introduction   *53*
2.2 Theory and Modeling   *54*
2.3 Dynamic Pulse Propagation   *60*
2.4 Symmetry and Asymmetry PANDA Ring Resonators   *67*

## vi Contents

2.5 Random Binary Code Generation  70
2.6 Binary Code Suppression and Recovery  77
2.7 Conclusion  80
References  80

### 3. Dark-Bright Soliton Conversion   83–110
3.1 Introduction  83
3.2 Operating Principle  85
3.3 Soliton Nonlinear Behaviors  87
3.4 Optical Soliton  90
3.5 Dark-Bright Soliton Conversion  93
3.6 Dark-Bright Soliton Conversion in Add/Drop Filter  95
3.7 Soliton Collision Management in a Microring Resonator  99
3.8 Soliton Collision Management  102
3.9 Conclusion  105
References  107

### 4. Dynamic Optical Tweezers   111–132
4.1 Introduction  111
4.2 The Add/Drop Optical Filter  113
4.3 Storage Trapping Tool  115
4.4 Dynamic Potential Well Generation  117
4.5 Dynamic Optical Tweezers via a Wavelength Router  122
4.6 Trapping Forces  123
4.7 Trapping Stability  124
4.8 Trapping and Transportation Mechanism  126
4.9 Atom/Molecule Transmission and Transportation via Wavelength Router  128
4.10 Conclusion  129
References  130

### 5. Hybrid Interferometer   133–146
5.1 Introduction  133
5.2 Theoretical Background  134
5.3 Hybrid Interferometer  138
5.4 Conclusions  144
References  144

### 6. Hybrid Transceiver   147–160
6.1 Introduction  147
6.2 Theory  148
6.3 Hybrid Transceiver and Repeater  153
  6.3.1 Hybrid Transceiver  154
  6.3.2 Hybrid Repeater  157
6.4 Conclusion  157
References  158

Contents **vii**

**7. Nanocommunication** 161–182
7.1 Introduction  161
7.2 Multi Variable Quantum Tweezers Generation and Modulation  162
7.3 Molecular Transporter Generation for Quantum-Molecular Transmission  169
  7.3.1 Transporter Generation  170
  7.3.2 Transporter Quantum State  174
  7.3.3 Multi Quantum-Molecular Transportation  176
7.4 Conclusion  177
 References  178

**8. Nanosensors** 183–222
8.1 Introduction  183
  8.1.1 Operating Principle  184
  8.1.2 Distributed Spatial Sensors  185
  8.1.3 Distributed Quantum Sensors  189
8.2 Network Sensors using a PANDA Ring Resonator Type  192
  8.2.1 General Review  192
  8.2.2 Principle and Method  193
  8.2.3 Distributed Network Sensors Using a Microring Sensing Transducer  196
8.3 Self-calibration in a Fiber Optic Sensing System  199
  8.3.1 General Review  199
  8.3.2 Operation Principle  200
  8.3.3 Entangled Photon States Walk-off Compensation  203
8.4 Conclusion  207
 References  208

**9. Optical and Quantum Computing** 213–244
9.1 Introduction  213
9.2 Quantum Controlled-NOT (CNOT) Gate  214
9.3 Quantum SWAP Gate  217
9.4 All-optical Logic Gate  218
9.5 Dark-Bright Soliton Conversion Mechanism  220
9.6 Optical XOR/XNOR Logic Gate Operation  221
9.7 Operation Principle of Simultaneous All-optical Logic Gates  225
9.8 OOK Generation  227
9.9 Conclusion  237
 References  238

**10. Drug Delivery** 245–258
10.1 Introduction  245
10.2 Optical Vortex Generation  247
10.3 Drug Trapping and Delivery  252
10.4 Conclusion  255
 References  255

## 11. Hybrid Transistor     259–298

11.1 Introduction    259
11.2 All-Optical Photonic Transistor    260
11.3 Single-Photon Transistor    268
    11.3.1 Single-Photon Transistor using Microtoroidal Resonators    268
    11.3.2 Single-Photon Transistor using Nanoscale Surface Plasmons    268
11.4 Single-Atom Transistor    273
11.5 Single-Electron Transistor    275
11.6 Single-Molecule Transistor    277
11.7 Photonic Transistor using PANDA Ring    279
11.8 Conclusion    285
*References*    285

## 12. Electron-Hole Pair Manipulation     291–300

12.1 Introduction    291
12.2 Single Electron-Hole Pair Generation    292
12.3 Multi Electron-Hole Pair Generation    296
12.4 Conclusion    299
*References*    299

## *Index*     301–302

# 1

# Linear and Nonlinear Ring Resonators

**CHAPTER OUTLINE**

- Introduction
- A Linear Microring Resonator
- A Linear Add/Drop Filter
- Vernier Effect
- All-Pass and Add/Drop Filter
- A Nonlinear Microring Resonator
- A Nonlinear Add/Drop Filter
- Conclusion
- References

## 1.1 Introduction

Optical devices have become one of the important components in the new era of merging research areas. The use of a small scale optical device has applications in the research area known as nanophotonics. Many research works have shown the potential of future applications which could be categorized as follows. Optical microcavities [1–8] or optical microring resonators have generated tremendous research progress in many aspects of optical science, such as all optical nonlinear switching [9–11], modulators [12–14], all-optical signal processing [15–17], biochemical sensing [18–20], slow-light structures or optical buffers [21–27], wavelength division-multiplexes (WDM), optical filters for optical networks and on-chip optical interconnects [28–35], ultralow-threshold microlasers [36–37], enhancement of Raman processes [38] as well as Raman lasing [39–41], and the fundamental experiments in quantum physics, e.g., cavity quantum electrodynamics (QED) [42]. All those applications are made possible by the strong light confinement in a small modal volume, i.e. small scale device. By adjusting the shape, size, and material composition, the microresonator can be designed to support a spectrum of optical modes with required polarization, frequency, and field patterns. The most fundamental form of optical resonator is known as the Fabry–Pérot (FP) etalon [43], which consists of a slab-shaped optical medium with highly parallel sides surrounded by a medium with a different refractive index. The build up of the intracavity intensity caused by reflections from the interfaces in this standing wave resonator, aided by constructive interference between the incoming and the intracavity fields, depends on the time spent by the light bouncing back and forth inside the cavity and is determined by the reflectivity of the interfaces. The resonances are periodically spaced in frequency with a free spectral range (FSR) that is inversely proportional to the optical cavity length (inter-mirror distance). Highly reflective mirrors in place of the interfaces result in resonances with high finesse, i.e., in high FSR to bandwidth (BW) ratio. The simple yet elegant concept of the FP resonator is the fundamental principle of operation found in all optical microresonators, both standing wave and traveling wave. In traveling wave resonators, the closed loop offers spatial separation between the waves impinging on and emerging from the couplers, preventing the formation of standing waves.

In principle, there are, so far two well-established mechanisms of light confinement and guidance inside the volume of an optical microresonator. The first is the conventional mechanism of total internal reflection (TIR) and the existence of evanescent waves, where the guiding medium must be optically denser, i.e., have a higher refractive index, than the surrounding one in order to achieve light confinement. The second is the photonic bandgap (PBG) found in artificial optical media having a spatial periodicity in one, two, or three dimensions, termed photonic crystal (PC) [44], which is a result of the phenomenon of Bragg reflection causing the formation of frequency bands where propagation of light is prohibited by the destructive interference of field harmonics inside the crystal. Highly confined optical modes can be achieved in these bands when certain defects are introduced in the otherwise perfectly periodic crystal. With PC defect modes [45–47], the light can be confined in a size comparable to its wavelength ($\lambda/n$), where $\lambda$ is the vacuum wavelength and $n$ is the medium refractive index. In this chapter, the behavior of light pulse propagation in either linear or nonlinear ring resonator is investigated, in which the nonlinear behavior of the simulation results such as chaos, bifurcation and bistability are also discussed in detail.

## 1.2 A Linear Microring Resonator

We start by considering a simple microring (MR) introducing its most basic characteristics. We shall use the ring resonator as an example but one can consider many alternative schemes for instance, Fabry–Pérot, photonic crystal, whispering gallery mode microdisk cavities. Figure 1.1 shows the schematic drawing of an all-pass single microring resonator (MRR), where $\kappa$ and $a$ are the amplitude coupling coefficient of the coupling region and roundtrip amplitude transmission coefficient, respectively. The transmittance of the complex field at the through port can be expressed as in Eq. (1.1).

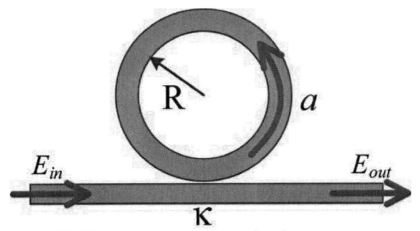

**FIGURE 1.1** A schematic diagram of single ring resonator.

$$T = \left|\frac{E_{out}}{E_{in}}\right|^2 = \left(\frac{r - ae^{j\phi}}{1 - are^{j\phi}}\right)\left(\frac{r - ae^{-j\phi}}{1 - are^{-j\phi}}\right) = \frac{r^2 - a^2 - 2ra\cos(\phi)}{1 + r^2a^2 - 2ra\cos(\phi)} \quad (1.1)$$

where $\phi$ is the roundtrip phase change of the ring, which is related to the optical frequency, and $r$ is the amplitude transmission coefficient of the coupling region which satisfies the relation $r^2 + \kappa^2 = 1$ for lossless coupling. The phase shift ($\Phi$) of the transmitted light can be derived as [48]

$$\Phi = \pi + \phi + \tan^{-1}\frac{r\sin(\phi)}{a - r\cos(\phi)} + \tan^{-1}\frac{ra\sin(\phi)}{1 - ra\cos(\pi)} \quad (1.2)$$

Figure 1.2 shows the transmission and phase at the through port with different coupling coefficients of Si microring resonator radius is 17.5 μm. At the over-coupling as shown in Figure 1.2(a)–(d), respectively, where the phase experiences a full $2\pi$ shift [49].

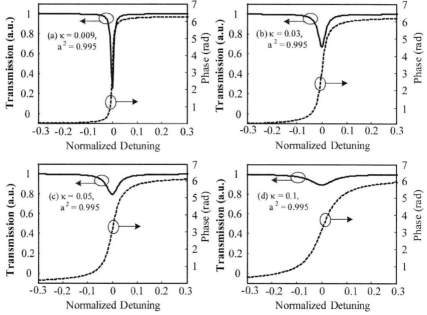

**FIGURE 1.2** Transmission and phase shift of the through port for the MRR under amplitude transmission coefficient for a round-trip along the ring, $a^2 = 0.995$ and different coupling conditions: (a) $\kappa = 0.009$, (b) $\kappa = 0.03$, (c) $\kappa = 0.05$, and (d) $\kappa = 0.1$.

## 1.3 A Linear Add/Drop Filter

### 1.3.1 Characteristics of Complementary Ring-resonator Add/drop [50]

In this section, the filtering characteristics of the two kinds of complementary ring-resonator add/drop filters are modeled and investigated. A graphical approach with signal flow graph is employed here for the analytical derivation of the optical transfer functions in Z-domain of filters. The characteristics of the complementary circuits including the transmittance and group delay of the drop port with respect to the input port are simulated. The present analysis is restricted to directional couplers characterized by two parameters, where the power coupling coefficient and power coupling loss is $j$ and $c$, respectively. Explicit expressions for the phase delay, full-width at half maximum, Q-factor and finesse are also given. Using the appropriate coupling coefficients, the filtering characteristics can be optimized.

Optical ring resonators (RR) have numerous applications in single mode lasers [51], biosensors [52], optical switching [53], add/drop filters [54], tunable lasers [55], signal processing [56] and dispersion compensators [57]. In any WDM system, optical filters are used for separating one optical channel from the combined signal without electronics. A filter referred to as an add/drop filter is required to separate the channel to be dropped from those that pass through filter unaffected [58]. Integrated photonic ring resonator filters have been demonstrated in a variety of material systems such as Si–SiO$_2$, GaAs–AlGaAs, InGaAsP–InP, etc. [59–61].

The basic building blocks, the single RR with two couplers in its two fundamental architectures as illustrated in Figures 1.3(a) and 1.4(a), are the simplest optical waveguide filters with a single pole response. The main performance characteristics of these resonators are the transmittance, free spectral range, finesse or Q-factor, and group delay, which have been demonstrated both theoretically and experimentally in many works [59, 62–64]. In addition, note that Figures 1.3(a) and 1.4(a) also show the two complementary architectures. In each figure, the circuits are called complementary because, as we shall see later in the results, they have the same performance when $\kappa$ and $1 - \kappa$ are interchanged.

All optical components of the circuits to be considered are passive, linear, time invariant, supporting a single mode and must be lumped (i.e., not distributed). The effects, such as the backscatter of light along the length of an optical waveguide, or the saturation of an optical amplifier, are therefore not considered here, because the former is a distributed phenomenon and the latter is a non-linear effect. It can be assumed that the two $2 \times 2$ couplers have no physical length so that $L$ is taken to represent the length of the entire ring resonator. Various techniques have been used for the analysis of fiber/integrated ring resonator. There are two main classes of such techniques. The first class is a set of analytical methods including the scattering matrix method [65], the transfer matrix/chain matrix algebraic method [66, 67] and the method of solving the field equations [68, 69]. The second class using a graphical approach is called the Signal Flow Graph (SFG) method proposed by Mason [70]. This method is originally used in the electrical circuits and not widely used in the analysis of optical circuits.

### 1.3.2 Graphical Representation of Photonic Circuits

The architectures of two complementary ring-resonator add/drop filters are illustrated in Figures 1.3(a) and 1.4(a), which is constructed by one ring resonator and two $2 \times 2$ optical couplers.

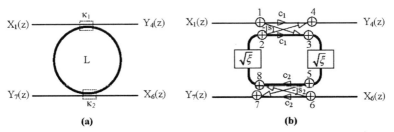

**FIGURE 1.3** The architecture of ring-resonator add/drop filter: (a) waveguide layout and (b) z-transform diagram (SFG).

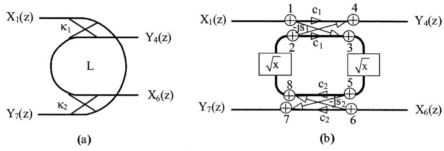

**FIGURE 1.4** The complementary architecture of Figure 1.3, where (a) waveguide layout and (b) z-transform diagram (SFG).

The two complementary optical circuits of ring-resonator add/drop filters can be represented in SFG diagrams as shown in Figure 1.3(b) and 1.4(b) according to Reference [71]. By taking into account the insertion loss of the coupler c and the coupling factor of the ith coupler, $\kappa_i$, when a coherent source is input into a device, the coupling intensity for the throughput path is denoted by $c_i = \sqrt{(1-\gamma)(1-\kappa_i)}$ and for the cross path is $-js_i = -j\sqrt{(1-\gamma)(\kappa_i)}$ were represents a $-(\pi/2)$ phase shift. In contrast, the coupling for the complementary circuit of Figure 1.4(a) is different from Figure 1.3(a), namely $-js_i = -j\sqrt{(1-\gamma)(\kappa_i)}$ for the throughput path and $c_i = \sqrt{(1-\gamma)(1-\kappa_i)}$ for the cross path [72, 73]. As to the transmission of light along the ring resonator (the closed path), we can written as $\xi = \exp(-\alpha L/2)z^{-1}$, where $\exp(-\alpha L/2)$ is the one round-trip losses coefficient, and the $z^{-1}$ is the z-transform parameter, which is defined as

$$z^{-1} = \exp(-j\beta L) \quad (1.3)$$

where $\beta = kn_{\text{eff}}$ is the propagation constant, $k = 2\pi/\lambda$ is the vacuum wave number, $n_{\text{eff}}$ is the effective refractive index of the waveguide and the circumference of the ring is $L = 2\pi R$, where $R$ is the radius of the ring.

### 1.3.3 Photonic Transfer Functions

#### 1.3.3.1 Mason's rule for optical circuits [70, 74]

A forward path is a connected sequence of directed links going from one node to another (along the link directions), encountering no node more than once. A loop is a forward path that begins and ends on the same node. The loop gain or path gain is the product of all the links along that loop or path, respectively. Two loops or paths are said to be non-touching, if they share no nodes in common. Mason's rule states that the transfer function or input–output transmittance relationship from node $X_1(z)$ to node $Y_n(z)$ in a signal flow graph is given by

$$H = \frac{1}{\Delta} \sum_{i=1}^{n} T_i \Delta_i \quad (1.4)$$

where $H$ is the network function relating an input and an output port, $T_i$ is the gain of the $i$th forward path from input to output port, and $n$ is the total number of forward paths from input to output.

The symbol $\Delta$ denotes the signal flow graph determinant, which is given as

$$\Delta = 1 - \sum_i L_i + \sum_{i,j} L_i L_j - \sum_{i,j,k} L_i L_j L_k + \cdots, \tag{1.5}$$

where $L_i$ is the transmittance gain of the $i$th loop. In each of the product summations in Eq. (1.5), only the products of non-touching loops are included. The term "non-touching" implies that the loops have no node in common, i.e., they the separated loops. A minus sign goes with a sum of products of an odd number of loop gains, and a plus sign with a sum of products of an even number of loop gains. The symbol $\Delta_i$ in Eq. (1.4) represents the determinant $\Delta$ after all loops that touch the path $T_i$ at any node which have been eliminated. It is noted-that the optical transmittance in this section is also given the same graphical representation.

### 1.3.3.2 Transfer functions of SRR add/drop filters

SFG for single RR add/drop filter is shown in Figure 1.3(b), in which $X_1(z)$ is the input node and $Y_7(z)$ is considered as the output node. The forward path transmittance for path $1 \to 3 \to 5 \to 7$ is given by

$$T_1 = -s_1 s_2 \sqrt{e^{-(\alpha/2)L}} z^{-1} \tag{1.6}$$

There is one individual loop gain for loop $3 \to 5 \to 8 \to 2 \to 3$, which is denoted as

$$L_1 = c_1 c_2 e^{-(\alpha/2)L} z^{-1} \tag{1.7}$$

From Mason's rule, the determinant of the SFG is given by

$$\Delta = 1 - L_1 = 1 - c_1 c_2 e^{-(\alpha/2)L} z^{-1} \tag{1.8}$$

Since the closed path touches the forward path, we have

$$\Delta_1 = 1 \tag{1.9}$$

By substituting Eqs (1.6)–(1.9) into Eq. (1.4), we obtain the transfer function in Figure 1.3(b), given by

$$\frac{Y_7(z)}{Y_1(z)} = H_{71}^1(z) = -\frac{s_1 s_2 \sqrt{e^{-(\alpha/2)L}} z^{-1}}{1 - c_1 c_2 e^{-(\alpha/2)L} z^{-1}} \tag{1.10}$$

Similarly referring to SFG for the complementary circuit in Figure 1.4(b), it's transfer function is represented by

$$\frac{Y_7(z)}{Y_1(z)} = H_{71}^2(z) = -\frac{c_1 c_2 \sqrt{e^{-(\alpha/2)L}} z^{-1}}{1 + s_1 s_2 e^{-(\alpha/2)L} z^{-1}} \tag{1.11}$$

Equations, new parameters will be used for simplification: $\phi = \beta L$ is the phase constant, $D = c_1 c_2 \exp[(-\alpha/2)L]$, when treating the SFG in

Figure 1.3(b), and we can use the same symbol with different definition for $D = s_1 s_2 \exp[(-\alpha/2)L]$ which is the complementary SFG in Figure 1.4(b). Thus, from Eq. (1.10), we obtain the output intensity

$$|H_{71}^2|^2 = -\frac{(s_1 s_2)^2 e^{-(\alpha/2)L}}{(1+D)^2 + 4D \sin^2\left(\frac{\phi}{2}\right)} \tag{1.12}$$

Equation (1.11) can be written as the output intensity expression similar to Eq. (1.12) except that $s_i$ is replaced by $c_i$ and there is a change of the sign in front of $D$. The two complementary ring resonator (RR) add/drop filters have the same periodic resonant responses in frequency domain with the free spectral range (FSR) between two resonance peaks given by [71]:

$$\text{FSR} = \Delta f = \frac{c}{n_g L} \tag{1.13}$$

where $n_g = n_{\text{eff}} + f_0(dn_{\text{eff}}/df)f_0$ is the group index of the ring waveguide and $f_0$ is the center frequency. The optical resonators resonate at a high order mode. At $f_0$, the perimeter of the ring is an integer number of guide wavelengths, and this integer $M_r$ is the order number of mode and $f_0 = M_r \text{FST}$. The maximum transmission for Figure 1.3(a) occurs at frequencies, where $\sin(\phi/2) = 0$, that is at $f = M_r c / n_{\text{eff}} L$; $c$ is the velocity of light in vacuum, while for Figure 1.4 ($\sin(\phi/2) = 1$). The full width at half maximum (FWHM, $\delta f$) in terms of frequency at the resonance peaks for both architectures has the same expression as [75].

$$\delta f = \frac{\text{FSR}}{\pi} \frac{1-D}{\sqrt{D}} \tag{1.14}$$

The finesse $F$ is defined as the ratio of FSR to FWHM

$$F = \frac{\pi \sqrt{D}}{1-D} \tag{1.15}$$

where $D$ in Eqs (1.14 and 1.15) has the different definition for the two complementary circuits.

The ability of the waveguide to confine the field is described by the quality factor $Q$. It is a measure of the sharpness of the transmission peak which is defined as the ratio of the resonance frequency to the full width at half maximum:

$$Q = \frac{f_0 F}{\text{FSR}} \tag{1.16}$$

### 1.3.3.3 Group delay of RR add/drop filters

The filter group delay is defined as the negative derivative for the phase of the transfer function with respect to the angular frequency given as follows [71]:

$$\tau_n = -\frac{\partial}{\partial \omega} \tan^{-1}\left[\frac{\text{Im}[H(z)]}{\text{Re}[H(z)]}\right] \tag{1.17}$$

where $\tau_n$ is normalized to the unit delay of the waveguide, $T$. Thus, from Eqs (1.10) and (1.11) we obtain equations for the phase delays of Figure 1.3(a) and 1.4(a), equations in terms of $D$ and $\phi$, given respectively as follows:

$$\Phi = \tan^{-1}\left[\frac{(D+1)}{(D-1)}\tan\left(\frac{\phi}{2}\right)\right] \tag{1.18}$$

and

$$\Phi = \tan^{-1}\left[\frac{(D-1)}{(D+1)}\tan\left(\frac{\phi}{2}\right)\right] \tag{1.19}$$

respectively.

Since the phase constant $\beta L$ is equal to $\omega T$ for the waveguide, then $Ld\beta = Td\omega$. Substituting Eqs (1.18) and (1.19) for Eq. (1.17), given that $d\tan^{-1}[g(x)]/dx = g'(x)/[1+g^2(x)]$ is the normalized group delay for the circuits of Figures 1.3(a) and 1.4(a), given by

$$\tau_r \frac{\tau_n}{T} = \frac{(1-D)^2/2}{[(D-1)\cos(\phi/2)]^2 + [(D+1)\sin(\phi/2)]^2} \tag{1.20}$$

$$\tau_r \frac{\tau_n}{T} = \frac{(1-D)^2/2}{[(D+1)\cos(\phi/2)]^2 + [(D-1)\sin(\phi/2)]^2} \tag{1.21}$$

respectively, where $\tau_r$ is the relative normalized group delay and $D$ is a different definition of the two complementary circuits.

### 1.3.4 Simulation Results

The transmission characteristics with lossless in RR ($\alpha = 0$), as shown in Figure 1.3, are the comparison of Figure 1.1(a) to Figure 1.4(a) response. The parameters of both circuits are identical except the symmetric coupling coefficients. In both cases the design frequency (wavelength) $f_0 = 193.1$ THz ($\lambda_0 = c/f_0 = 1552.52$ nm), $M_r = 1931$, $\gamma = 0.1\%$, FSR = 100 GHz and $n_g = 3.46$ (for the III–V semiconductor materials waveguide), which determines the circumference of the ring as $L = M_r\lambda_0/n_g = 0.86$ mm. Note that the output intensities of the two complementary circuits are the same when $\kappa_i$ and $1 - \kappa_i$ are interchanged in the relations. For example, $\kappa = 0.1, 0.15, 0.2$ for Figure 1.3(a) can obtain the same results from Figure 1.4(a) with $\kappa = 0.9$, 0.85, 0.8, respectively with merely a $\pi$ phase difference.

The FWHM of the transmission of circuit in Figure 1.3(a) decreases when $\kappa$ decreases as shown in Figure 1.5, for example, when the FWHM of pass band is 0.05 nm, $\kappa = 0.2$, while FWHM is degraded to 0.03 nm for $\kappa = 0.1$. The value of $\kappa$ in Figure 1.6(a) should be less than 0.2 in order to obtain a crosstalk lower than −20 dB to separate the channel from the other channels.

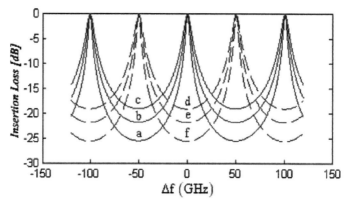

**FIGURE 1.5** Transmittance of the two complementary single RR add/drop filters compare different symmetric coupling coefficients $\kappa_{1,2} = \kappa$. Solid lines are for Figure 1.3 under different $\kappa$: (a) $\kappa = 0.1$, (b) $\kappa = 0.15$, (c) $\kappa = 0.2$. Dash lines are for Figure 1.4 under different $\kappa$: (d) $\kappa = 0.8$, (e) $\kappa = 0.85$ and (f) $\kappa = 0.9$.

The maximum transmission characteristic of the circuit in Figure 1.3(a) as a function of $\kappa_1$ and $\kappa_2$ is shown in Figure 1.6. The output intensities will be unity at resonance ($\beta L = 2M_r\pi$), which indicates that the resonance wavelength is fully extracted by the resonator, for identical symmetrical couplers $\kappa_1 = \kappa_2$, especially lossless in waveguide ($\alpha = 0$) and couplers ($\gamma = 0$). In accordance if $\kappa_i$ is replaced with $1 - \kappa_i$ and the other parameters are identical, the complementary output intensities at resonance, $\beta L = (2M_r + 1)\pi$ will be unity which is the same as in Figure 1.6.

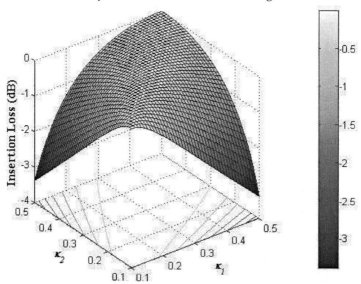

**FIGURE 1.6** The impact of $\kappa_1$ and $\kappa_2$ to transmittance characteristic at resonance for Figure 1.3(a), when $\alpha = 0$, $\gamma = 0$. Transmittance will be unity using symmetric coupling coefficient $\kappa_1 = \kappa_2$.

The finesse F and Q-factor of the circuit in Figure 1.3(a) are dependent on the symmetric coupling coefficients $\kappa_1 = \kappa_2 = \kappa$, and several values of attenuation coefficient $\alpha$ as shown in Figure 1.7. The other parameters are the same as those used for Figure 1.5. Note that both the finesse and Q-factor decrease to zero when $\kappa$ is increased until it approaches to 1, and it is also found that optical loss in ring resonator will result in deterioration of both the finesse and Q-factor. However, F- and Q-factor will saturate when $\kappa$ is decreased until it reaches zero. For example, if the finesse F = 60 and $\kappa$ = 0.05 and F is degraded to 33, when $\kappa$ = 0.05 and $\alpha$ = 1 dB/cm. For this case, the direction of increasing Q-factor values of the y-axis on the right is reversed from top to bottom. Note that when the $\kappa_i$ and $1 - \kappa_i$ are interchanged in Eqs (1.14) and (1.15), the complementary finesse and Q-factor characteristics for Figure 1.4(a) are the mirror image of Figure 1.7 over the $\kappa$ = 0.5 as shown in Figure 1.8.

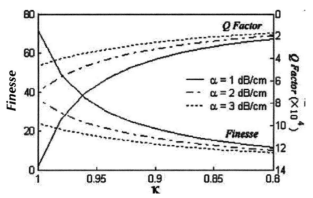

**FIGURE 1.7** The finesse and Q-factor of RR in Figure 1.3(a) as a function of the coupling coefficient for varying loss in waveguide, and assuming $\gamma = 0.1\%$.

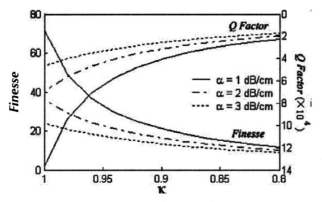

**FIGURE 1.8** The finesse and Q-factor of the complementary of Figure 1.4(a) as a function of the coupling coefficient for varying loss in the waveguide and assuming $\gamma = 0.1\%$.

Figure 1.9 illustrates Q-factor that is dependent on the finesse F of RR in Figure 1.3(a) when the radius $R$ = 100 µm, 150 µm, 200 µm and 250 µm at the center frequency. A high finesse (F > 100) and a low Q-factor (<$10^5$) is obtained, for example, a ring resonator with a low radius ($R$ = 100 µm). For laser applications a high Q-factor is required and this can be achieved by increasing ring radius, but this will lead to a decrease in FWHM.

The relative group delay responses of circuit in Figure 1.3(a) are plotted in Figure 1.10, assuming the circumference of ring $L$ = 0.86 mm and using symmetric coupling coefficients $\kappa_1 = \kappa_2 = \kappa$. The relative group delay response shows a period of 100 GHz, which is the same as the FSR of ring resonator and we found that as $\kappa$ is decreasing, it will become sharper and steeper. Similarly, if $\kappa_i$ is replaced with $1 - \kappa_i$ and the other parameters are identical, the complementary relative group delay can obtain the same results, with merely a $\pi$ phase difference as shown in Figure 1.11.

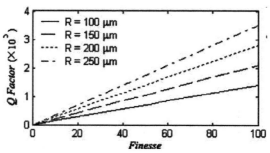

**FIGURE 1.9** Q-factor depending on the finesse for a specific radius $R$ of RR in Figure 1.3(a).

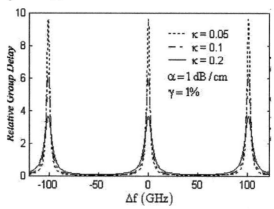

**FIGURE 1.10** The relative group delay responses of RR in Figure 1.3(a) comparing different symmetric coupling coefficients.

Similarly, as we can see, the group delay in Figure 1.10 is a periodic function of frequency, then from Eq. (1.18), the sharp peaks will occur when ($\beta L = 2M_r\pi$), namely $\sin(\phi/2) = 0$. Figure 1.12 shows the normalized group delay response as a function of $\kappa_1$ and $\kappa_2$ at resonance for Figure 1.3(a) with lossless. The surface descends, when the values of $\kappa_1$ and $\kappa_2$ are decreased.

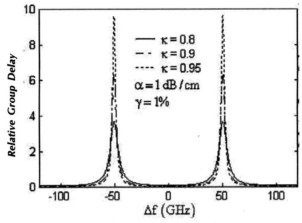

**FIGURE 1.11** The relative group delay responses of the complementary of Figure 1.4(a) comparing different symmetric coupling coefficients.

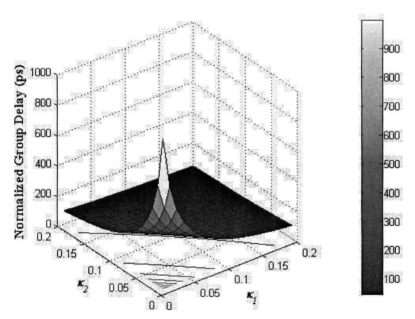

**FIGURE 1.12** The normalized group delay characteristic for Figure 1.3(a) at resonance as a function of $\kappa_1$ and $\kappa_2$ ($\alpha = 0$, $\gamma = 0$ and $L = 0.86$ mm).

## 1.4 Vernier Effect [76]

A dense wavelength-division multiplexing (DWDM) system requires optical channel filters with high selectivity (ability to separate two adjacent channels), so that they have a wide, free spectral range (FSR) to accommodate large channel counts. There are two ways to expand the FSR. The first employs a ring resonator constructed using a smaller ring waveguide.

The conventional single-ring resonator (SRR) has a disadvantage, that is, it cannot increase the FSR significantly. Since FSR is inversely proportional to the ring radius, the ring radius must be reduced in order to increase the FSR. However, the bending loss of a ring waveguide increases rapidly with decreasing ring radius. Therefore, there is a serious problem in using optical ring resonators as tunable filters. Thus, we investigate the second approach: Vernier operation of a multiple-ring resonator (MRR) with slightly different ring radii. This way, we expand the FSR without reducing the ring radius, by adding another ring waveguide. Investigations of these types of filters have been performed [77, 78].

Although Vernier operation is clearly advantageous for increasing the FSR, it has three major disadvantages: (1) small pass-band width, (2) decreasing rejection across the stop band with increasing integer ratios $N:M$, and (3) trade-offs between maximum pass-band transmission and stop-band rejection.

Optical filter design is typically approached with electromagnetic field equations, which are solved for the fields in the frequency or time domain. These techniques are required for characterizing ring resonator performance and directional couplers. However, they can become cumbersome and nonintuitive for filter design. Therefore, different analytical methods of signal processing, including the scattering-matrix method [3, 4] and the transfer-matrix–chain-matrix algebraic method [79–81] have been developed for determining optical filter transfer functions in the Z domain, considering the optical circuit to be linear and time-invariant. Another approach to analyse the complex photonic circuits and fast calculation of optical transfer functions is graphical approach, called the signal flow graph (SFG) method, proposed by Mason [82]. This method was originally used for electrical circuits, and is not yet used widely in the analysis of optical circuits.

Consider the architectures of ring-resonator add/drop filters as illustrated in Figures 1.13–1.15, which are constructed from $2 \times 2$ optical couplers. The $2 \times 2$ optical directional coupler can be represented in an SFG diagram according to Ref. 3. By taking into account the coupling factor $\kappa_i$ of the $i$th coupler ($i = 1, 2, \ldots, N$) and the insertion loss $\gamma$ for each coupler, the fraction of light passed through the throughput path is expressed as $c_i = [(1-\gamma)(1-\kappa)]^{1/2}$, and in contrast, the fraction passing though the cross path is expressed as $-js_i = -j[(1-\gamma)(\kappa_i)]^{1/2}$. The z-transform parameter $z^{-1}$ is defined as

$$z^{-1} = \exp(-j\beta L) \tag{1.22}$$

where $\beta = kn_{\text{eff}}$ is the propagation constant, $k = 2\pi/\lambda$ is the vacuum wave number, $n_{\text{eff}}$ is the effective refractive index of the waveguide, and the circumference of the ring is $L = 2\pi R$; here $R$ is the radius of the ring.

When all rings of MRR have the same circumference (we call such a device a *uniform* SMRR optical filter), the FSR of the device is determined by

$$\text{FSR} = \frac{c}{n_g L} \tag{1.23}$$

where $n_g = n_{eff} + f_0 (dn_{eff}/df)_{f_0}$ is the group refractive index of the ring, $n_{eff}$ is the effective refractive index, and $f_0$ is the design (center) frequency [79].

A *forward path* is a connecting sequence of directed links going from one node to another (along the link directions), encountering no node more than once. A *loop* is a forward path that begins and ends on the same node. The *loop gain* or *path gain* is the product of all the links along that loop or path, respectively. Two loops or paths are said to be *non-touching* if they have no nodes in common. *Mason's rule* states that the transfer function, or input-output transmittance relation, from the node $E_1(z)$ to the node $E_n(z)$ in a signal flow graph is given by

$$H = \frac{1}{\Delta} \sum_{i=1}^{n} T_i \Delta_i \quad (1.24)$$

where $H$ is the network function relating an input and an output port, $T_i$ is the gain of the $i$'th forward path from input to output port, and $n$ is the total number of forward paths from input to output. The symbol $\Delta$ denotes the signal flow graph determinant, which is given as

$$\Delta = 1 - \sum_i L_i + \sum_{i,j} L_i L_j - \sum_{i,j,k} L_i L_j L_k + \cdots \quad (1.25)$$

of the product summations in Eq. (1.25), only the products of non-touching loops are included. A minus sign goes with a sum of products of an odd number of loop gains, and a plus sign with a sum of products of an even number of loop gains. The symbol $\Delta i$ in Eq. (1.24) indicates the determinant $\Delta$ after all loops that touch the path $T_i$ at any node have been eliminated. It is noted that the optical transmittance is also given the same graphical representation as in the chapter.

### 1.4.1 Transfer Function of a Single-ring Resonator Filter

The optical transfer functions of the ring resonator filters at the drop port for an input port $E_1$ can be obtained by using Mason's rule. First the transfer function of an SRR filter is presented, followed by corresponding results on double- and triple-ring resonator Vernier filters. The SFG for an SRR filter is shown in Figure 1.13(b), in which the input nodes $E_1(z)$ and $E_8(z)$ are considered as the drop node.

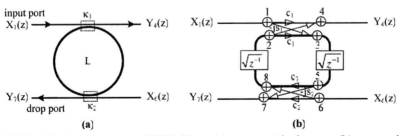

**FIGURE 1.13** Architecture of SRR filter: (a) waveguide layout; (b) z-transform diagram (SFG).

For the transfer function $E_8(z)/E_1(z)$, the individual loop gain of the SFG is denoted as

$$L_1^1 = c_1 c_2 x z^{-1} \tag{1.26}$$

There is only one forward path transmittance from node 1 to 8 for the drop port, and this forward path also touches the loop $L_1^1$. From Eq. (1.26), we have

$$T_{1d}^1 = -s_1 s_2 (xz^{-1})^{1/2}$$

$$\Delta_1 = 1 \tag{1.27}$$

From Eq. (1.4), the determinant of the SFG from Mason's rule is given by

$$\Delta = 1 - L_1^1 = 1 - c_1 c_2 x z^{-1} \tag{1.28}$$

Substituting Eqs (1.27) and (1.28) into Eq. (1.24), we obtain the transfer function for the drop port of the SFG in Figure 1.13(b) as

$$\frac{E_8(z)}{E_1(z)} = H_d^1 = -\frac{s_1 s_2 \sqrt{e^{(\alpha/2)L}} z^{-1}}{1 - c_1 c_2 e^{(\alpha/2)L} z^{-1}} \tag{1.29}$$

## 1.4.2 Transfer Function of Double-ring Resonator Vernier Filter

The multiple-ring resonator filter opens the possibility of expanding the FSR to the least common multiple of the FSRs of the individual ring resonators. This is done by choosing different radii in the MRR, which is called Vernier operation. In the case of different radii, light passing through the MRR is launched from the drop port when the resonant conditions of the multiple single-ring resonators are satisfied. In this article, the two- and three-ring serially coupled Vernier filters are investigated.

The architecture of a double-ring resonator (DRR) Vernier filter is shown in Figure 1.14(a) and (b) shows its SFG. Here the input node is $E_1(z)$, and $E_{12}(z)$ is the drop node. The FSR of a DRR with two different radii is expressed by

$$\text{FSR} = N \bullet \text{FSR}_1 = M \bullet \text{FSR}_2 \tag{1.30}$$

where $N$ and $M$ are coprime natural numbers ($M > N$) and have to be chosen, carefully.

For the transfer function $E_{12}(z)/E_1(z)$, there are three individual loop gains of the SFG to be obtained. They are expressed as

$$L_1^2 = c_1 c_2 z^{-N} \tag{1.31}$$

$$L_2^2 = c_2 c_3 x_2 z^{-M} \tag{1.32}$$

$$L_3^2 = c_1 (x_1 z^{-N})^{1/2} (-js_2)(x_2 z^{-M})^{1/2} c_3 (x_2 z^{-M})^{1/2}) - (js_2)(x_1 z^{-N})^{1/2}$$
$$= -c_1 c_3 s_2^2 x_1 x_2 z^{-(N+M)} \tag{1.33}$$

where $x_i = \exp(-\alpha L_i/2)(i=1,2)$ are the ring losses of ring 1 and ring 2, respectively.

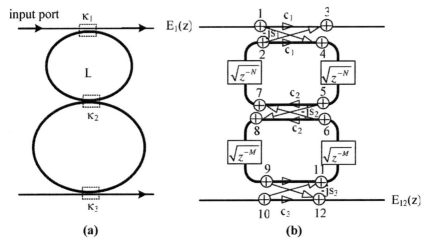

**FIGURE 1.14** Architecture of DRR Vernier filter: (a) waveguide layout; (b) z-transform diagram (SFG).

There is one possible product of transmittances of two non-touching loops, resulting from the separation of loops $L_1^2$ and $L_2^2$, and given by

$$L_{12}^2 = c_1 c_2^2 c_3 x_1 x_2 z^{-(N+M)} \tag{1.34}$$

There is only one forward path transmittance from node 1 to 12 for the drop port, since all loops touch this forward path,

$$T_{1d}^2 = (-js_1)(x_1 z^{-N})^{1/2}(-js_2)(x_2 z^{-M})^{1/2}(-js_3)$$
$$= -js_1 s_2 s_3 \xi (x_1 x_2 z^{-(N+M)})^{1/2} \tag{1.35}$$
$$\Delta_1 = 1$$

From Eq. (1.25), by using the relation $s_2^2 + c_2^2 = 1$, the determinant of the SFG from Mason's rule is given by

$$\Delta = 1 - [L_1^2 + L_2^2 + L_3^2) + L_{12}^2 = 1 - c_1 c_2 \xi - c_2 c_3 \xi + c_1 c_3 \xi \tag{1.36}$$

Substituting Eqs (1.35) and (1.36) into (1.24), we get the transfer function of the SDRR Vernier filter of the SFG in Figure 1.14(b) at the drop port as

$$\frac{E_{12}(z)}{E_1(z)} = H_d^2 = \frac{js_1 s_2 s_3 (x_1 x_2)^{1/2} (z^{-(N+M)})^{1/2}}{1 - c_1 c_2 x_1 z^{-N} - c_2 c_3 x_2 z^{-M} + c_1 c_3 x_1 x_2 z^{-(N+M)}} \tag{1.37}$$

Normally, the coupling coefficients of the couplers adjacent to the bus waveguides ($\kappa_1$ and $\kappa_N$) are larger than those of the innermost couplers.

### 1.4.3 Transfer Function of a Triple-ring Resonator Vernier Filter

The architecture of triple-ring resonator (TRR) Vernier filter is shown in Figures 1.15(a) and (b) shows its SFG. Here the input node is $E_1(z)$, and $E_{16}(z)$ is the drop node. The FSR of the TRR Vernier filter with three different radii is expressed by

$$\text{FSR} = N \bullet \text{FSR}_1 = M \bullet \text{FSR}_2 = L \bullet \text{FSR}_3 \quad (1.38)$$

where $N$, $M$, and $L$ are resonant numbers for each ring resonator and all are integers ($L > M > N$).

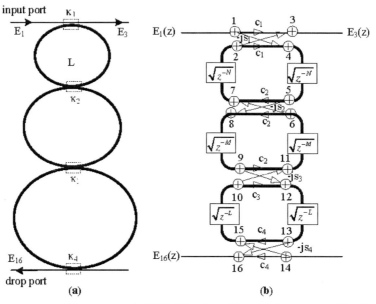

**FIGURE 1.15** Architecture of TRR Vernier filter: (a) waveguide layout; (b) z-transform diagram (SFG).

There are six individual loop gains of the SFG to be obtained, for the transfer function $E_{16}(z)/E_1(z)$, and they are expressed as

$$L_1^3 = c_1(x_1 z^{-N})^{1/2} c_2 (x_1 z^{-N})^{1/2} = c_1 c_2 x_1 z^{-N} \quad (1.39)$$

$$L_2^3 = c_2(x_2 z^{-M})^{1/2} c_3 (x_2 z^{-M})^{1/2} = c_2 c_3 x_2 z^{-M} \quad (1.40)$$

$$L_3^3 = c_3(x_3 z^{-L})^{1/2} c_4 (x_3 z^{-L})^{1/2} = c_3 c_4 x_3 z^{-L} \quad (1.41)$$

$$\begin{aligned} L_4^3 &= c_1(x_1 z^{-N})^{1/2}(-js_2)(x_2 z^{-M})^{1/2} c_3 (x_2 z^{-M})^{1/2} (-js_2)(x_1 z^{-N})^{1/2} \\ &= -c_1 c_3 s_2^2 x_1 x_2 z^{-(N+M)} \end{aligned} \quad (1.42)$$

$$\begin{aligned} L_5^3 &= c_2(x_2 z^{-M})^{1/2}(-js_3)(x_3 z^{-L})^{1/2} c_4 (x_3 z^{-L})^{1/2} = (-js_2)(x_2 z^{-M})^{1/2} \\ &= -c_2 c_4 s_3^2 x_2 x_3 z^{-(M+L)} \end{aligned} \quad (1.43)$$

$$\begin{aligned} L_6^3 &= c_1(x_1 z^{-N})^{1/2}(-js_2)(x_2 z^{-M})^{1/2} c_4 (x_3 z^{-L})^{1/2}(-js_3)(x_2 z^{-M})^{1/2}(-js_2)(x_1 z^{-N})^{1/2} \\ &= c_1 c_4 s_2^2 s_3^2 x_1 x_2 x_3 z^{-(N+M+L)} \end{aligned} \quad (1.44)$$

where $x_i = \exp(-\alpha L_i/2)(i = 1, 2, 3)$ are the ring losses of rings 1, 2, and 3 from top to bottom.

There are five possible products of transmittances of two non-touching loops, given by

$$L_{12}^3 = c_1 c_2^2 c_3 x_1 x_2 z^{-(N+M)} \tag{1.45}$$

$$L_{23}^3 = c_1 c_3^2 c_4 x_2 x_3 z^{-(M+L)} \tag{1.46}$$

$$L_{13}^3 = c_1 c_2 c_3 c_4 x_1 x_3 z^{-(N+L)} \tag{1.47}$$

$$L_{34}^3 = -c_1 c_3^2 c_4 s_2^2 x_1 x_2 x_3 z^{-(N+M+L)} \tag{1.48}$$

$$L_{15}^3 = -c_1 c_2^2 c_4 s_3^2 x_1 x_2 x_3 z^{-(N+M+L)} \tag{1.49}$$

There is one possible product of transmittance of three non-touching loops, given by

$$L_{123}^3 c_1 c_2^2 c_3^2 c_4 x_1 x_2 x_3 z^{-(N+M+L)} \tag{1.50}$$

There is only one forward path transmittance from node 1 to 16 for the drop port, and all loops also touch this forward path. We have

$$T_{1d}^3 = (-js_1)(x_1 z^{-N})^{1/2}(-js_2)(x_2 z^{-M})^{1/2}(-js_3)(x_3 z^{-L})^{1/2}(-js_4)$$
$$= s_1 s_2 s_3 s_4 (x_1 x_2 x_3)^{1/2} (z^{-(N+M+L)})^{1/2} \tag{1.51}$$

$$\Delta_1 = 1$$

The loop determinant $\Delta$ of the SFG is given by Eq. (1.25), and by using the relation $s_i^2 + c_i^2 = 1$ ($i = 2, 3$), the resulting expression can be simplified as

$$\Delta = 1 - c_1 c_2 \xi - c_2 c_3 \xi + c_3 c_4 \xi + c_1 c_3 \xi^2 + c_2 c_4 \xi^2 - c_2 c_4 \xi^3 + c_1 c_2 c_3 c_4 \xi^2 \tag{1.52}$$

Substituting Eqs (1.51) and (1.52) into Eq. (1.24), the transfer function of the TRR Vernier filter of the SFG in Figure 1.15(b) at the drop port is expressed by

$$\frac{E_{16}(z)}{E_1(z)} = H_d^3 \frac{s_1 s_2 s_3 s_4 (x_1 x_2 x_3)^{1/2} (z^{-(N+M+L)})^{1/2}}{\begin{pmatrix} 1 - c_1 c_2 x_1 z^{-N} - c_2 c_3 x_2 z^{-M} - c_3 c_4 x_3 z^{-L} + c_1 c_3 x_1 x_2 z^{-(N+M)} \\ + c_2 c_4 x_2 x_3 z^{-(M+L)} + c_1 c_2 c_3 c_4 x_1 x_3 z^{-(N+L)} - c_1 c_4 x_1 x_2 x_3 z^{-(N+M+L)} \end{pmatrix}} \tag{1.53}$$

### 1.4.4 Simulation Results

First we present the Vernier effect of the DRR filter, and then the corresponding result on TRR filter. In this study, we choose the DRR Vernier filter with $R_1 = 273$ μm, $R_2 = 341$ μm and we use symmetric coupling coefficients of $\kappa_{1,3} = 0.5$ for the outer couplers and $\kappa_2 = 0.13$ for the coupler at the center, with lossless waveguides ($\alpha = 0$) and couplers ($\gamma = 0$).

According to Eq. (1.23), the FSR for single-ring resonators are 50 GHz for resonator 1 and 40 GHz for resonator 2, and the FSR of the DRR is calculated to be 200 GHz, as shown in Figure 1.16. Therefore, the resonant numbers for each ring resonator are taken as $N = 4$, $M = 5$. The interstitial resonance suppression of the drop port is approximately 5.2 dB.

The suppression of interstitial resonances of the DRR is as shown in Figure 1.17. This can be improved by decreasing the coupling coefficient of the center coupler to $\kappa_2 = 0.08$ or $0.05$, where the other parameters are set identically to those in Figure 1.16. Interstitial resonance suppressions of 6.9 dB (solid line) and 8.8 dB (dashed line) are respectively achieved. We found that the decrease of $\kappa$ in the center coupler produces a small deterioration of the resonant loss at resonance peaks. By setting the coupling coefficients $\kappa_1 = \kappa_3 = 0.5$ for the outer couplers and within the tolerance range of $\kappa_2 = 0.09$ to $0.13$ for the center coupler, a side mode suppression of more than 5 dB with no loss at the resonance peaks can be obtained.

Figure 1.18 shows a comparison of the calculated magnitude responses for FSR ratios $N : M$ of $1 : 2$, $4 : 5$, $7 : 8$, and $11 : 12$ based on the same parameters. These ratios are for the same FSR, which is equal to 200 GHz. Note the increasing interstitial resonances for the ratios of larger integer values. The passband width of the resonant transmission peak is dominated by the largest ring. Consequently, the response can be made sharper only at the expense of reducing the passband width. For this reason, the best FSR of the DRR Vernier of Figure 1.6 is a resonance ratio $4 : 5$, which is suitable for designing the FSR expansion. A similar result can also be obtained when using a resonance ratio $1 : 2$; however, that leads to a bending loss due to the decreasing ring radius.

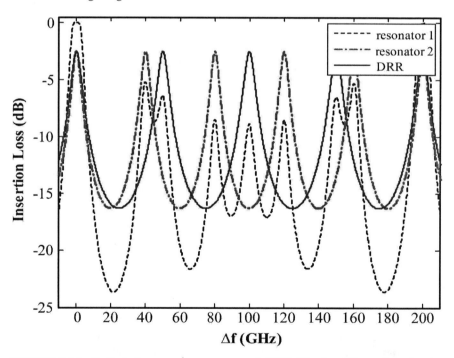

**FIGURE 1.16** Optical frequency response of DRR Vernier filter of each ring resonator (resonator 1 with $FSR_1 = 50$ GHz, resonator 2 with $FSR_2 = 40$ GHz), and of the DRR with FSR = 200 GHz.

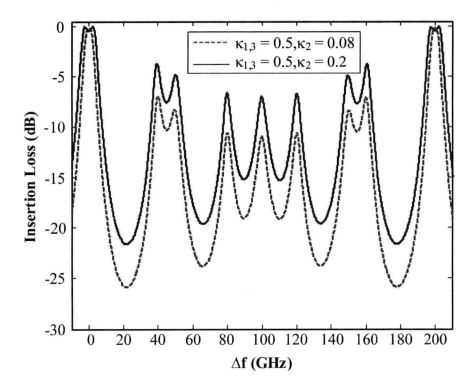

**FIGURE 1.17** The improvement in suppression of interstitial resonances of the DRR Vernier filter.

A DRR Vernier filter does not offer sufficient suppression of the interstitial resonances, while those that consist of more than four rings are too complex and expensive to fabricate. Moreover they fail to provide the needed improvement in suppression. In order to improve the suppression of interstitial resonances, three- and four-ring series-coupled Vernier filters are used. In this section, the next filter simulation was mainly focused on the TRR Vernier effect. Another advantage of TRR Vernier filters is that they can be designed to reduce the unit delay length to half that of the DRR Vernier filter, so that twice the FSR of the DRR Vernier filter can be achieved. The frequency responses of the TRR Vernier filter with $R_1 = 273$ μm, $R_2 = 341$ μm, and $R_3 = 511$ μm using symmetric coupling coefficients of $\kappa_{1,4} = 0.5$ and $\kappa_{2,3} = 0.01$ are shown in Figure 1.19. The FSR of resonator 1 is 50 GHz, the FSR of resonator 2 is 40 GHz, and the FSR of resonator 3 is 26.67 GHz as shown in Figure 1.19(a) from top to bottom. The FSR of the TRR Vernier is calculated using Eq. (1.23) to be 400 GHz as shown in Figure 1.19(b); it is twice that of the DRR Vernier. Here, the resonant numbers for each ring resonator are chosen $N = 8$, $M = 10$, and $L = 15$. The interstitial resonance suppression of the drop port is noted to be 18.5 dB. In order to suppress the interstitial resonances, lower coupling coefficients in

**22** Nanoscale Nonlinear PANDA Ring Resonator

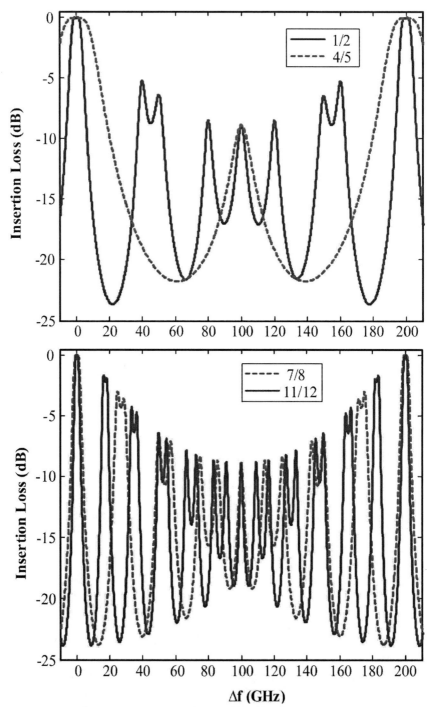

**FIGURE 1.18** Comparison of DRR Vernier filter response with different FSR ratios $N:M = 1:2$, $4:5$, $7:8$, and $11:12$.

the center couplers are chosen as shown in Figure 1.8. Here, the parameters of the TRR Vernier filter are identical to those in Figure 1.19. The interstitial resonance suppression of the drop port with $\kappa_1 = \kappa_4 = 0.5$, $\kappa_2 = \kappa_3 = 0.008$ is approximately 22 dB, and with $\kappa_1 = \kappa_4 = 0.5$, $\kappa_2 = \kappa_3 = 0.005$ is 26.5 dB, as shown in Figures 1.20(a) and (b), respectively. However, we found that the response of the TRR Vernier filter is reduced in the pass-band width, while it is sharper at the resonance peaks.

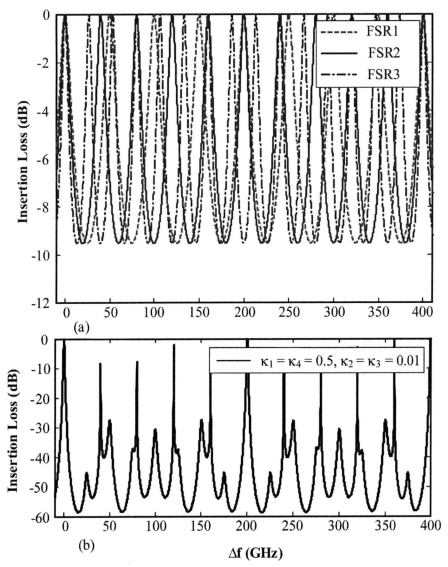

**FIGURE 1.19** Optical frequency response of TRR Vernier filter of each ring resonator: (a) resonator 1 with $FSR_1 = 50$ GHz, resonator 2 with $FSR_2 = 40$ GHz, resonator 3 with $FSR_3 = 26.67$ GHz; (b) TRR with FSR = 400 GHz.

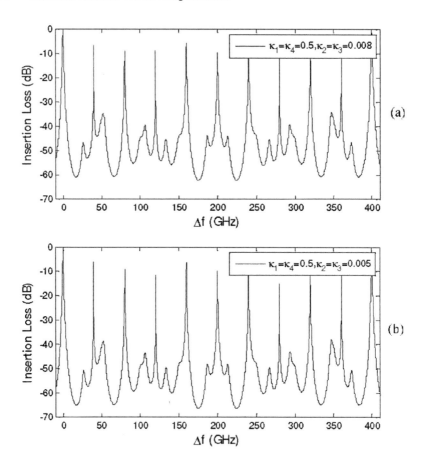

**FIGURE 1.20** Improvement in suppression of interstitial resonances of TRR Vernier filter: (a) $\kappa_{1,4} = 0.5$, $\kappa_{2,3} = 0.008$; (b) $\kappa_{1,4} = 0.5$; $\kappa_{2,3} = 0.005$.

## 1.5 All-Pass and Add/Drop Filter

Consider the architecture of a double coupler ring resonator that is sometimes called an add/drop filter (ADF), as illustrated in Figure 1.21, which is often constructed by the 2 × 2 optical couplers.

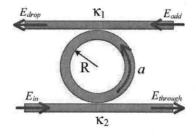

**FIGURE 1.21** A schematic diagram of single ADF.

Similarly, the optical transfer functions of the ring resonator filters at the throughput port and drop port for an input port ($E_{i1}$) can be given by Eqs (1.54) and (1.55).

$$E_{t1} = \sqrt{1-\gamma_1}\left[j\sqrt{\kappa_1}E_4 + \sqrt{1-\kappa_1}E_{i1}\right] \tag{1.54}$$

$$E_1 = \sqrt{1-\gamma_1}\left[j\sqrt{\kappa_1}E_{i1} + \sqrt{1-\kappa_1}E_4\right] \tag{1.55}$$

where $\gamma$ and $\kappa_1$ are the loss and the coupling coefficients, respectively. The incoming light of $E_{i1}$ and $E_4$ are coupled through the first coupler to the output light $E_{t1}$ and $E_1$, and the output light $E_1$ is transmitted through the ring becomes output light $E_2$. According to light transmission theory in linear optical systems, we obtain the following relation between $E_1$ and $E_2$, which is given by

$$E_2 = E_1 e^{\frac{\alpha L}{22} jk_n \frac{L}{2}} \tag{1.56}$$

where the transmission line length is $L/2$. The relationship of the second coupler ($\kappa_2$) is given by

$$E_{t2} = jE_1\sqrt{1-\gamma_2}\sqrt{\kappa_2}e^{\frac{\alpha L}{22} jk_n \frac{L}{2}} \quad \text{at } E_{i2} = 0 \tag{1.57}$$

$$E_3 = E_1\sqrt{1-\gamma_2}\sqrt{1-\kappa_2}e^{\frac{\alpha L}{22} jk_n \frac{L}{2}} \tag{1.58}$$

Using the transmission theory, we obtain $E_4$ in terms of $E_3$, yields

$$E_4 = E_3 e^{\frac{\alpha L}{22} jk_n \frac{L}{2}} \tag{1.59}$$

$$E_1 = \frac{jE_{i1}\sqrt{1-\gamma_1}\sqrt{\kappa_1}}{1 - \sqrt{1-\gamma_1}\sqrt{1-\gamma_2}\sqrt{1-\kappa_1}\sqrt{1-\kappa_2}e^{\frac{\alpha}{2}L - jk_n L}} \tag{1.60}$$

$$E_4 = \frac{jE_{i1}\sqrt{1-\gamma_1}\sqrt{\kappa_1}\sqrt{1-\gamma_2}\sqrt{1-\kappa_2}e^{\frac{\alpha}{2}L - jk_n L}}{1 - \sqrt{1-\gamma_1}\sqrt{1-\gamma_2}\sqrt{1-\kappa_1}\sqrt{1-\kappa_2}e^{\frac{\alpha}{2}L - jk_n L}} \tag{1.61}$$

By using these equations, the transfer function for throughput and drop ports in Figure 1.21 can thus be expressed as the following:
Throughput port,

$$\frac{E_{t1}}{E_{i1}} = \frac{\sqrt{1-\gamma_1}\sqrt{1-\kappa_1} - (1-\gamma_1)\kappa_1\sqrt{1-\kappa_2}e^{\frac{\alpha}{2}L - jk_n L} - (1-\gamma_1)(1-\kappa_1)\sqrt{1-\gamma_2}\sqrt{1-\kappa_2}e^{\frac{\alpha}{2}L - jk_n L}}{1 - \sqrt{1-\gamma_1}\sqrt{1-\gamma_2}\sqrt{1-\kappa_1}\sqrt{1-\kappa_2}e^{\frac{\alpha}{2}L - jk_n L}} \tag{1.62}$$

Drop port,

$$\frac{E_{t2}}{E_{i1}} = \frac{-\sqrt{1-\gamma_1}\sqrt{1-\gamma_2}\sqrt{\kappa_1\kappa_2}e^{\frac{\alpha L}{22}jk_n\frac{L}{2}}}{1-\sqrt{1-\gamma_1}\sqrt{1-\gamma_2}\sqrt{1-\kappa_1}\sqrt{1-\kappa_2}e^{\frac{\alpha}{2}L-jk_nL}} \quad (1.63)$$

The intensity relations for the throughput and drop ports can be obtained by normalizing the transfer functions in Eqs (1.62) and (1.63), which are given by

$$\frac{I_{t1}}{I_{i1}} = \left|\frac{E_{t1}}{E_{i1}}\right|^2$$

$$= \frac{1-(1-\gamma_1)\kappa_1-2\sqrt{1-\gamma_1}\sqrt{1-\kappa_1}\sqrt{1-\gamma_2}\sqrt{1-\kappa_2}e^{\frac{\alpha}{2}L}\cos(k_nL)-(1-\gamma_2)(1-\kappa_2)e^{-\alpha L}}{1+(1-\gamma_1)(1-\gamma_2)(1-\kappa_1)(1-\kappa_2)e^{-\alpha L}-2\sqrt{1-\gamma_1}\sqrt{1-\gamma_2}\sqrt{1-\kappa_1}\sqrt{1-\kappa_2}e^{\frac{\alpha}{2}L}\cos(k_nL)} \quad (1.64)$$

$$\frac{I_{t2}}{I_{i1}} = \left|\frac{E_{t2}}{E_{i1}}\right|^2$$

$$= \frac{(1-\gamma_1)(1-\gamma_2)\kappa_1\kappa_2 e^{\frac{\alpha}{2}L}}{1+(1-\gamma_1)(1-\gamma_2)(1-\kappa_1)(1-\kappa_2)e^{-\alpha L} - 2\sqrt{1-\gamma_1}\sqrt{1-\gamma_2}\sqrt{1-\kappa_1}\sqrt{1-\kappa_2}e^{-\frac{\alpha}{2}L}\cos(k_nL)} \quad (1.65)$$

For simplification, the calculation of the intensity relation does not take into account the coupling losses ($\gamma$) and it is neglected. For simplify, we given by

$$x = \exp\left(-\frac{\alpha}{2}L\right)$$
$$c_1 = \sqrt{1-\kappa_1} \quad (1.66)$$
$$c_1 = \sqrt{1-\kappa_2}$$

The intensity relations Eqs (1.64) and (1.65) are then given by

$$\frac{I_{t1}}{I_{i1}}(\phi) = \left|\frac{E_{t1}}{E_{i1}}\right|^2 = 1 - \frac{(1-c_1^2)(1-c_2^2 x^2)}{(1-c_1c_2x)^2 + 4c_1c_2x\sin^2\left(\frac{\phi}{2}\right)} \quad (1.67)$$

$$\frac{I_{t2}}{I_{i1}}(\phi) = \left|\frac{E_{t2}}{E_{i1}}\right|^2 = 1 - \frac{(1-c_1^2)(1-c_2^2)x}{(1-c_1c_2x)^2 + 4c_1c_2x\sin^2\left(\frac{\phi}{2}\right)} \quad (1.68)$$

Figure 1.22 shows the transmission spectrum of add/drop with the radius as 136 μm (Si waveguide), two coupling coefficients, $\kappa_1 = \kappa_2 = 0.1$, $A_{eff} = 1$ μm$^2$, and free spectral range, FSR = 1 THz.

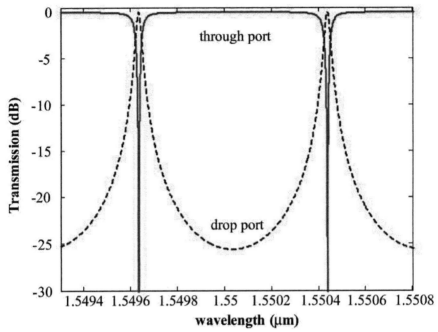

**FIGURE 1.22** Transmission spectrum and wavelength of single ADF, when the ADF radius, $R = 136$ μm, two coupling coefficients, $\kappa_1 = \kappa_2 = 0.1$, and free spectral range, FSR = 1 THz.

A ring resonator can be used in two configurations, an all-pass filter (APF) [83] when it is coupled into a single optical input–output bus as shown in Figure 1.23(a), or with two coupled waveguides making it essentially an add–drop filter (ADF) [84, 85] as shown in Figure 1.23(b). The resonator

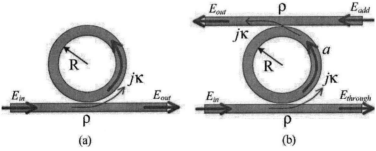

**FIGURE 1.23** Shows (a) an all-pass filter (APF) based on a single-microring resonator; (b) an add-drop filter (ADF) based on a single-microring resonator.

is characterized by a circumference ($L$), effective index ($n_{\text{eff}}$), group index ($n_g = n_{\text{eff}} + \omega dn_{\text{eff}}/d\omega$), and an amplitude coupling coefficient ($\kappa$). Disregarding the loss, the transmission function of the APF is given by

$$\frac{E_{\text{out}}(\omega)}{E_{\text{in}}(\omega)} = \frac{\rho - e^{-j(\omega - \omega_0)\tau}}{1 - \rho e^{-j(\omega - \omega_0)\tau}} \equiv e^{-j\Phi(\omega)} \tag{1.69}$$

where the resonant angular frequency ($\omega_0$) is determined from the condition $n_{\text{eff}}\omega_0 L/c = 2m\pi$, $\rho^2 = 1 - \kappa^2$ and $\tau = Ln_g/c$ is the ring round-trip time. As the name "all pass" implies, the amplitude transmission of the loss-free MR is unity for all frequencies and only the phase is affected as

$$\tan \Phi(\omega) = \frac{\kappa^2 \sin[(\omega - \omega_0)\tau]}{(1 + \rho)^2 \cos[(\omega - \omega_0)\tau] - 2\rho} \tag{1.70}$$

The group delay of the APF is

$$T_d(\omega) = \frac{\partial \Phi(\omega)}{\partial \omega} = \frac{(1 - \rho^2)\tau}{1 = \rho^2 - 2\rho\cos(\omega - \omega_0)\tau}$$

$$= T_d^{(0)} + T_4^{(2)}(\omega - \omega_0)^2 + \cdots \tag{1.71}$$

where the group delay at the resonant frequency is

$$T_d^{(0)} = \tau \frac{1 + \rho}{1 - \rho} \tag{1.72}$$

and the group delay dispersion (GDD) is

$$T_d^{(2)} = -\tau^3 \frac{\rho(1 + \rho)}{(1 - \rho)^3} = -[T_d^{(0)}]^3 \frac{\rho}{(1 + \rho)^2} \tag{1.73}$$

When a small loss per ring round-trip $\alpha_R$ is included, which results from a combination of absorption, scattering, and bending losses in the ring, the expression for the transmission of the APF becomes

$$\left|\frac{E_{\text{out}}(\omega)}{E_{\text{in}}(\omega)}\right|^2 = \frac{1 - \alpha_R + \rho^2 - 2\sqrt{1 - \alpha_R}\,\rho\cos(\omega - \omega_0)\tau}{1 + (1 - \alpha_R)\rho^2 - 2\sqrt{1 - \alpha_R}\,\rho\cos(\omega - \omega_0)\tau}$$

$$= 1 - \frac{\alpha_R(1 - \rho^2)}{1 + (1 - \alpha_R)\rho^2 - 2\sqrt{1 - \alpha_R}\,\rho\cos(\omega - \omega_0)\tau} \tag{1.74}$$

For an APF with small loss we can write from Eqs (1.23) and (1.20)

$$\left|\frac{E_{\text{out}}(\omega)}{E_{\text{in}}(\omega)}\right|^2 = 1 - \alpha_R \frac{T_d(\omega)}{\tau} \tag{1.75}$$

which makes perfect sense indicating that the total loss is equal to the per ring round-trip loss multiplied by the average number of round trips the light makes inside the MR.

Figure 1.24(a) and (b) shows the phase and group delay characteristics of an APF with circumference $L = 36$ µm, effective index $n_g = 4.16$ ($\tau = 500$ fs) and $\kappa = 0.3$ [86, 87], while Figure 1.24(c) shows the transmission characteristic of the same APF with a 0.04 dB round-trip loss. As one can see, large group delays (exceeding $\tau$ by a factor of 40) can be achieved in a relatively narrow bandwidth near $\omega_0$. In fact, one can estimate the full width at half maximum (FWHM) of the transmission and group delay characteristics as

$$\Delta\omega_{APF} = \frac{2}{\tau}\frac{1-\rho}{\rho^{1/2}} = \frac{2}{T_d^{(0)}}\frac{1+\rho}{\rho^{1/2}} \quad (1.76)$$

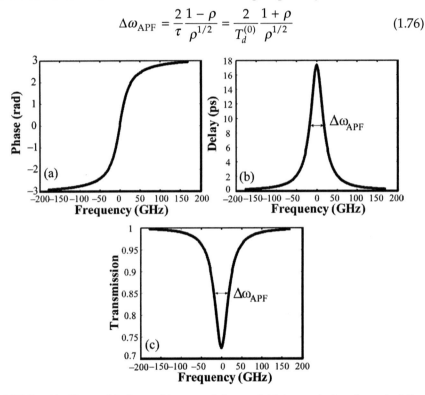

**FIGURE 1.24** Shows (a) phase, (b) group delay, and (c) transmission characteristics of the APF of Figure 1.23(a)

This provides the simple relation between the group delay and the width of the resonance

$$\Delta\omega_{APF} T_d^{(0)} = 2\frac{1+\rho}{\rho^{1/2}} \quad (1.77)$$

or simply $\Delta f_{APF} T_d(0) \approx 2/\pi \approx 0.64$ for resonators with small coupling coefficient $\kappa$. Note that using simple power series expansion in Eq. (1.71) one can obtain the expression for the FWHM of APF as

$$\Delta\omega_{APF} \approx \left(\frac{2T_d^{(0)}}{-T_d^{(2)}}\right)^{1/2} = \frac{2^{1/2}}{T_d^{(0)}}\frac{1+\rho}{\rho^{1/2}} \quad (1.78)$$

which differs from the exact equation only by a factor of $2^{1/2}$. Now, intuitively the strength of any optical effect, linear or nonlinear, increases with the interaction time, i.e., the time the light spends inside the resonator. Therefore the resonance enhancement in the MR can only come at the expense of the bandwidth, with the product of the two staying constant.

Let us now turn our attention to the ADF of Figure 1.23(b) whose transmission coefficient is

$$\frac{E_{out}(\omega)}{E_{in}(\omega)} = \frac{-\kappa^2 e^{-j(\omega - \omega_0)\tau/2}}{1 - \rho^2 e^{-j(\omega - \omega_0)\tau}} \quad (1.79)$$

and has the Fabry–Pérot-like intensity spectrum shown in Figure 1.25(a). The corresponding FWHM is given by

$$\Delta\omega_{ADF} = 2\kappa^2 \tau^{-1} \quad (1.80)$$

The delay time of the signal is

$$T_d(\omega) = \frac{\partial \Phi(\omega)}{\partial \omega} = \frac{(1 - \rho^2)\rho^2}{1 - 2\rho^2 \cos(\omega - \omega_0)\tau + \rho^4}, \quad (1.81)$$

i.e., the time spent inside the resonator is shown in Figure 1.25(b) and reaches to maximum at resonance:

$$T_d^{(0)} = \tau \frac{\rho^2}{1 - \rho^2} = \tau \frac{1 - \kappa^2}{\kappa^2} \quad (1.82)$$

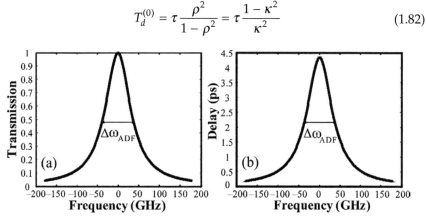

**FIGURE 1.25** Shows (a) transmission characteristic and (b) group delay of the ADF of Figure 1.23(b).

From Eqs (1.29) and (1.31) we obtain the simple relationship between delay and pass-band width, which is given by

$$T_d^{(0)} \Delta\omega_{ADF} = 2(1 - \kappa^2) \quad (1.83)$$

For weakly coupled resonators $\Delta f_{ADF} T_d^{(0)} \approx 1/\pi \approx 0.32$, i.e., one-half of the delay-bandwidth of the APF. No matter what the application of MR is, the longer the time the light spends inside the resonator, the stronger is the effect one can impose on the light. Therefore Eq. (1.79) would lead to a gain-bandwidth product for any application.

## 1.6 A Nonlinear Microring Resonator

The non-linear response indicated in this section is seen only when the wavelength is closed to a microring resonance and on the long-wavelength side of the resonance. The response can be understood by considering how the gain saturation mechanism interacts with the ring resonances. Consider the case when the input wavelength is slightly red-shifted from the resonance peak. Since it is closed to the resonance, the optical intensity in the microring resonator builds up. The microring is biased such that the quantum wells are in inversion and there is optical gain. Consequently the built-up intensity in the resonator undergoes optical amplification by stimulated recombination as it travels around the microring. The effect of the stimulated recombination is to decrease carrier density in the quantum wells, which in turn affects the refractive index. The refractive index increases due to (a) change in the plasma frequency of the electron/hole gas with carrier density (b) change in imaginary part of the refractive index (absorption/gain) and consequent effect on real part through the Kramers-Kroenig relation [88]. The rising refractive index causes the microring resonances to shift towards the longer wavelength side or closer to the input wavelength. This in turn means that the intensity in the ring builds up even further and there is a net positive feedback. The positive feedback results in the resonance moving all the way to the input wavelength and past, till the input now lies on the shorter wavelength side of the resonance.

Once the resonance moves past the input wavelength, any further increase in input intensity only causes the resonance to move further away and the intensity amplification factor in the resonator drops. This 'snapping' of the resonance to the right is what causes the first sharp transition in the transfer response. When the input intensity begins to drop, the resonance peak begins moving back towards the shorter wavelengths and hence closer to the input wavelength. This has the effect of higher intensity amplification within the microring and hence a negative feedback. Consequently, the intensity in the microring remains more or less constant as the input intensity decreases. Below a certain threshold input intensity, when the resonance is aligned exactly with the input wavelength, the negative feedback effect no longer occurs and the resonance 'snaps' back due to positive feedback. This 'snapping' to the left is what causes the second sharp transition in the output waveform. To summarize, the positive feedback occurs when the input wavelength is red-shifted with respect to the resonance and negative feedback occurs when blue-shifted. The whole phenomenon is inverted when absorption saturation is used as the nonlinear effect instead of gain-saturation. Consequently, the non-linear transfer response is seen when the input wavelength is on the shorter wavelength side of the microring resonance. The rest of this section deals with modeling the steady-state non-linear response from first principles and recreating the non-linear transfer function and bistability seen in the experiments [89]. The starting point for the model is the microring optical intensity relations reproduced here for convenience, we obtain

$$I_t = I_{in} \cdot \frac{\tau^2 + a^2 - 2a\tau\cos(\phi)}{1 + a^2\tau^2 - 2a\tau\cos(\phi)} \quad (1.84)$$

$$I_R = I_{in} \cdot \frac{\kappa^2}{1 + a^2\tau^2 - 2a\tau\cos(\phi)} \quad (1.85)$$

Here, $I_R$, $I_t$, and $I_{in}$ are optical intensities in the microring, at the through-port and input respectively. In a linear device, the round-trip gain ($a$) and phase ($\phi$) are independent of the intensity within the microring. However, for the non-linear system, the round-trip gain is given by

$$a = \exp\left[-\frac{(\Gamma g L - \alpha_{loss})}{2}\right] \quad (1.86)$$

The optical gain in the quantum wells ($g$) is dependent on the optical intensity in the microring and is reproduced here,

$$g = \frac{g_0}{1 + I_R/I_{sat}} \quad (1.87)$$

Here, $g_0$ is the unsaturated gain in the quantum well, when there is zero optical intensity. The round trip phase ($\phi$) is also dependent upon intensity in the ring and is given by

$$\phi = \phi_0 + \frac{2\pi}{\lambda}\Gamma \cdot \Delta n(I_R) L \quad (1.88)$$

Here $\phi_0$ is the initial detuning from resonance. $\Delta n(I_R)$ is the intensity induced refractive index change in the quantum wells and is given by the relation for gain saturation induced index shift and reproduced here, which is given by

$$\Delta n(I_R) = \frac{\alpha_H \gamma g_0}{4\pi} \cdot \frac{\frac{I_g}{I_{sat}}}{1 + \frac{I_R}{I_{sat}}} \quad (1.89)$$

A system of non-linear equations that needs to be solved numerically for a given input optical intensity ($I_{in}$), microring parameters ($\kappa$, $\phi_0$, $\Gamma$), material parameters ($\alpha_H$, $g_0$) and wavelength ($\lambda$). Once the solution for intensity in the microring, roundtrip gain and phase is obtained, the output power can be calculated. This was implemented to various parameters such as intensity in microring, output intensity, phase-change and round-trip gain plotted as a function of the input intensity. Figure 1.26 shows the results from such a calculation. The various parameters used in the calculation are indicated in Table 1.1.

**Table 1.1** Parameters used in the simulation of optical nonlinearity and bistability shown in Figure 1.26

| Parameters | Values |
|---|---|
| Wavelength ($\lambda$) | 1550 nm |
| Microring (microdisk) radius ($R$) | 20 μm |
| Coupling coefficient ($\kappa$) | 0.6 |
| Loss in ring ($\alpha$) | 15 cm$^{-1}$ |
| Linewidth enhancement factor ($\alpha_H$) | 3 |
| Unsaturated gain ($g_0$) | 1000 cm$^{-1}$ |
| Overlap factor ($\Gamma$) | 0.05 |
| Initial detuning from resonance ($\phi_0$) | −0.2 rad |

All the intensities in the graphs in Figure 1.26 are normalized to the saturation intensity ($I_{sat}$), the estimated value of about 64 kW/cm$^2$ is seen in Table 1.1. Figure 1.26(a) shows the intensity in the ring as a function of the input intensity. Figure 1.26(b) shows the round-trip gain ($a$) as a function of the input intensity. The lasing threshold is indicated by a dashed line. Figure 1.26(c) shows the round-trip phase. It indicates the relative position of the resonance with respect to the input wavelength. They are coincident when $\phi = 0$ (indicated by a dashed line). The movement of the resonance is seen to corroborate the physical process described earlier in this section. The last Figure 1.26(d) shows the output intensity as a function of the input. The blue curves denote the path taken when the input intensity is increasing and the red curve denotes the path taken on the way down as indicated by the arrows.

The simulation was repeated with lower gain. The parameters used are listed in Table 1.2. Figure 1.27 shows the plot of the various parameters as a function of input intensity for the case of lower gain. As previously stated, Figure 1.27(a) shows the intensity in the microring. The bistable

**Table 1.2** Parameters used in the simulation of optical nonlinearity and bistability shown in Figure 1.27

| Parameters | Values |
|---|---|
| Wavelength ($\lambda$) | 1550 nm |
| Microring (microdisk) radius ($R$) | 20 μm |
| Coupling coefficient ($\kappa$) | 0.4 |
| Loss in ring | 15 cm$^{-1}$ |
| Linewidth enhancement factor ($\alpha_H$) | 3 |
| Unsaturated gain ($g_0$) | 1000 cm$^{-1}$ |
| Overlap factor ($\Gamma$) | 0.05 |
| Initial detuning from resonance ($\phi_0$) | −0.3 rad |

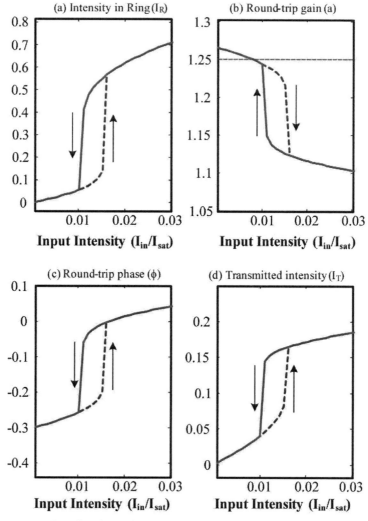

**FIGURE 1.26** Simulated non-linear response in a microring (microdisk) using parameters from Table 1.1. All intensities are normalized with respect to the saturation intensity ($I_{sat}$), where (a) intensity in disk/ring, (b) round-trip gain, (c) round-trip phase and (d) transmitted intensity [90].

operation and direction have not changed from what they were earlier. Figure 1.27(b) shows the round-trip gain as a function of the input intensity. Notice how the gain drops from above unity to below indicating net loss in the round-trip. Figure 1.27(c) and (d) shows the round-trip phase and transmitted intensity respectively. The transmitted intensity is seen to drop with increasing input intensity beyond a threshold due to a corresponding drop in the round-trip gain. Also, the direction of bistability is seen to change to clockwise as was observed in the experiment [90].

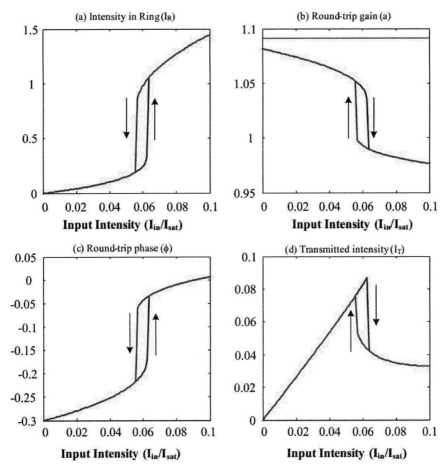

**FIGURE 1.27** Simulated non-linear response in a microring (microdisk) using parameters from Table 1.2. All intensities are normalized with respect to the saturation intensity ($I_{sat}$), in applications (a) intensity in disk/ring, (b) roundtrip gain, (c) round-trip phase and (d) transmitted intensity [90].

## 1.7 A Nonlinear Add/Drop Filter

Modern optical processing systems require the design and development of fast, small and compact all-optical functional devices. Silicon-based photonic devices with a high index contrast promise to meet this criterion [91–92]. The nonlinear switching in a microring resonator has been experimentally demonstrated by several research groups [93–97], because it can be used for a wide variety of photonic functions [98–103]. However, a unified analytical theory for optical bistability inside microring resonator has recently developed [104]. When light enters the silicon material, it can be absorbed through two-photon absorption (TPA) and results in a change in the refractive index effect proportional to the light intensity (Kerr effect). Through the TPA process, free carriers are excited which result in additional absorption (Free Carrier Absorption, FCA) and an associated index change (Free Carrier Dispersion, FCD). After a while, these carriers will recombine, resulting in a thermal refractive index change (Thermo-Optic effect, TOE).

In this section, we theoretically analyze the impact of linear loss, TPA, FCA, FCD, and TOE effects simultaneously on the nonlinear switching in the double-coupler nonlinear silicon-on-insulator (SOI) ring resonator. We start from general equations describing evolution of the electric fields associated with the continuous-wave (CW) propagating inside the SOI ring resonator. We then solve analytically the input-output correlation function for the case in which the effect of TPA and TOE are negligible. We subsequently consider the progress of electric field when TPA and TOE cannot be ignored. Finally, we present a new switching scheme exploiting the impact of the free carrier lifetime on the input-output relation. This scheme is appropriate for all-optical switching with fixed input intensity.

A schematic diagram of a nonlinear ring resonator and details of the notation used is shown in Figure 1.28. Let us consider a CW signal at an angular frequency $\omega$ propagating inside a straight SOI waveguide coupled laterally to a silicon ring of radius $R$. The evolution of electric field $E(z)$ associated with this optical wave is given by [104]

$$\frac{1}{A}\frac{dA}{dz} = -\frac{\alpha}{2} - \left(\frac{\beta}{2} - i\gamma\right)|A|^2 - \left(\frac{\xi_r}{2} + i\xi_i\right)|A|^4 \quad (1.90)$$

where $A(z)$ is the complex amplitude (with $|A^2|$ representing intensity) related to the electric field as $E(z) = \bar{\omega}A(z)\exp(ib_0 z)$; $\bar{\omega} = (\mu_0/\varepsilon_0)_{1/4}(2n_0)^{1/2}$; $\mu_0$ and $\varepsilon_0$ are, respectively, the intrinsic permeability and permittivity of vacuum; $n_0$ is the linear refractive index, $\beta_0 = n_0 k$ is the propagation constant with $k = \omega/c$; and $c$ is the speed of light in vacuum. The used parameters in Eq. (1.39) are $\alpha$, $\beta$, $\gamma = kn_2$ linear losses, TPA, and the Kerr effect, where effects $n_2$ being the nonlinear Kerr parameter. The parameters $\xi_r$ and $\xi_i$ are accounted for the free-carrier effects, which they are defined as

$$\xi_r = (1.45 \times 10^{-21})\left(\frac{\lambda}{1.55 \times 10^{-6}}\right)^2 \frac{\tau\beta\lambda}{2hc} \quad (1.91a)$$

$$\xi_i = (5.3 \times 10^{-27})\left(\frac{\lambda}{1.55 \times 10^{-6}}\right)^2 \frac{\tau\beta}{2hc} - \frac{k\beta\kappa\vartheta}{C\rho} \quad (1.91b)$$

where $\kappa$, $\vartheta$, $C$, and $\rho$ are the thermo-optic coefficient, thermal dissipation time, thermal capacity, and density of silicon, respectively. Here, the second term of Eq. (1.91b) vanishes, when the TOE is absence.

The evolution of the electric field of Eq. (1.90) with $\beta = 0$ in the form $A(z) = \sqrt{I(z)}\exp[i\phi(z)]$, along both the straight and ring waveguide in Figure 1.28, is given by

$$I(z) = \frac{I_0 \exp(-\alpha z)}{\sqrt{1 + I_0^2(\xi_r/\alpha)[1 - \exp(-2\alpha z)]}} \quad (1.92a)$$

$$\phi(z) = \phi_0 + \gamma I_0 L_{\text{eff}}(z) - \frac{\xi_i}{\xi_r}\left(\ln\frac{I_0}{I(z)} - \alpha z\right) \quad (1.92b)$$

$$L_{\text{eff}}(z) = \frac{\tan^{-1}\left[I_0\sqrt{\xi_r/\alpha}\right] - \tan^{-1}\left[I(z)\sqrt{\xi_r/\alpha}\right]}{I_0\sqrt{\alpha\xi_r}} \quad (1.92c)$$

where $I_0$ and $\phi_0$ are the values of intensity and phase at $z = 0$, respectively and $L_{\text{eff}}$ is the generalized effective length.

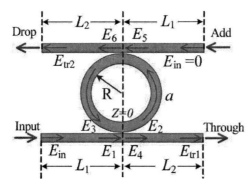

**FIGURE 1.28** A schematic diagram of the silicon ring resonator and details of the notation used.

The coupling between optical waves propagating in the ring and straight waveguides is assumed to be combined together at point $z = 0$ (see Figure 1.28), then the electric field relations can be expressed by

$$E_4 = rE_1 + itE_3 \quad (1.93)$$

$$E_2 = itE_1 + rE_3 \quad (1.94)$$

$$E_3 = rE(2\pi R) \quad (1.95)$$

$$E_6 = itE(\pi R) \quad (1.96)$$

$$E_5 = 0 \quad (1.97)$$

where $t = \sqrt{1 - r^2}$, and $r^2$ is the fraction of power remaining in the straight waveguide after the coupler. Without loss of generality, we assume that the coupling constant of both couplers are the same values. Substituting Eq. (1.95) into Eqs (1.93) and (1.94) and eliminating $E_1$, we can express the intensity $I_4 = |E_4|^2/\omega^2$ in terms of $I_0 = |E^2|2/\omega^2$ as

$$I_4 = \left[ r^2[I_0 + I(2\pi R) - 2\sqrt{I_0 I(2\pi R)} \cos \Delta\phi]/(1-r^2) \right] \quad (1.98)$$

where $\Delta\phi = 2\pi\beta_0 R + \phi(2\pi R) - \phi_0$ is the phase shift acquired during one round trip within the ring.

A similar relation for the intensity $I_6$ can be obtained by using Eqs (1.96) and (1.97)

$$I_6 = t^2 I(\pi R) = (1 - r^2) I(\pi R) \quad (1.99)$$

From energy conservation law, we obtain

$$I_1 = I_0 + I_4 - [r^2 I(2\pi R)] \quad (1.100)$$

If $L_1$ and $L_2$ are lengths of the straight waveguides before and after the coupling points (see Figure 1.28), the intensities at the input and output ends of the straight waveguides can be related to $I_1$ and $I_4$, which is given by

$$I_{in}(I_0) = \frac{I_1(I_0)\exp(\alpha L_1)}{\sqrt{1 + I_1^2(I_0)(\xi_r/\alpha)[1 - \exp(2\alpha L_1)]}} \quad (1.101a)$$

$$I_{tr1}(I_0) = \frac{I_4(I_0)\exp(-\alpha L_2)}{\sqrt{1 + I_4^2(I_0)(\xi_r/\alpha)[1 - \exp(-2\alpha L_2)]}} \quad (1.101b)$$

$$I_{tr2}(I_0) = \frac{I_6(I_0)\exp(-\alpha L_2)}{\sqrt{1 + I_6^2(I_0)(\xi_r/\alpha)[1 - \exp(-2\alpha L_2)]}} \quad (1.101c)$$

Equation (1.101) gives the evolution of the input–output relationship in a parametric form, where the parameter $I_0$ lies in the interval $[0, +\infty]$. This parametric formulation offers the multiple input–output value $I_{in}(I_{tr})$ possibility for a fixed value of $I_0$.

In the presence of TPA, the solutions of Eq. (1.90) in the implicit form is given by [104]

$$2\alpha z + \beta I_0 L_{eff}(z) = \ln \frac{\alpha I^{-2}(z) + \beta I^{-1}(z) + \xi_r}{\alpha I_0^{-2} + \beta I_0^{-2} + \xi_r} \quad (1.102a)$$

$$\phi(Z) = \phi_0 + \gamma I_0 L_{eff}(z) - \frac{\xi_i}{\xi_r}\left[\ln \frac{I_0}{I(z)} - \alpha z - \frac{\beta}{2\gamma} L_{eff}(z)\right] \quad (1.102b)$$

where

$$L_{eff}(z) = \frac{q}{\beta I_0 \ln}\left[\frac{qK(z)+1}{qK(z)-1}\frac{qK(0)+1}{qK(0)-1}\right] \quad (1.102c)$$

with $q = (1 - 4\alpha\xi_r/\beta^2)^{-1/2}$, and $K(z) = 1 + 2(\xi_r/\beta)I(z)$.

## 1.7.1 The effect of TPA

Figure 1.29 shows the bistable behavior of the throughput port [Figure 1.29(a))] and the drop port [Figure 1.29(b)] of a silicon ring resonator with 5 μm radius and set $L_1 = L_2 = 10$ μm, $n_0 = 3.484$, $\alpha = 1$ dB/cm, $\beta = 0.5$ cm/GW, $n_2 = 6 \times 10^{-5}$ cm$^2$/GW, and $\tau = 1$ ns. This figure demonstrates agreement between the analytical (black circles) and the exact numerical results (circles). In this case, the contribution of TPA cannot be ignored, numerical simulations do not reveal any qualitative difference between the true bistable response of the ring resonators and that predicted by Eq. (1.101). Therefore, all the findings based on results are valid in the presence of TPA, and can thus be considered justification under general conditions. One can see that the output intensity at the drop port is a bit higher than throughput port, the drop port has wider hysteresis loop than the throughput port. Input part of the curve with negative slope represents unstable branch. Black arrows show the abrupt changes occurring in the outputs at the boundaries of unstable regions A(B).

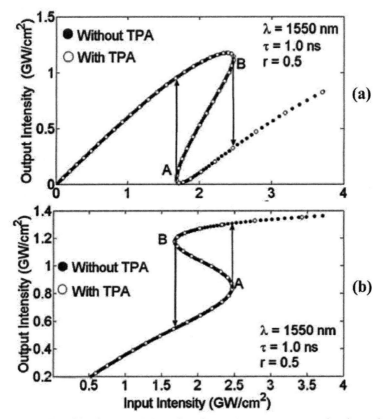

**FIGURE 1.29** Bistable characteristics of a silicon ring resonator at the throughput channel (a) and at the drop channel and (b) in the absence of TOE effect. Other parameters are shown in the text.

## 1.7.2 The Impact of Coupling Coefficient

The evolutions of the transmitted intensity versus input intensity by varying the coupling coefficient $r$ in the absence of TOE, are illustrated in Figure 1.30. This figure shows the bistable behavior of a silicon ring resonator with 5 μm radius and set the $L_1 = L_2 = 10$ μm, $n_0 = 3.484$, $\alpha = 1$ dB/cm, $\beta = 0.5$ cm/GW, $n_2 = 6 \times 10^{-5}$ cm$^2$/GW, and $\tau = 1$ ns. One can see that, optical bistability is unfeasible for the coupling coefficient $r$ below 0.40 approximately. As the coupling coefficient $r$ is increased, the switching power increases, and also increasing the hysteresis loops, in which a good discrimination between the high and low output intensity levels is achieved. It is shown that the transmission intensity at the drop port is higher than throughput port, in which the drop port has wider hysteresis than the throughput port.

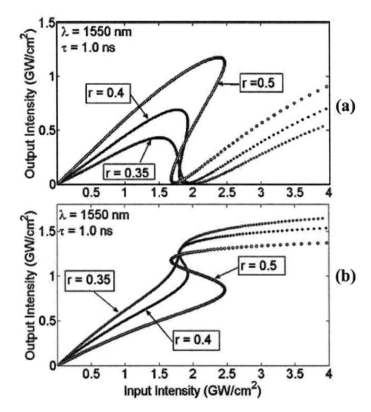

**FIGURE 1.30** The effect of coupling coefficients on the input–output characteristics at throughput port (a) and drop port (b). Other parameters are the same as in Figure 1.29.

In case, the shape of the bistability curve and the hysteresis loop of the throughput port are similar to a single-couple ring resonator [104] and that of the drop port are similar to those Fabry–Pérot resonators [105].

### 1.7.3 The Impact of Wavelength

The bistable behavior of silicon ring resonators depend dramatically on the operating wavelength $\lambda$. As illustrated in Figure 1.31, the decreases in $\lambda$ from 1550 to 1545 nm nearly triples the hysteresis width, but further decreasing of $\lambda$ up to 1540 nm results in the collapse of the bistability phenomenon. This situation is similar to a single couple ring resonator, where the reduction of wavelength results in a collapse of bistability.

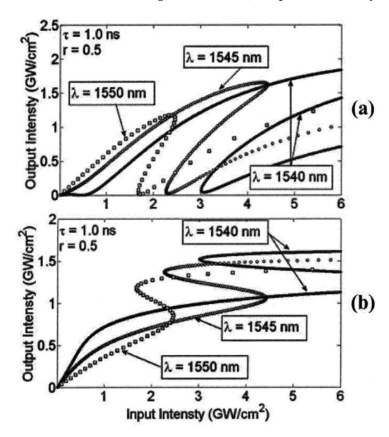

**FIGURE 1.31** Bistable curves at three operating wavelengths of the throughput port (a) and drop port (b). Other parameters are the same as in Figure 1.29.

### 1.7.4 The impact of TOE

Next, we consider the impact of the TOE which excluded from the increase in the refractive index resulting from heat generation due to linear losses. Figure 1.32 shows $I_{tr}$ as a function of $I_{in}$ in the presence of TOE. The presence of the TOE reduces the input intensities at which bistable switching occurs and results in optical multistability.

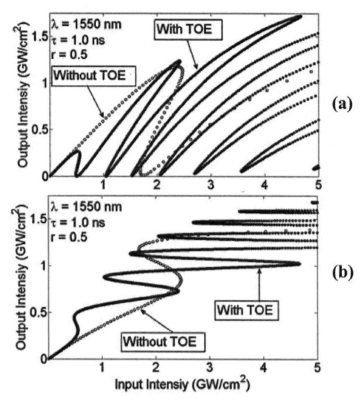

**FIGURE 1.32** The influence of TOE on the bistable behaviors at the throughput (a) and drop port. Other parameters are the same as in Figure 1.29.

### 1.7.5 The Impact of Photon Lifetime

Finally, we present a scheme for optical switching of the output ports at a fixed input intensity by varying free carrier lifetime $\tau$. If the system in Figure 1.33(a) is initially in the state $a$ for $I_{in} \approx 1.7$ GW/cm$^2$, changing $\tau$ from 2.15 ns → 0.975 ns → 2.15 ns will switch the system up and then back to state $a$ again through the transition path $a \to b \to c \to d \to a$. Similarly, by changing $\tau$ from 0.975 ns → 0.45 ns → 0.975 ns will switch the system from $e$ and then back to initial state $e$ again through the path $e \to f \to g \to h \to e$. Figure 1.33(b) shows the process to switch the system up and then return to its initial state using the same values of $\tau$ as in Figure 1.33(a). If the system in Figure 1.33(b) is initially in the state $a$ for $I_{in} \approx 1.7$ GW/cm$^2$, changing $\tau$ from 2.15 ns → 0.975 ns → 2.15 ns will switch the system up and then back to state $a$ again through the transition path $a \to b \to c \to a$. Similarly, changing $\tau$ from 0.975 ns → 0.45 ns → 0.975 ns will switch the system from $d$ and then back to initial state $d$ again through the path $d \to e \to f \to d$. This switching scheme may lead to optical logic gates, and memories.

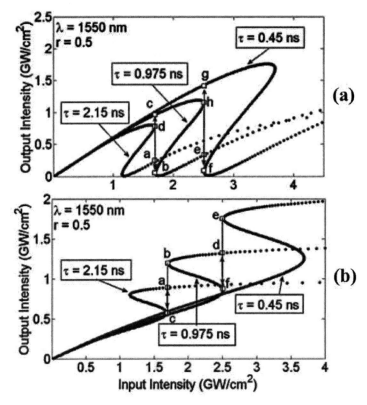

**FIGURE 1.33** The optical switching can be obtained by varying photon lifetime. Bistable curves the throughput (a) and drop port (b). Other parameters are the same as in Figure 1.29.

## 1.8 Conclusion

A novel simulated characteristics and the comparison between the two complementary RR add/drop filters have been presented. The graphical approach is used in the analysis for derivation of the optical transfer functions. The treatment of the complementary circuits covers the analysis, both the transmission and the group delay. It also includes the finesse, the resonant transmittance and the Q-factor. All characteristics are characterized by these parameters, two types of losses: waveguide ring loss and coupler loss and the coupling coefficient and the ring radius. The optimum characteristics, such as a cross talk lower than 20 dB, a high finesse and Q-factor and group delay can be obtained with symmetric coupling coefficients $\kappa_1 = \kappa_2 < 0.2$ for Figure 1.3 and $\kappa_1 = \kappa_2 > 0.8$ for Figure 1.4. In addition, the influence of the optical loss in the ring resonators is also analyzed and found that it deteriorates all characteristics. Finally, the characteristics of the two complementary circuits have the same performances when $\kappa_i$ and $1 - \kappa_i$ are interchanged.

We have proposed a new attempt to employ a graphical approach in the analytical derivation of the optical transfer functions of MRR Vernier filters. The graphical approach with an SFG is used in our analysis for fast derivation of the optical transfer functions. MRR filters that exploit the Vernier effect were investigated which opens the possibility to realize a larger FSR than would be achieved using only a single-ring resonator. The suppression of interstitial resonances of the DRR Vernier was improved by decreasing the coupling coefficient of the center coupler. We have noted increasing interstitial resonances for the ratios of larger integer values. The DRR Vernier filter does not adequately suppress the interstitial resonances that lie between the main resonances separated by the FSR. A solution to this problem can be obtained by using a TRR Vernier filter. In addition, the TRR Vernier filter can be designed to obtain twice the FSR of the DRR Vernier filter.

We have studied the performance of MRs as optical switches operating at high bit rates. We have considered two alternative schemes. One is a single-resonator optical switch where the resonance frequency is shifted causing the change in transmission. The other is a "slow-light" scheme where the phase shift is accumulated as the light passes through a coupled resonator structure. The nonlinear switching features of silicon-based double coupler ring resonator in the presence of the linear losses, TPA, FCA, and FCD effects and compared their behaviors with those of single coupler ring resonator. It was shown that the general features of the nonlinear switching are similar to a single-couple ring resonator. We found that the domain of the nonlinear switching can be controlled by the several parameters which provide multiple options for observing bistable behavior in silicon ring resonators and ensures their broad applications. We believe the present analysis will be useful in a future experimental verification of optical switching in SOI ring resonators and in the device design.

# REFERENCES

[1] H. Rokhsari, T.J. Kippenberg, T. Carmon, and K.J. Vahala, "Theoretical and experimental study of radiation pressure-induced mechanical oscillations (parametric instability) in optical microcavities," *IEEE J. Sel. Top. Quantum Electron.*, **12**(1), 96–107 (Jan. 2006).

[2] T.J. Kippenberg, S.M. Spillane, B.Min, and K.J. Vahala, "Theoretical and experimental study of stimulated and cascaded Raman scattering in ultrahigh-Q optical microcavities," *IEEE J. Sel. Top. Quantum Electron.*, **10**(5), 1219–1228 (Sept. 2004).

[3] T.J. Kippenberg, H. Rokhsari, T. Carmon, A. Scherer, and K.J. Vahala, "Analysis of radiation-pressure induced mechanical oscillation of an optical microcavity," *Phys. Rev. Lett.*, **95**, 033901 (15 July 2005).

[4] P.D. Haye, A. Schliesser, O. Arcizet, T. Wilken, R. Holzwarth, and T.J. Kippenberg, "Optical frequency comb generation from a monolithic microresonator," *Nature*, **450**, 1214–1217 (Dec. 2007).

[5] G. Anetsberger, R. Rivière, A. Schliesser, O. Arcizet, and T.J. Kippenberg, "Ultralow-dissipation optomechanical resonators on a chip," *Nat. Photon.*, **2**, 627–633 (Oct. 2008).

[6] T. Carmon, T.J. Kippenberg, L. Yang, H. Rokhsari, S. Spillane, and K.J. Vahala, "Feedback control of ultra-high-Q microcavities: application to micro-Raman lasers and microparametric oscillators," *Opt. Express*, **13**(9), 3558–3566 (May 2005).

[7] P.D. Haye, O. Arcizet, A. Schliesser, R. Holzwarth, and T.J. Kippenberg, "Full stabilization of a microresonator-based optical frequency comb," *Phys. Rev. Lett.*, **101**, 053903 (1 Aug. 2008).

[8] S.M. Spillane, T.J. Kippenberg, K.J. Vahala, K.W. Goh, E. Wilcut, and H.J. Kimble, "Ultrahigh-Q toroidal microresonators for cavity quantum electrodynamics," *Phys. Rev. A*, **71**, 013817 (2005).

[9] F. Michael, N. Jan, P. Tobias, B. Jens, W. Thorsten, M. Christian, and K. Heinrich, "High-speed all-optical switching in ion-implanted silicon-on-insulator microring resonators," *Opt. Lett.* **32**, 2046–2048 (2007).

[10] V.R. Almeida and M. Lipson, "Optical bistability on a silicon chip," *Opt. Lett.* **29**, 2387–2389 (2004).

[11] Y. Dumeige and P. Féron, "Dispersive tristability in microring resonator," *Phys. Rev. E.* **72**, 066609 (2005).

[12] P. Rabiei, W.H. Steier, C. Zhang, and L.R. Dalton, "Polymer microring filters and modulators," *J. Lightw. Technol.*, **20**, 1968–1975 (2002).

[13] Q. Xu, B. Schimdt, S. Pradhan, and M. Lipson, "Micrometre-scale silicon electro-optic modulator," *Nature* **435**, 325–327 (2005).

[14] H. Tazawa, Y.-H. Kuo, I. Dunayevskiy, J. Luo, A.K.-Y. Jen, H.R. Fetterman, and W.H. Steier, "Ring resonator-based electro-optic polymer traveling-wave modulator," *J. Lightw. Technol.* **24**, 3514–3519 (2006).

[15] T.A. Ibrahim, K. Amarnath, L.C. Kuo, R. Grover, V. Van, and P.-T. Ho, "Photonic logic NOR gate based on two symmetric microring resonators," *Opt. Lett.* **29**, 2779–2781 (2004).

[16] T. A. Ibrahim, R. Grover, L.-C. Kuo, S. Kanakaraju, L.C. Calhoun, and P.-T. Ho, "All-optical AND/NAND logic gates using semiconductor microresonators," *IEEE Photon. Technol. Lett.* **15**, 1422–1424 (2003).

[17] V. Van, T.A. Ibrahim, P.P. Absil, F.G. Johnson, R. Grover, and P.-T. Ho, "Optical signal processing using nonlinear semiconductor microring resonators," *IEEE J. Sel. Top. Quant. Electron.* **8**, 705–713 (2002).

[18] C.-Y. Chao and L.J. Guo, "Biochemical sensors based on polymer microrings with sharp asymmetrical resonance," *Appl. Phys. Lett.* **83**, 1527–1529 (2003).

[19] A. Yalc, K.C. Popat, J.C. Aldridge, T.A. Desai, J. Hryniewicz, N. Chbouki, B.E. Little, O. King, V. Van, S. Chu, D. Gill, M.A. Washburn, M. Selim, and B.B. Goldberg, "Optical sensing of biomolecules using microring resonators," *IEEE J. Sel. Top. Quantum Electron.* **12**, 148–154 (2006).

[20] K.D. Vos, I. Bartolozzi, E. Schacht, P. Bienstman, and R. Baets, "Silicon-on-Insulator microring resonator for sensitive and label-free biosensing," *Opt. Express* **15**, 7610–7615 (2007).

[21] F. Xia, L. Sekaric, and Y. Vlasov, "Ultracompact optical buffers on a silicon chip," *Nat. Photon.*, **1**, 65–71 (2007).

[22] J.K.S. Poon, J. Scheuer, Y. Xu, and A. Yariv, "Designing coupled-resonator optical delay lines," *J. Opt. Soc. B* **21**, 1665–1673 (2004).

[23] F. Morichetti, A. Melloni, A. Breda, A. Canciamilla, C. Ferrari, and M. Martinelli, "A reconfigurable architecture for continuously variable optical slow-wave delay lines," *Opt. Express* **15**, 17273–17282 (2007).

[24] A. Melloni, F. Morichetti, C. Ferrari, and M. Martinelli, "Continuously tunable 1 byte delay in coupled-resonator optical waveguides," *Opt. Lett.* **33**, 2389–2391 (2008).

[25] F. Morichetti, A. Melloni, C. Ferrari, and M. Martinelli, "Error-free continuously-tunable delay at 10 Gbit/s in a reconfigurable on-chip delay-line," *Opt. Express* **16**, 8395–8405 (2008).

[26] J.B. Khurgin, "Optical buffers based on slow light in electromagnetically induced transparent media and coupled resonator structures: comparative analysis," *J. Opt. Soc. Am. B* **22**, 1062–1074 (2005).

[27] Y.M. Landobasa and M.K. Chin, "Optical buffer with higher delay-bandwidth product in a two-ring system," *Opt. Express* **16**, 1796–1807 (2008).

[28] J.V. Hryniewicz, P.P. Absil, B.E. Little, R.A. Wilson, and P.-T. Ho, "Higher order filter response in coupled microring resonators," *IEEE Photon. Technol. Lett.* **12**, 320–322 (2000).

[29] A. Melloni, "Synthesis of a parallel-coupled ring-resonator filter," *Opt. Lett.* **26**, 917–919 (2001).

[30] B.E. Little, S.T. Chu, P.P. Absil, J.V. Hryniewicz, F.G. Johnson, F. Seiferth, D. Gill, V. Van, O. King, and M. Trakalo, "Very high-order microring resonator filters for WDM applications," *IEEE Photon. Technol. Lett.* **16**, 2263–2265 (2004).

[31] T. Barwicz, M.A. Popovic, M.R. Watts, P.T. Rakich, E.P. Ippen, and H.I. Smith, "Fabrication of add/drop filters based on frequency-matched microring resonators," *J. Lightw. Technol.* **24**, 2207–2218 (2006).

[32] F. Xia, M. Rooks, L. Sekaric, and Y. Vlasov, "Ultra-compact high order ring resonator filters using submicron silicon photonic wires for on-chip optical interconnects," *Opt. Express* **15**, 11934–11941 (2007).

[33] S. Xiao, M. Khan, H. Shen, and M. Qi, "Silicon-on-Insulator microring add-drop filters with free spectral ranges over 30 nm," *J. Lightw. Technol.* **26**, 228–236 (2008).

[34] S. Xiao, M. Khan, H. Shen, and M. Qi, "A highly compact third-order silicon microring add-drop filter with a very large free spectral range, a flat pass-band and a low dispersion," *Opt. Express* **15**, 14765–14771 (2007).

[35] Y.M.T. Landobasa, P. Dumon, R. Baets, and M.K. Chin, "Boxlike filter response based on complementary photonic bandgap in two-dimensional microresonator arrays," *Opt. Lett.* **33**, 2512–2514 (2008).

[36] L. Yang, T. Carmon, B. Min, S.M. Spillane, and K.J. Vahala, "Erbium-doped and Raman microlasers on a silicon chip fabricated by the sol-gel process," *Appl. Phys. Lett.* **86**, 091114 (2005).

[37] L. Yang, D.K. Armani, and K.J. Vahala, "Fiber-coupled Erbium microlasers on a chip," *Appl. Phys. Lett.* **83**, 825–826 (2003).

[38] H.-B. Lin, and A.J. Campillo, "CW nonlinear optics in droplet microcavities displaying enhanced gain," *Phys. Rev. Lett.* **73**, 2440–2443 (1994).

[39] S. Spillane, T. Kippenberg, and K.J. Vahala, "Ultralow-threshold Raman laser using a spherical dielectric microcavity," *Nature* **415**, 621–623 (2002).

[40] B. Min, L. Yang, and K. Vahala, "Controlled transition between parametric and Raman oscillations in ultrahigh-Q silica toroidal microcavities," *Appl. Phys. Lett.* **87**, 181109 (2005).

[41] T.J. Kippenberg, S.M. Spillane, D.K. Armani, and K.J. Vahala, "Ultralow-threshold microcavity Raman laser on a microelectronic chip," *Opt. Lett.* **29**, 1224–1227 (2004).

[42] K.J. Vahala, "Optical microcavities," *Nature* **424**, 839–846 (2003).

[43] C. Fabry and A. Pérot, "Théorie et applications d'une nouvelle méthode de spectroscopie interférentielle," *Ann. Chim. Phys.* **16**, 115 (1899).

[44] J.D. Joannopoulos, S.G. Johnson, J.N. Winn, and R.D.M. Joannopoulos, *Photonic Crystals: Molding the Flow of Light*, second edition, Princeton: Princeton University Press (2008).

[45] J. Vûckovíc, O. Painter, Y. Xu, A. Yariv, and A. Scherer, "Finite-difference time-domain calculation of the spontaneous emission coupling factor in optical microcavities," *IEEE J. Quant. Electron.* **35**, 1168–1175 (1999).

[46] O. Painter, J. Vuckovic, and A. Scherer, "Defect modes of a two-dimensional photonic crystal in an optically thin dielectric slab," *J. Opt. Soc. Am. B* **16**, 275–285 (1999).

[47] Y. Xu, J.S. Vuckovic, R.K. Lee, O.J. Painter, A. Scherer, and A. Yariv, "Finite-difference time-domain calculation of spontaneous emission lifetime in a microcavity," *J. Opt. Soc. Am. B* **16**, 465–474 (1999).

[48] M. Pu, L. Liu, W. Xue, Y. Ding, L.H. Frandsen, H. Ou, K. Yvind, and J.M. Hvam, "Tunable microwave phase shifter based on silicon-on-insulator microring resonator," *IEEE Photon. Technol. Lett.*, **22**(12), 869–871 (June 15, 2010).

[49] J. Heebner, A.V. Wong, A. Schweinsberg, R.W. Boyd, and D.J. Jackson, "Optical transmission characteristics of fiber ring resonators," *IEEE J. Quantum Electron.* **40**(6), 726–730 (2004).

[50] P.P. Yupapin, P. Saeung, and C. Li, "Characteristics of complementary ring-resonator add/drop filters modeling by using graphical approach," *Opt. Comm.*, **272**, 81–86 (2007).

[51] S. Park, S.-S. Kim, L. Wang, and S.-T. Ho, "Single-mode lasing operation using a microring resonator as a wavelength selector," *IEEE J. Quantum Electron.* **38**, 270–273 (2002).

[52] R.W. Boyd and J.E. Heebner, "Sensitive Disk Resonator Photonic Biosensor," *Appl. Opt.* **40**, 5742–5747 (2001).

[53] K. Djordjev, S.-J. Choi, and P.D. Dapkus, "Vertically coupled InP microdisk switching devices with electroabsorptive active regions," *IEEE Photon. Technol. Lett.* **14**, 1115–1117 (2002).

[54] S. Suzuki, Y. Hatakeyama, Y. Kokubun, and S.T. Chu, "Precise control of wavelength channel Spacing of microring resonator add–drop filter array," *J. Lightw. Technol.* **20**, 745–750 (2002).

[55] B. Liu, A. Shakouri, and J.E. Bowers, "Wide tunable double ring resonator coupled lasers," *IEEE Photon. Technol. Lett.* **14**, 600–602 (2002).

[56] J. Azana and L.R. Chen, "Multiwavelength optical signal processing using multistage ring resonators," *IEEE Photon. Technol. Lett.* **14**, 654–656 (2002).

[57] H. Takahashi, R. Inohara, M. Hattori, K. Nishimura, and M. Usami, "Expansion of compensation bandwidth of tunable dispersion compensator based on ring resonators," *Electron. Lett.* **40**, 1014–1015 (2004).

[58] B.E. Little, S.T. Chu, H.A. Haus, J.S. Foresi, and J.-P. Laine, "Microring resonator channel dropping filters," *IEEE J. Lightwave Technol.* **15**, 998–1005 (1997).

[59] B.E. Little, J.S. Foresi, G. Steinmeyer, E.R. Thoen, S.T. Chu, H.A. Haus, E.P. Ippen, L.C. Kimerling, and W. Greene, "Ultra-compact Si–SiO$_2$ microring resonator optical channel dropping filters," *IEEE Photon. Technol. Lett.* **10**, 549–551 (1998).

[60] V. Van, P.P. Absil, J.V. Hryniewicz, and P.-T. Ho, "Propagation loss in single-mode GaAs–AlGaAs microring resonators: measurement and model," *J. Lightwave Technol.* **19**, 1734–1739 (2001).

[61] R. Grover, P.P. Absil, V. Van, J.V. Hryniewicz, B.E. Little, O. King, L.C. Calhoun, F.G. Johnson, and P.-T. Ho, "Vertically coupled GaInAsP-InP microring resonators," *Opt. Lett.* **26**, 506–508 (2001).

[62] C.K. Madsen and J.H. Zhao, "A general planar waveguide autoregressive optical filter," *J. Lightw. Technol.* **14**, 437–447 (1996).

[63] M.K. Chin and S.T. Ho, "Design and modeling of waveguide-coupled single-mode microring resonators," *J. Lightwave Technol.* **16**, 1433 (1998).

[64] A. Vorckel, M. Monster, W. Henscel, P.H. Bolivar, and H. Kurz, "Asymmetrically coupled silicon-on-insulator microring resonators for compact add–drop multiplexers," *IEEE Photon. Technol. Lett.* **15**, 921–923 (2003).

[65] O. Schwelb, "Generalized analysis for a class of linear interferometric networks—Part II: simulations," *IEEE Trans. Microwave Theory Tech.* **46**, 1409–1418 (1998).

[66] J. Capmany, M.A. Muriel, S. Sales, J.J. Rubio, and D. Pastor, "Microwave V-I Transmission Matrix Formalism for the Analysis of Photonic Circuits: Application to Fiber Bragg Gratings," *J. Lightw. Technol.* **21**, 3125–3134 (2003).

[67] B. Moslehi, J.W. Goodman, M. Tur, and H.W. Shaw, "Fiber-optic lattice signal processing," *Proc. IEEE* **72**, 909–930 (1984)..

[68] Y.H. Ja, "Generalized theory of optical fiber loop and ring resonators with multiple couplers. 2: General characteristics," *Appl. Opt.* **29**, 3524–3529 (1990).

[69] A. Rostami and G. Rostami, "Full-optical realization of tunable low pass, high pass and band pass optical filters using ring resonators," *Opt. Comm.* **240**, 133–151 (2004).

[70] S.J. Mason, "Feedback Theory: Further Properties of Signal Flow Graphs," *Proc. IRE*, **44**, 920–926 (1956).

[71] C.K. Madsen and J.H. Zhao, *Optical Filter Design and Analysis: A Signal Processing Approach*, Wiley & Sons, Inc. (1999).

[72] K. Ogusu, "Dynamic behavior of reflection optical bistability in a nonlinear fiber ring resonator," *IEEE J. Quantum Electron.* **32**, 1537–1543 (1996).

[73] Y. Imai and T. Tamura, "Coherence effect on nonlinear dynamics in fiber-optic ring resonator," *Opt. Comm.* **195**, 259–265 (2001).

[74] L.N. Binh, N.Q. Ngo, and S.F. Luk, "Graphical representation and analysis of the Z-shaped double-coupler optical resonator," *J. Lightw. Technol.*, **11**, 1782–1792 (Nov. 1993).

[75] C. Vazquez, S.E. Vargas, and J.M.S. Pena, "Sagnac loop in ring resonators for tunable optical filters" *J. Lightwave Technol.* **23**, 2555–2567 (2005).

[76] P. Saeung and P.P. Yupapin, "Vernier effect of multiple-ring resonator filters modeling by a graphical approach," *Opt. Eng.*, **46**(7), 075005 (July 2007).

[77] S. Suzuki, O. Kazuhiro, and H. Yoshinori, "Integrated-optic double-ring resonators with a wide free spectral range of 100 GHz," *IEEE J. Lightwave Technol.* **13**, 1766–1771 (1995).

[78] K. Oda, N. Takato, and H. Toba, "A wide-FSR waveguide double-ring resonator for optical FDM transmission systems," *IEEE J. Lightwave Technol.* **9**, 728–736 (1991).

[79] C.K. Madsen and J.H. Zhao, *Optical Filter Design and Analysis: A Signal Processing Approach*, New York: John Wiley & Sons, Inc. (1999).

[80] O. Schwelb, "Generalized analysis for a class of linear interferometric networks. Part I: analysis," *IEEE Trans. Microwave Theory Tech.* **46**, 1399–1408 (1998).

[81] J. Capmany and M.A. Muriel, "A new transfer matrix for the analysis of fiber ring resonators: compound coupled structures for FDMA demultiplexing," *IEEE J. Lightwave Technol.* **8**, 1904–1919 (1990).

[82] S.J. Mason, "Feedback theory—further properties of signal flow graphs," *Proc. IRE* **44**, 920–926 (1956).

[83] C.K. Madsen and G. Lenz, "Optical all-pass filters for phase response design with applications for dispersion compensation," *IEEE Photon. Technol. Lett.* **10**, 994–996 (1998).

[84] S. Xiao, M.H. Khan, H. Shen, and M. Qi, "Multiple-channel silicon micro-resonator based filters for WDM applications," *Opt. Express* **15**, 7489–7498 (2007).

[85] S.N. Magdalena, L. Tao, W. Xuan, and R.P. Roberto, "Tunable silicon microring resonator with wide free spectral range," *Appl. Phys. Lett.* **89**, 071110 (2006).

[86] F. Xia, L. Sekaric, and Y. Vlasov, "Ultra-compact optical buffers on a silicon chip," *Nat. Photon.* **1**, 65–71 (2006).

[87] J.K. Poon, L. Zhu, G.A. DeRose, and A. Yariv, "Transmission and group delay of microring coupled-resonator optical waveguides," *Opt. Lett.* **31**, 456–458 (2006).

[88] B.R. Bennett, R.A. Soref, and J.A. Del Alamo, "Carrier-induced change in refractive index of InP, GaAs and InGaAsP," *IEEE J. Quantum Electron.*, **26**, 113–22 (1990).

[89] M.J. Adams, H.J. Westlake, M.J. O'Mahony, and I.D. Henning, "A comparison of active and passive optical bistability in semiconductors," *IEEE J. Quantum Electron.*, **QE-21**, 498–504 (1985).

[90] K. Amarnath, "Active microring and microdisk optical resonators on indium phosphide" Ph. D. Thesis, University of Maryland (2006).

[91] L. Pavesi and D.J. Lockwood, *Silicon Photonics*, New York: Springer (2004).

[92] R.A. Soref, "The Past, Present, and Future of Silicon Photonics," *IEEE J. Sel. Top. Quantum Electron.*, **12**, 1678–1687 (2006).

[93] T.A. Ibrahim, R. Grover, L.C. Kuo, S. Kanakaraju, L.C. Calhoun, and P.T. Ho, "All-optical AND/NAND logic gates using semiconductor microresonators," *IEEE Photon. Technol. Lett.*, **15**, 1422–1424 (2003).

[94] C.Y. Chao, W. Fung, and L.J. Guo, "Polymer Microring Resonators for Biochemical Sensing Applications," *IEEE J. Sel. Top. Quantum Electron.*, **12**, 134–142 (2006).

[95] Q. Xu and M. Lipson, "Carrier-induced optical bistability in silicon ring resonators," *Opt. Lett.*, **31**, 341–343 (2006).

[96] M. Forst, J. Niehusmann, T. Plotzing, J. Bolten, T. Wahlbrink, C. Moormann, and H. Kurz, "High-speed all-optical switching in ion-implanted silicon-on-insulator microring resonators," *Opt. Lett.*, **32**, 2046–2048 (2007).

[97] Q. Xu, D. Fattal, and R.G. Beausoleil, "Silicon microring resonators with 1.5-µm radius," *Opt. Express*, **16**, 4309-4315 (2008).

[98] T.A. Ibrahim, K. Amarnath, L.C. Kuo, R. Grover, V. Van, and P.T. Ho, "Photonic logic NOR gate based on two symmetric microring resonators," *Opt. Lett.*, **29**, 2779-2781 (2004).

[99] G. Priem, P. Dumon, W. Bogaerts, D.V. Thourhout, G. Morthier, and R. Baets, "Optical bistability and pulsating behaviour in silicon-on-insulator ring resonator structures," *Opt. Express*, **13**, 9623–9628 (2005).

[100] H. Ma, X. Zhang, Z. Jin, and C. Ding, "Waveguide-type optical passive ring resonator gyro using phase modulation spectroscopy technique," *Opt. Eng.*, **45**, 080506 (2006).

[101] Q. Xu and M. Lipson, "All-optical logic based on silicon microring resonators" *Opt. Express*, **15**, 924–929 (2007).

[102] A.C. Turner, M.A. Foster, A.L. Gaeda, and M. Lipson, "Ultra-low power frequency conversion in silicon microring resonators," *Proc. Conf. Lasers Electro-Optics* (OSA, Washington, DC, paper CPDA3, (2007).

[103] N. Pornsuwancharoen, S. Chaiyasoonthorn, and P.P. Yupapin, "Fast and slow light generation using chaotic signals in nonlinear microring resonators for communication security," *Opt. Eng.*, **48**, 015002 (2009).

[104] I.D. Rukhlenko, M. Premaratne, and G.P. Agrawal, "Analytical study of optical bistability in silicon ring resonators,"*Opt. Lett.*, **35**, 55–57 (2010).

[105] I.D. Rukhlenko, M. Premaratne, and G.P. Agrawal, "Analytical study of optical bistability in silicon-waveguide resonators,"*Opt. Express*, **17**, 22124–22137 (2009).

# 2

# A PANDA Ring Resonator

**CHAPTER OUTLINE**

- Introduction
- Theory and Modeling
- Dynamic Pulse Propagation
- Symmetry and Asymmetry PANDA Ring Resonators
- Random Binary Code Generation
- Binary Code Suppression and Recovery
- Conclusion
- References

## 2.1 Introduction

In this chapter, we have derived and presented the dynamic behavior of dark-bright soliton collision within the modified add/drop filter, which it is known as PANDA ring resonator. By using the dark-bright soliton conversion control, the obtained outputs of the dynamic states can be used to form the random binary codes, which can be available for communication security application. Results obtained have shown that the random binary codes can be formed by using the polarized light components. The retrieved (decoded) codes can be obtained by using the dark-bright soliton conversion signals.

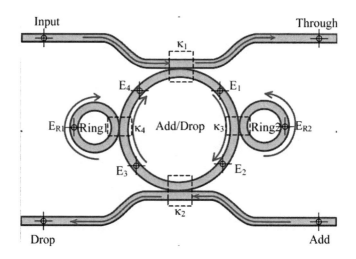

**FIGURE 2.1** A schematic diagram of single-PANDA ring resonator

## 2.2 Theory and Modeling

The proposed system consists of an add/drop filter and double nanoring resonators known as a **PANDA** ring resonator as shown in Figures 2.1 and 2.2. To perform the dark-bright soliton conversion, dark and bright solitons are input into the add/drop optical filter system. The input optical field ($E_{i1}$) and the control port optical field ($E_{con}$) of the dark-bright solitons pulses are given by [1]

$$E_{i1}(t) = A\tanh\left[\frac{T}{T_0}\right]\exp\left[\left(\frac{x}{2L_D}\right) - i\phi(t)\right] \quad (2.1a)$$

$$E_{con}(t) = A\operatorname{sech}\left[\frac{T}{T_0}\right]\exp\left[\left(\frac{x}{2L_D}\right) - i\phi(t)\right] \quad (2.1b)$$

in which $A$ and $z$ are the optical field amplitude and propagation distance, respectively.

$$\phi(t) = \phi_0 + \phi_{NL} = \phi_0 + \frac{2\pi n_2 L}{A_{eff}\lambda}|E_0(t)|^2$$

is the random phase term related to the temporal coherence function of the input light, $\phi_0$ is the linear phase shift, $\phi_{NL}$ is the nonlinear phase shift, $n_2$ is the nonlinear refractive index of InGaAsP/InP waveguide. The effective mode core area of the device is given by $A_{eff}$, $L = 2\pi R_{ad}$, $R_{ad}$ is the radius of device, $\lambda$ is the input wavelength light field and $E_0(t)$ is the circulated field within nanoring coupled to the right and left add/drop optical filter system as shown in Figure 2.1. $T$ is a soliton pulse propagation time in a frame moving at the group velocity, $T = t - \beta_1 z$, where $\beta_1$ and $\beta_2$ are the coefficients of the linear and second-order terms of Taylor expansion of the propagation constant. $L_D=T_0^2/|\beta_2|$ is the dispersion length of the soliton pulse. $T_0$ in equation is a soliton pulse propagation time at initial input (or soliton pulse width), where $t$ is the soliton phase shift time, and the frequency shift of the soliton is $\omega_0$. This solution describes a pulse that keeps its temporal width invariance as it propagates, and is thus called a temporal soliton. When a soliton peak intensity $(|\beta_2/\Gamma T_0^2|)$ is given, the value of propagation time for initial input $T_0$ is known. For the soliton pulse in the microring device, a balance should be achieved between the dispersion length ($L_D$) and the nonlinear length ($L_{NL}=1/\Gamma\phi_{NL}$). $\Gamma = n_2k_n$ is the length scale over which dispersive or nonlinear effects makes the beam become wider or narrower. For a soliton pulse, there is a balance between dispersion and nonlinear lengths, hence $L_D = L_{NL}$.

When light propagates within the nonlinear medium, the refractive index ($n$) of light within the medium is given by

$$n = n_0 + n_2 I = n_0 + \frac{n_2}{A_{eff}} P \qquad (2.2)$$

with $n_0$ and $n_2$ as the linear and nonlinear refractive indexes, respectively. $I$ and $P$ are the optical intensity and the power, respectively. The effective mode core area of the device is given by $A_{eff}$. For the add/drop optical filter design, the effective mode core areas range from 0.50 to 0.10 μm². Parameters were obtained by using the related practical material parameters (InGaAsP/InP) [2–4]. When a dark soliton pulse is input and propagated within a add/drop optical filter as shown in Figure 2.1, the resonant output is formed.

Figure 2.1 consists of add/drop optical multiplexing used for generated random binary coded light pulse and add/drop optical filter device for decoded binary code signal. The resonator output field, $E_{t1}$ and $E_1$ consists of the transmitted and circulated components within the add/drop optical multiplexing system, which can perform the driven force to photon/molecule/atom.

When the input light pulse passes through the first coupling device of the add/drop optical multiplexing system, the transmitted and circulated components can be written as

$$E_{t1} = \sqrt{1-\gamma_1}\left[\sqrt{1-\kappa_1}E_{i1} + j\sqrt{\kappa_1}E_4\right] \qquad (2.3)$$

$$E_1 = \sqrt{1-\gamma_1}\left[\sqrt{1-\kappa_1}E_4 + j\sqrt{\kappa_1}E_{i1}\right] \qquad (2.4)$$

$$E_2 = E_{R2}e^{-\frac{\alpha}{2}\frac{L}{2}-jk_n\frac{L}{2}} \qquad (2.5)$$

where $\kappa_1$ and $\gamma_1$ are the intensity coupling coefficient and the fractional coupler intensity loss of the add/drop optical filter, respectively. $\alpha$ is the attenuation coefficient, $k_n = 2\pi/\lambda$ is the wave propagation number, $\lambda$ is the input wavelength light field and $L = 2\pi R_{ad}$, $R_{ad}$, is the radius of add/drop device.

For the second coupler of the add/drop optical multiplexing system,

$$E_{t2} = \sqrt{1-\gamma_2}\left[\sqrt{1-\kappa_2}\,E_{i2} + j\sqrt{\kappa_2}\,E_2\right] \quad (2.6)$$

$$E_3 = \sqrt{1-\gamma_2}\left[\sqrt{1-\kappa_2}\,E_2 + j\sqrt{\kappa_2}\,E_{i2}\right] \quad (2.7)$$

$$E_4 = E_{R1}e^{-\frac{\alpha L}{2 2}-jk_n\frac{L}{2}} \quad (2.8)$$

where $\kappa_2$ is the intensity coupling coefficient, $\gamma_2$ is the fractional coupler intensity loss. The circulated light fields, $E_{R1}$ and $E_{R2}$ are the light field circulated components of the nanoring radii, $R_1$ and $R_2$ which coupled into the left and right sides of the add/drop optical multiplexing system, respectively. The light field transmitted and circulated components in the right nanoring, $R_2$, are given by

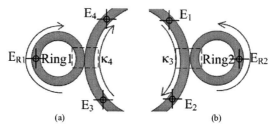

**FIGURE 2.2** A schematic diagram of PANDA ring, where (a) the left nanoring and (b) the right nanoring.

$$E_{R2} = \sqrt{1-\gamma_3}\left[\sqrt{1-\kappa_3}\,E_1 + j\sqrt{\kappa_3}\,E_{r2}\right] \quad (2.9)$$

$$E_{r1} = \sqrt{1-\gamma_3}\left[\sqrt{1-\kappa_3}\,E_{r2} + j\sqrt{\kappa_3}\,E_1\right] \quad (2.10)$$

$$E_{r2} = E_{r1}e^{-\frac{\alpha}{2}L_2-jk_nL_2} \quad (2.11)$$

where $\kappa_3$ and $\gamma_3$ are the intensity coupling coefficient and the fractional coupler intensity loss of the right nanoring, respectively. $\alpha$ is the attenuation coefficient, $k_n = 2\pi/\lambda$ is the wave propagation number, $\lambda$ is the input wavelength light field and $L_2 = 2\pi R_2$, $R_2$ is the radius of right nanoring.

From Eqs (2.9)–(2.11), the circulated roundtrip light fields of the right nanoring radii, $R_2$, are given in Eqs (2.12) and (2.13), respectively.

$$E_{r1} = \frac{j\sqrt{1-\gamma_3}\sqrt{\kappa_3}\,E_1}{1-\sqrt{1-\gamma_3}\sqrt{1-\kappa_3}\,e^{-\frac{\alpha}{2}L_2-jk_nL_2}} \quad (2.12)$$

$$E_{r2} = \frac{j\sqrt{1-\gamma_3}\sqrt{\kappa_3}\,E_1 e^{-\frac{\alpha}{2}L_2-jk_nL_2}}{1-\sqrt{1-\gamma_3}\sqrt{1-\kappa_3}\,e^{-\frac{\alpha}{2}L_2-jk_nL_2}} \quad (2.13)$$

Thus, the output circulated light field, $E_{R2}$, for the right nanoring is given by

$$E_{R2} = E_1 \left\{ \frac{\sqrt{(1-\gamma_3)(1-\kappa_3)} - (1-\gamma_3)e^{-\frac{\alpha}{2}L_2 - jk_n L_2}}{1 - \sqrt{(1-\gamma_3)(1-\kappa_3)}e^{-\frac{\alpha}{2}L_2 - jk_n L_2}} \right\} \quad (2.14)$$

Similarly, the output circulated light field, $E_{R1}$, for the left nanoring at the left side of the add/drop optical multiplexing system is given by

$$E_{R1} = E_3 \left\{ \frac{\sqrt{(1-\gamma_4)(1-\kappa_4)} - (1-\gamma_4)e^{-\frac{\alpha}{2}L_1 - jk_n L_1}}{1 - \sqrt{(1-\gamma_4)(1-\kappa_4)}e^{-\frac{\alpha}{2}L_1 - jk_n L_1}} \right\} \quad (2.15)$$

where $\kappa_4$ is the intensity coupling coefficient, $\gamma_4$ is the fractional coupler intensity loss, $\alpha$ is the attenuation coefficient, $k_n = 2\pi/\lambda$ is the wave propagation number, $\lambda$ is the input wavelength light field and $L_1 = 2\pi R_1$, $R_1$ is the radius of left nanoring.

From Eqs (2.3)–(2.15), the circulated light fields, $E_1$, $E_3$ and $E_4$ are defined by given $x_1 = (1-\gamma_1)^{1/2}$, $x_2 = (1-\gamma_2)^{1/2}$, $y_1 = (1-\kappa_1)^{1/2}$, and $y_2 = (1-\kappa_2)^{1/2}$.

$$E_1 = \frac{jx_1\sqrt{\kappa_1}E_{i1} + jx_1x_2y_1\sqrt{\kappa_2}E_{R1}E_{i2}e^{-\frac{\alpha L}{44} - jk_n\frac{L}{4}}}{1 - x_1x_2y_1y_2E_{R2}E_{R1}e^{-\frac{\alpha}{2}L - jk_n L}} \quad (2.16)$$

$$E_3 = x_2y_2E_{R2}E_1 e^{-\frac{\alpha L}{42} - jk_n\frac{L}{2}} + jx_2\sqrt{\kappa_2}E_{i2} \quad (2.17)$$

$$E_4 = x_2y_2E_{R2}E_{R1}E_1 e^{-\frac{\alpha}{2}L - jk_n L} + jx_2\sqrt{\kappa_2}E_{R1}E_{i2}e^{-\frac{\alpha L}{44} - jk_n\frac{L}{4}} \quad (2.18)$$

Thus, from Eqs (2.3), (2.5), (2.16)–(2.18), the output optical field of the through port ($E_{t1}$) is expressed by

$$E_{t1} = x_1y_1E_{i1} + \left(jx_1x_2y_2\sqrt{\kappa_1}E_{R2}E_{R1}E_1 - x_1x_2\sqrt{\kappa_1\kappa_2}E_{R1}E_{i2}\right)e^{-\frac{\alpha L}{44} - jk_n\frac{L}{4}} \quad (2.19)$$

The power output of the through port ($P_{t1}$) is written by

$$P_{t1} = (E_{t1}) \cdot (E_{t1})^*$$

$$= \left| x_1y_1E_{i1} + \left(jx_1x_2y_2\sqrt{\kappa_1}E_{R2}E_{R1}E_1 - x_1x_2\sqrt{\kappa_1\kappa_2}E_{R1}E_{i2}\right)e^{-\frac{\alpha L}{44} - jk_n\frac{L}{4}} \right|^2 \quad (2.20)$$

Similarly, from Eqs (2.5), (2.6), (2.16)–(2.18), the output optical field of the drop port ($E_{t2}$) is given by

$$E_{t2} = x_2y_2E_{i2} + jx_2\sqrt{\kappa_2}E_{R2}E_1 e^{-\frac{\alpha L}{44} - jk_n\frac{L}{4}} \quad (2.21)$$

The power output of the drop port ($P_{t2}$) is expressed by

$$P_{t2} = (E_{t2}) \cdot (E_{t2})^* = \left| x_2 y_2 E_{i2} + jx_2\sqrt{\kappa_2} E_{R2} E_1 e^{-\frac{\alpha L}{44} - jk_n \frac{L}{4}} \right|^2 \quad (2.22)$$

In order to retrieve the required signals, we propose to use the add/drop optical multiplexing device with the appropriate parameters. This is given in the following details. The optical circuits of a PANDA ring resonator for the through port and drop port can be given by Eqs (2.20) and (2.22), respectively. The chaotic noise cancellation can be managed by using the specific parameters of the add/drop multiplexing device. The required signals can be retrieved by the specific users. $\kappa_1$ and $\kappa_2$ are the coupling coefficients of the add/drop filters, $k_n = 2\pi/\lambda$ is the wave propagation number in a vacuum, and the waveguide (ring resonator) loss is $\alpha = 5 \times 10^{-5}$ dB mm$^{-1}$. The fractional coupler intensity loss is $\gamma = 0.01$. In the case of the add/drop multiplexing device, the nonlinear refractive index is neglected. Figure 2.3 shows the a schematic diagram of single-PANDA ring by using OptiFDTD commercial software. The dynamic pulse train is generated in z-direction of InGaAsP/InP waveguide with $n_0 = 3.34$ by using OptiFDTD. As shown in Figure 2.4, (a) z = 0, (b) z = 0.88 µm, (c) z = 1.62 µm, (d) z = 2.74 µm, (e) z = 3.30 µm, (f) z = 3.44 µm, (g) z = 5.72 µm, (h) z = 8.10 µm, (i) z = 9.06 µm, (j) z = 10.7 µm, (k) z = 10.0 µm, and (l) z = 11.0 µm. The results are obtained by the through (Th) and drop

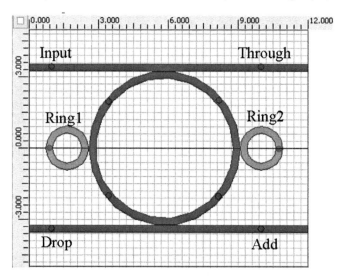

**FIGURE 2.3** A schematic diagram of single-PANDA ring with a 10 × 12 µm² size, which is drawn by using the OptiFDTD commercial software.

A PANDA Ring Resonator **59**

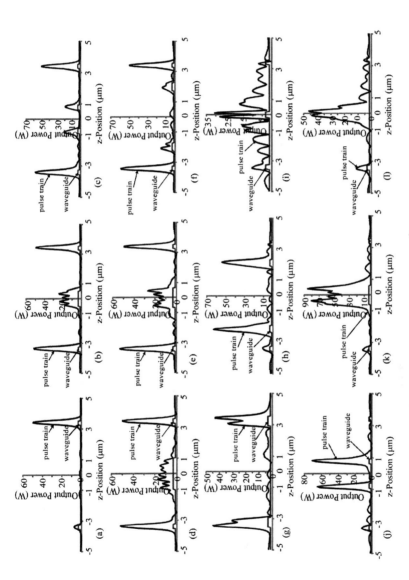

**FIGURE 2.4** Dynamic pulse in z-direction of PANDA ring size $10 \times 12$ μm$^2$ by using OptiFDTD, where (a) $z = 0$, (b) $z = 0.88$ μm, (c) $z = 1.62$ μm, (d) $z = 2.74$ μm, (e) $z = 3.30$ μm, (f) $z = 3.44$ μm, (g) $z = 5.72$ μm, (h) $z = 8.10$ μm, (i) $z = 9.06$ μm, (j) $z = 10.7$ μm, (k) $z = 10.0$ μm and (l) $z = 11.0$ μm.

(Dr) ports as shown in Figure 2.5. We found that the output power at the drop port is higher than the through port, which means that the required and the transmitted signals are obtained, in which the delay signals within left and right nanoring are seen as shown in Figure 2.6.

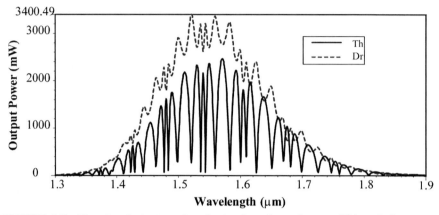

**FIGURE 2.5** The simulation results obtained at through port (Th) and drop port (Dr).

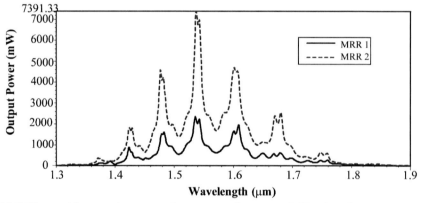

**FIGURE 2.6** The simulation results generated at ring1 (MRR1) and ring2 (MRR2).

## 2.3 Dynamic Pulse Propagation

A schematic diagram of the double PANDA ring resonator system for dynamic pulse observation is designed and shown in Figure 2.7 [5]. In operation, to form the amplification part, a nanoring resonator is embedded within the add/drop optical filter in the system. The modulated Gaussian continuous wave (CW) with center wavelength ($\lambda_0$) at 1.55 μm, peak power at 50 mW is input into the system. The combination between the input and reflected light output can be seen as shown in Figure 2.8(a). The suitable ring parameters are used, for instance, ring radii $R_1 = R_2 = R_3 = 0.80$ μm and the add/drop $R_{ad} = 3.2$ μm. To make the system associate with the practical device [6, 7], selected parameters of the system are fixed to $n_0 = 3.34$ (InGaAsP/InP), $A_{eff} = 0.50$ and $0.1~\mu m^2$ for a microring (add/drop filter) and the nanoring, respectively, $\alpha = 0.5~dB~mm^{-1}$, $\gamma = 0.1$. In this investigation, the coupling coefficient ($j$) of the microring

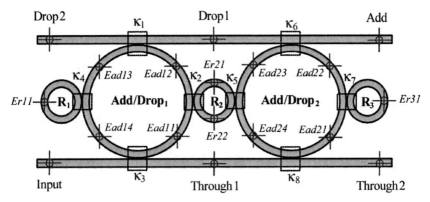

**FIGURE 2.7** A schematic diagram of a double PANDA ring resonators with the dynamic locations $10 \times 20~\mu m^2$ [5].

resonator is 0.8. The nonlinear refractive index of the microring used is $n_2 = 2.2 \times 10^{-17}~m^2/W$. In this case, the attenuation of light propagates within the system (i.e., wave guided) used is $0.5~dB~mm^{-1}$. The wafer (device) length ($z$) is $20~\mu m$ and width $10~\mu m$. Figure 2.8 shows the dynamics pulse train within double PANDA ring (see Figure 2.7) in $z$-direction, where (a) $z = 0$, (b) $z = 0.84~\mu m$, (c) $z = 1.67~\mu m$, (d) $z = 2.60~\mu m$, (e) $z = 3.395~\mu m$, (f) $z = 5.72~\mu m$, (g) $z = 8.18~\mu m$, (h) $z = 9.21~\mu m$, (i) $z = 9.95~\mu m$, (j) $z = 11.06~\mu m$, (k) $z = 11.86~\mu m$, (l) $z = 14.0~\mu m$, (m) $z = 16.65~\mu m$, (n) $z = 17.58~\mu m$, (o) $z = 18.37~\mu m$, (p) $z = 19.069~\mu m$ and (q) $z = 20.0~\mu m$.

The dynamic locations can be configured as follows; $E_{ad}$: pulse propagation in the add/drop device and $E_r$: pulse propagation in the nanoring resonator, which has been located in Figure 2.7. The through and drop ports are also identified by the add/drop filter structure. In practice, the required output signals are obtained and seen at the drop and through ports. Figures 2.8–2.13 show the results in the different locations in the double PANDA ring resonator system. The maximum power of 120 mW is obtained as shown in Figure 2.9(b) that is in the nanoring ($E_{r1}$). Maximum number of peaks of 18 is seen in Figure 2.11(b). The three-dimensional (3D) image of the dynamic modulated Gaussian CW is as shown in Figure 2.11, whereas the maximum power within the nanoring of 120 mW is obtained. The 3-D dynamic pulse propagation is as shown in Figure 2.13. Figure 2.14 shows the reflection intensity of double PANDA ring (see Figure 2.7) with two frequency center input 193.5 THz and 229 THz. We found that the reflection intensity is about $19.788~W/m^2$ seen, the resonances peak about 44 peaks as shown in Figures 2.15–2.19.

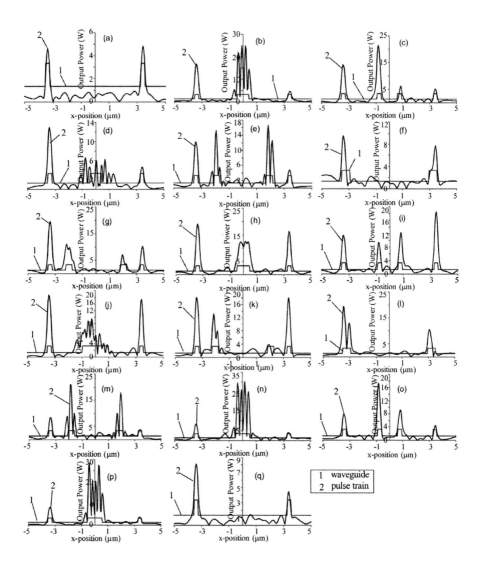

**FIGURE 2.8** Results of dynamic pulse train in z-direction using the OptiFDTD software [9], where (a) z = 0, (b) z = 0.84 μm, (c) z = 1.67 μm, (d) z = 2.60 μm, (e) z = 3.395 μm, (f) z = 5.72 μm, (g) z = 8.18 μm, (h) z = 9.21 μm, (i) z = 9.95 μm, (j) z = 11.06 μm, (k) z = 11.86 μm, (l) z = 14.0 μm, (m) z = 16.65 μm, (n) z = 17.58 μm, (o) z = 18.37 μm, (p) z = 19.069 μm and (q) z = 20.0 μm.

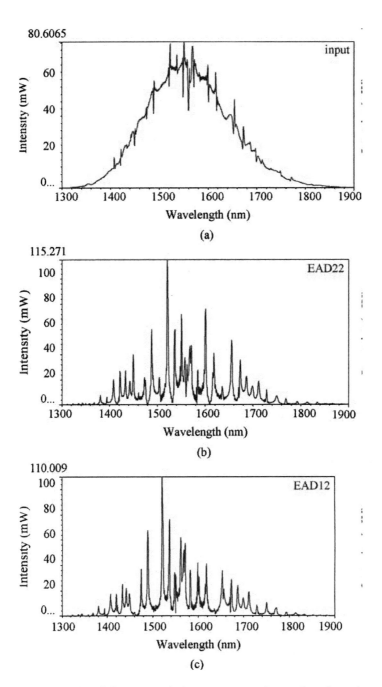

**FIGURE 2.9** Results of the output light intensity and wavelength at the certain location at (a) an input pulse, (b) EAD22: (Ead22), and (c) EAD12: (Ead12) [5].

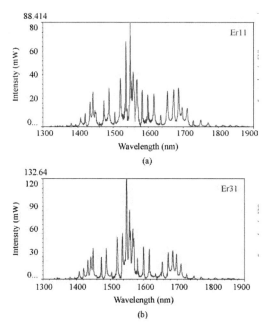

**FIGURE 2.10** Results of the output light intensity and wavelength at the certain location at (a) Er11 and (b) Er31 [5].

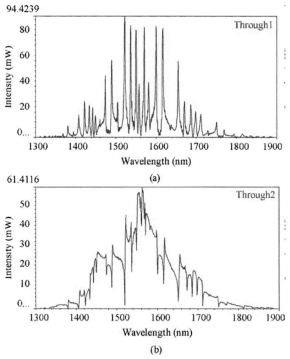

**FIGURE 2.11** Results of the output light intensity and wavelength at the certain location at (a) Through1 and (b) Through2 [5].

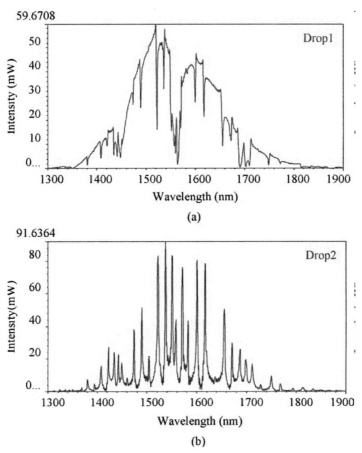

**FIGURE 2.12** Results of the output light intensity and wavelength at the certain location at (a) Drop1 and (b) Drop2 [5].

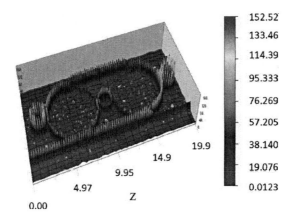

**FIGURE 2.13** The 3D dynamic graphic results obtained using the OPTIWAVE PROGRAMMING [5].

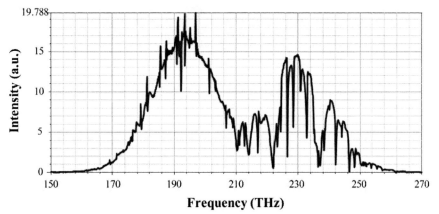

**FIGURE 2.14** Reflection intensity of double-PANDA ring for input frequency center 193.5 THz and 229 THz.

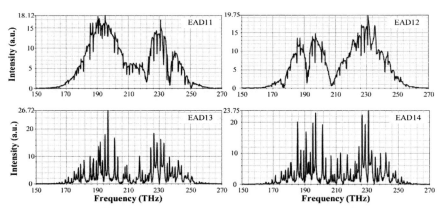

**FIGURE 2.15** Dynamic intensity traveling within the first-PANDA ring, with the input frequency center 193.5 THz and 229 THz.

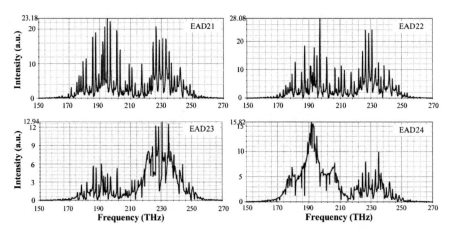

**FIGURE 2.16** Dynamic intensity traveling within the second-PANDA ring for input frequency center 193.5 THz and 229 THz.

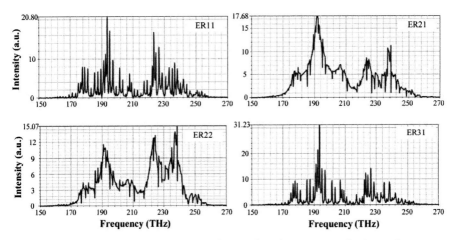

**FIGURE 2.17** Dynamic intensity traveling within the nanoring for input frequency center 193.5 THz and 229 THz.

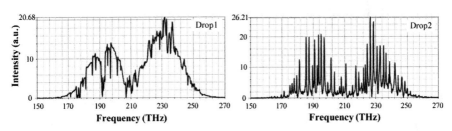

**FIGURE 2.18** Output intensity measured at drop ports of first- and second-PANDA ring.

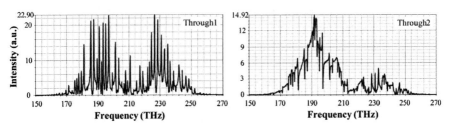

**FIGURE 2.19** Output intensity obtained at the through ports of the first- and second-PANDA rings.

## 2.4 Symmetry and Asymmetry PANDA Ring Resonators

PANDA ring resonator [5], [9] has become a ubiquitous component of photonic technology. Similarly, the coupled resonator optical waveguide (CROW) has modified versions to serve various signal processing functions have also been described and proposed [10–12]. The defect-assisted CROW (DACROW) [13], the subject of this communication, is yet another variant consisting of two subsidiary CROW arms bracketing a perturbed resonator, closely simulating the performance of a Fabry–Pérot (F–P) device and some recently investigated one-dimensional, or slab-type, photonic crystal structures incorporating a defect [14]. Here the perturbed resonator, which is but fractionally different from the others, represents the 'cavity' and the CROW arms on either side are the reflectors. Although defect modes in one-dimensional periodic structures such as coupled resonator arrays have already been described in the literature [15, 16], a detailed account of the properties of the DACROW characteristics, in particular its group delay dispersion and loss sensitivity. In this section, the occurring strong and weak perturbations of the light pulse superposition within PANDA ring are investigated. The strong perturbation is occurred when series connect in symmetry with PANDA ring whereas asymmetry with PANDA ring gives the weak perturbation is seen.

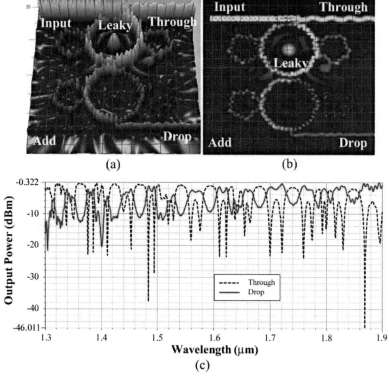

**FIGURE 2.20** Result of a strongest perturbation when a symmetry two-PANDA ring series connection is used. The parameters are $W = 300$ nm, depth = 500 nm, $R_{ad} = 1.56$ μm, $R_1 = R_2 = 0.775$ μm, gap coupling = 0, input power 50 W, $T_0 = 4 \times 10^{-14}$ s, $hw = 1.5 \times 10^{-14}$ s with wavelength center at 1.55 μm, where (a) 3D view, (b) 2D view and (c) output signal.

All numerical simulation in this section is used OptiWave FDTD [8] with Gaussian modulated continuous wave input pulse, which is given by

$$E(x) = A \exp\left[-\frac{(t-t_0)^2}{2T^2}\right] \cdot \sin(\omega t) \quad (2.23)$$

where $A$ is field amplitude, $t_0$ is the time offset, $T$ is the pulse width, $\omega = (2\pi/\lambda)c$ is the frequency of the input wave.

The strongest perturbation is generated by using the double PANDA ring series connection as shown in Figure 2.20. We found that the transmission signal is –46.011 dBm at the through port. The stronger field has the higher leaky output at the center first PANDA ring in the circuit. Figure 2.21 shows the strong perturbation of symmetry triple PANDA ring

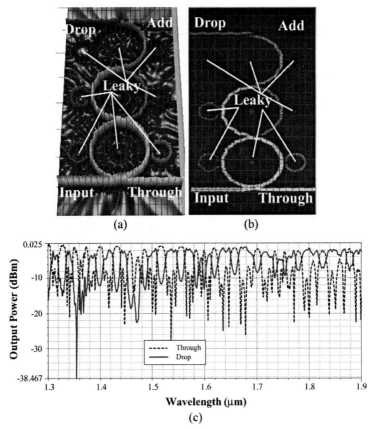

**FIGURE 2.21** Result of the strong perturbation when the symmetry three-PANDA ring series connection is used. The parameters are $W = 300$ nm, depth = 500 nm, $R_{ad} = 2.325$ μm, $R_1 = R_2 = 0.775$ μm, gap coupling = 0, input power 50 W, $T_0 = 4 \times 10^{-14}$ s, $hw = 1.5 \times 10^{-14}$ s with wavelength center 1.55 μm, where (a) 3D view, (b) 2D view and (c) output signal.

series connection. In the drop port, the storage or memory peak signal is −38.467 dBm. Three small peaks resonance of each peak (Box-like shapes) as shown in Figure 2.22 is seen. The weak perturbation occurs when the asymmetry triple PANDA ring series connection is operated as shown in Figure 2.23.

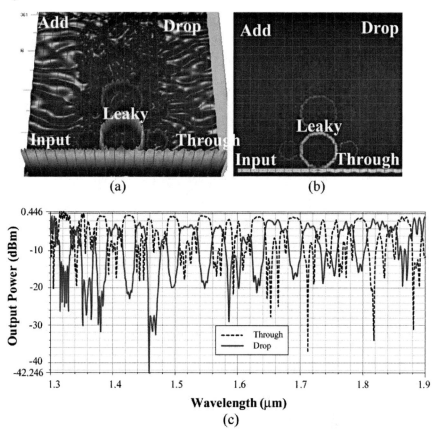

FIGURE 2.22 Result of the strong perturbation when the symmetry four-PANDA ring series connection is used. The parameters are $W = 300$ nm, $R_{ad} = 1.55$ μm, $R_1 = R_3 = 0.775$ μm, gap coupling $= 0$, input power 50 W, $T_0 = 4 \times 10^{-14}$ s, $hw = 1.5 \times 10^{-14}$ s with wavelength center 1.55 μm, where (a) 3D view, (b) 2D view and (c) output signal.

## 2.5 Random Binary Code Generation

All optical devices have become important equipments for advanced optical technology. The increasing demand for processing speed in various applications such as multimedia and telemedicine has generated significant interest in developing all-optical signal processing technologies for future photonic networks. In operation, optical signal processing functions [17], including pulse repetition rate multiplication (PRRM) and arbitrary

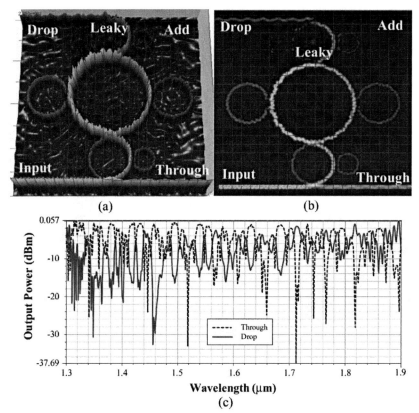

**FIGURE 2.23** Result of the weak perturbation when the asymmetry PANDA ring is used. The parameters are $W = 300$ nm, $R_{ad} = 5$ µm, $R_1 = 1.55$ µm, $R_2 = 0.775$ µm, gap coupling = 0, input power 50 W, $T_0 = 4 \times 10^{-14}$ s, $hw = 1.5 \times 10^{-14}$ s with wavelength center 1.55 µm, where (a) 3D view, (b) 2D view and (c) output signal.

waveform generation (AWG), are useful to generate ultrafast pulse trains with binary amplitude code patterns for ultra-wideband applications. It can also be used as codes for labels in label-switched networks and optical code-division multiple access. There are evidences of growing interest in either theory or experiment of photonic integrated circuits (PICs) for high-speed all-optical signal processing development. Xia et al. [18] have demonstrated that the 10-40 GHz pulse repetition rate multiplication with amplitude control using a 4-stage lattice-form Mach–Zehnder interferometer (LF-MZI) was fabricated in silica-on-silicon planar lightwave circuit (PLC) technology. Jain et al. [19] have reported the design and analysis of a 6-stage tunable LF-MZI for arbitrary binary code generation at 40 GHz by used the direct temporal domain approach. It can be used to determine the filter parameters which could be tuned via the thermo-optic effect in silica, using numerical simulations. It shows that the 6-stage device could generate any 4-bit binary amplitude or phase pattern at 40 GHz from a uniform 10 GHz input pulse train. Ou et al. [20] have demonstrated a novel optical generation approach for binary phase-coded, direct sequence ultra-wideband (UWB) signals. A system consists of a laser array, a polarization modulator (PolM), a fiber Bragg grating (FBG), a length of single mode fiber, and a photo detector (PD). The FBG, designed based on the superimposed, chirped grating, is used as the multi-channel frequency discriminator. The input electronic Gaussian pulse is modulated on the optical carrier by the PolM and then converted into UWB monocycle or doublet pulses sequence by the multi-channel frequency discriminator. The PolM is used to select the desired binary phase code pattern by adjusting the polarization state of each laser, rather than tuning the laser wavelengths. The desired UWB shape, monocycle or doublet, could be selected by tuning the FBG. An optical coding scheme using optical interconnection for a photonic analog-to-digital conversion has been proposed and demonstrated [21]. It allows us to convert a multi-power level signal into a multiple-bit binary code to detect it in a bit-parallel format by binary photodiode array. The proposed optical coding is executed after optical quantization using self-frequency shift. Optical interconnection based on a binary conversion table generates a multiple-bit binary code by appropriate allocation of a level identification signal which is provided as a result of optical quantization.

Recently, an experimentally demonstration of a novel technique to implement bipolar UWB pulse coding has been reported [22]. Multichannel chirped FBG was used, in combination with a dispersive fiber, to produce a multichannel frequency discriminator with a step-increased group-delay response. Binary monocycle or doublet sequences with the desired phase coding patterns were generated by applying a phase modulated Gaussian pulse train to the multichannel frequency discriminator. The time delay difference between adjacent UWB pulses remains unchanged when the wavelengths are tuned. Different phase coding patterns were generated by simply tuning the states of polarization of the wavelengths sent to the PolM. In both cases, the dispersion of the dispersive fiber was compensated by the multichannel chirped FBG. Therefore the distortion of the generated UWB pulses due to the fiber dispersion was eliminated. A novel and simple method for all-optically generating UWB pulses is by using a NOLM-based optical switch. Both Gaussian monocycle, doublet pulses and their polarity reversed pulses have been proposed [23]. The spectra accord with Federal Communications Commission (FCC) standard, are generated experimentally and distributed by the SMF, which acts as both dispersive and transmission media. A tunable laser adopted in the scheme ensures that the different length of SMF has no effects on the generated UWB pulses. Furthermore, pulse shape modulation at a high speed can easily be achieved by employing an intensity modulator. The authors in reference [6] have reported a Barker binary phase code for speckle reduction in line scan laser projectors, and a speckle contrast factor decrease down to 13%. Barker-like binary phase codes of lengths longer than 13 are used at an intermediate image plane. It is shown by theoretical calculation that a much better speckle reduction with a speckle contrast factor up to 6% can be achieved by using longer binary phase codes other than the Barker code.

From the previous reports, the searching for new optical encoding technique remains. Therefore in this chapter we propose the use of dark-bright optical soliton conversion and control within a tiny ring resonator system [7, 24]. It is formed by using an all optical devices, which consists of an add/drop optical filter known as a PANDA ring resonator [5]. The binary codes can be formed and retrieved by using the random polarized light and dark-bright soliton conversion, respectively. The advantage of this technique is that the random binary codes can be formed by controlling one of the input solitons into the PANDA ring resonator ports, which can be available for the use in information security requirement. Moreover the device dimension can be used to form an array of large photonic circuits, whereas the coding capacity can be increased.

The electric field detected by photodetector $D_3$ is given by [25]

$$E_{D_3} = E_{t1} \frac{-\sqrt{1-\kappa_4}\, e^{-\frac{\alpha}{2}L_b - jk_n L_b} + \sqrt{1-\kappa_4}}{1 - \sqrt{1-\kappa_4}\sqrt{1-\kappa_5}\, e^{-\frac{\alpha}{2}L_b - jk_n L_b}} \qquad (2.24)$$

where $L_b = 2\pi R_b$, $R_b$ is radius of add/drop optical filter decoded as shown in Figure 2.24. The light pulse output power detected by photodetector $D_3$ is defined as

$$P_{D_3} = \left|\frac{E_{D_3}}{E_{t1}}\right|^2$$

$$= \frac{\left(1 - \kappa_4 - 2\sqrt{1-\kappa_4}\sqrt{1-\kappa_5}\, e^{-\frac{\alpha}{2}L_b}\cos(k_n L_b) + (1-\kappa_4)e^{-\alpha L_b}\right)}{\left(1 + (1-\kappa_4)(1-\kappa_5)e^{-\alpha L_b} - 2\sqrt{1-\kappa_4}\sqrt{1-\kappa_5}\, e^{-\frac{\alpha}{2}L_b}\cos(k_n L_b)\right)} \qquad (2.25)$$

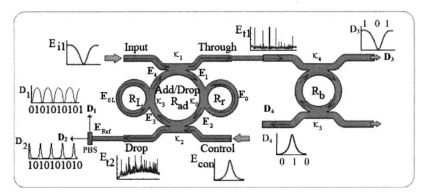

**FIGURE 2.24** A schematic diagram of random binary code generation using dark-bright soliton conversion. $D_n$: photodetectors, $\kappa_n$: coupling coefficient of couplers, $E_{Ref}$: references filed, $E_n$: electric fields, PBS: Polarized Beam Splitter.

The electric field detected by photodetector $D_4$ is given by

$$E_{D_4} = E_{t1} \frac{-\sqrt{\kappa_4 \kappa_5} e^{-\frac{\alpha L_b}{2 2} - jk_n \frac{L_b}{2}}}{1 - \sqrt{1-\kappa_4}\sqrt{1-\kappa_5} e^{-\frac{\alpha}{2}L_b - jk_n L_b}} \quad (2.26)$$

The light pulse output power detected by photodetector $D_4$ is defined as

$$P_{D_4} = \left|\frac{E_{D_4}}{E_{t1}}\right|^2 = \frac{\kappa_4 \kappa_5 e^{-\frac{\alpha}{2}L_b}}{\left(1 + (1-\kappa_4)(1-\kappa_5)e^{-\alpha L_b} - 2\sqrt{1-\kappa_4}\sqrt{1-\kappa_5}e^{-\frac{\alpha}{2}L_b}\cos(k_n L_b)\right)} \quad (2.27)$$

For the random binary code generation, we used the add/drop optical filter for decoded binary code signal as shown in Figure 2.24. When the light pulse signal passes through the coupler with coupling coefficient, $\kappa_4 = 0.5$, the light has split into two ways, one passes through photodetector, $D_3$ and detected output binary code signal and other circulated within the add/drop optical filter and passes through the coupler which coupling coefficient, $\kappa_5 = 0.5$, to photodetector, $D_4$ and detected once output binary code signal.

In simulation, the used parameters of the first add/drop filter (optical multiplexer, PANDA ring resonator) are fixed to be $\kappa_0 = 0.1$, $\kappa_0 = 0.35$, $\kappa_2 = 0.1$, and $\kappa_3 = 0.2$ respectively. The ring radii are $R_{ad} = 300$ nm, $R_r = 30$ nm, and $R_l = 15$ nm. $A_{eff}$ are 0.50, 0.25 and 0.25 mm$^2$ [5, 9] for the add/drop optical multiplexer, right and left nanoring resonators, respectively. The parameters of the second add/drop optical filter are fixed to be $\kappa_4 = \kappa_5 = 0.5$, $R_b = 200$ nm and $A_{eff} = 0.25$ mm$^2$, respectively. Simulation results of the random binary code generation with center wavelengths are at $\lambda_0 = 1.50$ μm. For random binary code generation as shown in Figure 2.25, the simulation result of the light pulse generated within the add/drop optical filter system at center wavelength $\lambda_0 = 1.50$ mm for random binary code generation: (a) $|E_1|^2$, (b) $|E_2|^2$, (c) $|E_3|^2$, (d) $|E_4|^2$ and (e) are the reflected outputs from the throughput port, and (f) is the output at the drop port. Here: (a) $E_1$, (b) $E_2$, (c) $E_3$ and (d) $E_4$ (e) are the through port and (f) drop port signals. When a dark soliton light pulse with 1 W peak power is input into the input port traveled and passes through the first coupler, $\kappa_1$, which one part split into to through port and other to arc at $E_1$ of add/drop optical multiplexer. The cross (X) phase modulation (XPM) output power is as shown in Figure 2.25(a). Then the light pulse entered and circulated in the right nanoring resonator radii, $R_r$, which it passes to the add/drop optical multiplexer at $E_2$. The output power has been amplified as shown in Figure 2.25(b). To use the control function, a bright soliton with 1 W peak power is input into the control port, then it passes through the second coupler, $\kappa_2$, and it is multiplexed with the light pulse from $E_2$. The output light is then split two ways: one part is split into drop port, $E_{t2}$, the power output as shown in Figure 2.25(e) and the other is split into the arc of add/drop at position $E_3$, the output power as shown in Figure 2.25(c). Finally, the light pulse travels and enters the left nanoring resonator radii, $R_L$, that passes to the add/drop optical multiplexer at position $E_4$. The output power is amplified again as shown in Figure 2.25(d) and traveled into the first coupler, $\kappa_1$, then entering into through port, $E_{t1}$, and $E_1$. The amplitude is more amplified compared to the one as shown in Figure 2.25(f).

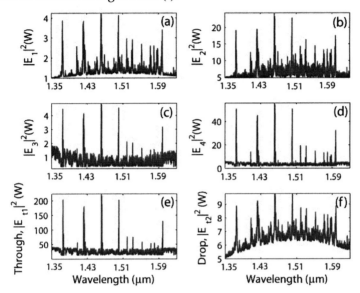

FIGURE 2.25 Simulation result of the light pulse generated within the add/drop optical filter system at center wavelength $\lambda_0 = 1.50$ mm for random binary code generation, where (a) $|E_1|^2$, (b) $|E_2|^2$, (c) $|E_3|^2$, (d) $|E_4|^2$, (e) are through port and (f) drop port signals, where $R_r = 15$ nm, $R_L = 30$ nm, $R_{ad} = 300$ nm and $a = 5 \times 10^{-5}$ dBmm$^{-1}$.

In operation, the random binary codes are generated within the add/drop optical multiplexer, which are seen at the drop port (Figure 25(f)). Then, it passes through the PBS, where the polarization phase shift of the two components is 90°. The random polarization states of two components can be used to form the random binary code patterns and the binary code signals. It can be observed by photodetectors $D_1$ and $D_2$. The referencing binary code patterns are set as shown in Figure 2.26(a) and (b), where the binary codes detected by $D_1$ are '10101010101010101010101010101010101010 101010101010' and '10101010101010101010 1010101010101010101010101010' patterns. The binary codes detected by $D_2$ are '10101010101010101010101 01010101010101010101010' patterns. In this chapter, the obtained pulse switching time of 12.8 ns is notes. In operation, the random binary codes can be retrieved in the form of dark-bright soliton conversion within the add/drop optical filter system as shown in Figure 2.27. The pattern of dark soliton conversion is '101', and the bright soliton is '010'. In Figure 2.27(a)–(b) shows that the random binary codes are formed because dark-bright soliton conversion detected by photodetectors $D_3$ and $D_4$ does not show the certain pattern.–They depend on the light pulse generated within add/drop optical multiplexer and input and control input signals.

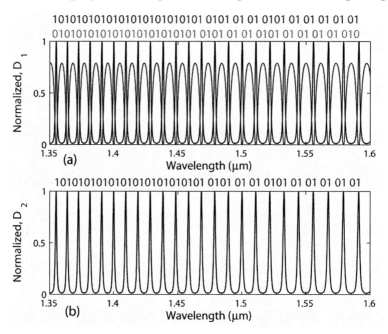

**FIGURE 2.26** Random binary code generation for reference output signal, $E_{Ref}$, passes through PBS at the drop port, which (a) detected by $D_1$ and (b) detected by $D_2$.

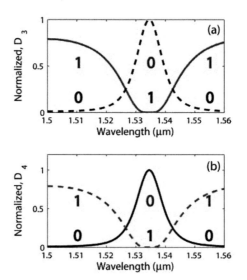

**FIGURE 2.27** Binary code signal form dark-bright solitons conversion, where (a) dark and (b) bright, and $\kappa_4 = \kappa_5 = 0.5$, $R_b = 200$ nm.

## 2.6 Binary Code Suppression and Recovery

The proposed system consists of a suppression system and recovery system as shown in Figures 2.28 and 2.29. The binary code suppression is designed by using the add/drop filter and double microring resonators, which is known as PANDA ring resonator. The binary code recovery is obtained by using add/drop filter. In simulation, the parameters of PANDA ring resonator are fixed to be $\kappa_0 = 0.1$, $\kappa_1 = 0.2$, $\kappa_2 = 0.2$, and $\kappa_3 = 0.1$ respectively. The ring radii are $R_{ad} = 200$ μm, $R_r = 15$ μm, and $R_l = 15$ μm. $A_{eff}$ are 0.50, 0.25 and 0.25 μm² [2, 4, 9] for PANDA ring resonator, right and left microring resonators, respectively. For the binary code recovery, the parameters of the add/drop optical filter are fixed to be $\kappa_4 = \kappa_5 = 0.2$, $R_b = 50$ μm and $A_{eff} = 0.25$ μm², respectively. Moreover, our binary suppression and recovery system support to the possible fabricated devices. Simulation result of the binary code suppression with center wavelength is presented at $\lambda_0 = 1.50$ μm. The binary code suppression result is as shown in Figure 2.30. The binary code recovery result is as shown in Figure 2.31.

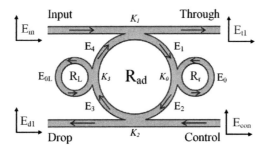

**FIGURE 2.28** A schematic diagram of binary code suppression system.

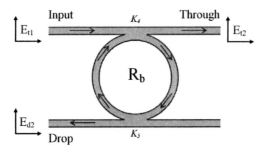

**FIGURE 2.29** A schematic diagram of binary code recovery system.

Figure 2.30 shows the simulation result of the light pulse generated at center wavelength $\lambda_0 = 1.50$ μm for binary code suppression, where (a) $|E_{in}|^2$, (b) $|E_{con}|^2$, (c) $|E_{t1}|^2$, (d) $|E_{d1}|^2$ and (e) $|E_{t1}|^2$, where $R_r = 15$ μm, $R_L = 15$ μm, $R_{ad} = 200$ μm and $\alpha = 5 \times 10^{-5}$ dBmm$^{-1}$. Here (a) is the input port for binary code suppression. By using bright soliton, the light pulse with 1 W peak power is input into the input port. Figure 2.30(b) is the control port that uses a dark soliton light pulse with 1 W peak power. The power output of the drop port for binary code suppression is as shown in Figure 2.30(d). Then, Figure 2.30(c) and (e) shows power output of the through port for binary code suppression which is the binary code suppression signal that can be transmitted to the receiver for high security communication. The reference signal can be formed for reference binary code in communication. Moreover, the peak power output from through and drop ports are 2.3 and 3.2 W, respectively which is larger than the input light pulse.

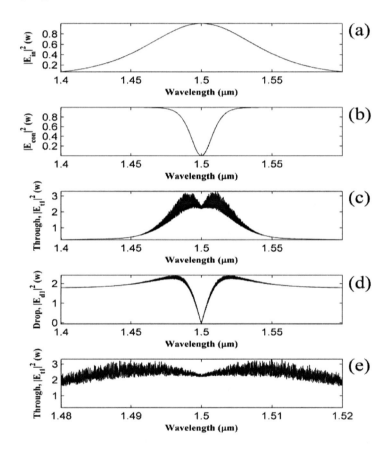

**FIGURE 2.30** Simulation result of the light pulse generated at center wavelength $\lambda_0 = 1.50$ μm for binary code suppression, where (a) $|E_{in}|^2$, (b) $|E_{con}|^2$, (c) $|E_{t1}|^2$, (d) $|E_{d1}|^2$ and (e) $|E_{t1}|^2$, where $R_r = 15$ μm, $R_L = 15$ μm, $R_{ad} = 200$ μm and $\alpha = 5 \times 10^{-5}$ dBmm$^{-1}$.

Then, Figure 2.31 shows the simulation result of the light pulse generated at center wavelength $\lambda_0 = 1.50$ μm for binary code recovery, where (a) $|E_{t1}|^2$, (b) $|E_{t2}|^2$, (c) $|E_{d2}|^2$ and (d) compare $|E_{t2}|^2$ and $|E_{d2}|^2$, where $R_b = 50$ μm and $\alpha = 5 \times 10^{-5}$ dBmm$^{-1}$. Here (a) is the input port for binary code recovery.

The binary code suppression signal which looks like noisy signal is input into this port. It shows that this technique of communication is highly secured. Figure 2.31(b) and (c) shows the power output of the through port and drop port for binary code recovery respectively. The power output from drop port represents the binary code, which is shown in Figure 2.32. Figure 2.31(d) shows the comparing signals between the power outputs of though and drop ports which have shown the signals relationship.

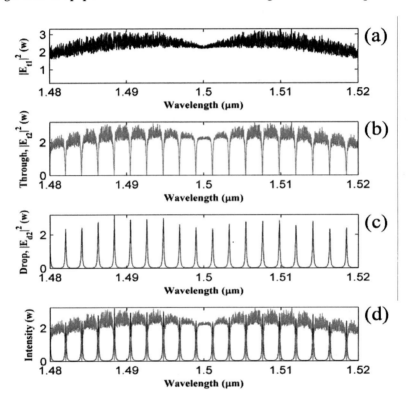

**FIGURE 2.31** Simulation result of the light pulse generated at center wavelength $\lambda_0 = 1.50$ μm for binary code recovery, where (a) $|E_{t1}|^2$, (b) $|E_{t2}|^2$, (c) $|E_{d2}|^2$ and (d) compare $|E_{t2}|^2$ and $|E_{d2}|^2$, where $R_b = 50$ μm and $\alpha = 5 \times 10^{-5}$ dBmm$^{-1}$.

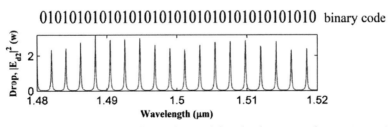

**FIGURE 2.32** The binary code is obtained by the binary code recovery signals (Figure 2.31(c)).

## 2.7 Conclusion

The authors have analyzed and described dark-bright soliton collision behavior within a PANDA ring resonator system. There is an interesting aspect called dark-bright soliton conversion that can be useful for random binary code application. We have also presented the use of the proposed system to generate the random codes, where the random binary codes using dark-bright optical soliton conversion within add/drop optical multiplexer can be formed. In order to achieve the required binary codes, first we set the dark and bight soliton pulses to be '0' and '1', respectively. By using the dark-bright soliton pulse trains, the dark and bright solitons can be separated by using PBS, which can be used to set the referencing codes. Finally, the required codes can be relatively obtained to the referencing codes. The advantage of the device is that the different binary code patterns can be generated randomly by changing the setting parameters, which can be available for security code applications.

In this chapter, the authors also propose a new design for security application in optical communication by using the binary code suppression and recovery methods. By using the dark-bright soliton pair within a PANDA ring resonator. The binary code suppression is designed and the recovery code obtained. The designed binary code suppression and recovery system is supported to the possible creation. Moreover, the simulation results obtained have shown that the proposed system can indeed be achieved. The binary code suppression and recovery, in which high security can be available for optical communication security. In future, it deserves to consider extending this work to real data suppression and recovery for realistic optical communication. Moreover, the development of the proposed technique for high security optical network can also be extended.

## REFERENCES

[1] G.P. Agrawal, *Nonlinear Fiber Optics*, 4th edition, New York: Academic Press (2007).

[2] F.M. Lee, C.L. Tsai, C.W. Hu, F.Y. Cheng, M.C. Wu, and C.C. Lin, "High-reliable and high-speed 1.3 µm complex-coupled distributed feedback buried-heterostructure laser diodes with Fe-doped InGaAsP/InP hybrid grating layers grown by MOCVD," *IEEE Trans. on Elect. Devices*, vol. 55, no. 2, pp. 540–546 (2008).

[3] S. Tomofuji, S. Matsuo, T. Kakitsuka, and K. Kitayama, "Dynamic switching characteristics of InGaAsP/InP multimode interference optical waveguide switch," *Opt. Express*, vol. 17, no. 26, pp. 23380–23388 (Dec. 2009).

[4] Y. Kokubun, Y. Hatakeyama, M. Ogata, S. Suzuki, and N. Zaizen, "Fabrication technologies for vertically coupled microring resonator with multilevel crossing busline and ultracompact-ring radius," *IEEE J. Sel. Top. Quantum Electron.*, vol. 11, pp. 4–10 (2005).

[5] K. Uomwech, K. Sarapat, and P.P. Yupapin, "Dynamic modulated Gaussian pulse propagation within the double PANDA ring resonator system," *Microw. Opt. Techn. Lett.*, vol. 52, no. 8, pp. 1818–1821 (Aug. 2010).

[6] M.N. Akram, V. Kartashov, and Z. Tong, "Speckle reduction in line-scan laser projectors using binary phase codes," *Opt. Lett.*, vol. 35, no. 3, pp. 444–446 (Feb. 2010).

[7] K. Sarapat, N. Sangwara, K. Srinuanjan, P.P. Yupapin and N. Pornsuwancharoen, "Novel dark-bright optical solitons conversion system and power amplification," *Opt. Eng.*, vol. 48, pp. 045004 (2009).

[8] OptiFDTD by OptiWave Corp. ©, ver. 8.0, single license (KMITL) (2008).

[9] T. Phatharaworamet, C. Teeka, R. Jomtarak, S. Mitatha, and P.P. Yupapin, "Random Binary Code Generation Using Dark-Bright Soliton Conversion Control Within a PANDA Ring Resonator," *J. Lightw. Techn.*, vol. 28, no. 19, pp. 2804–2809 (2010).

[10] Y. Chen and S. Blair, "Nonlinearity enhancement in finite coupled-resonator slow-light waveguides," *Opt. Express*, vol. 12, pp. 3353–3366 (2004).

[11] A. Melloni, F. Morichetti, C. Ferrari, and M. Martinelli, "Continuously tunable 1 byte delay in coupled-resonator optical waveguides," *Opt. Lett.*, vol. 33, pp. 2389–2391 (2008).

[12] M. Sumetsky, "Vertically-stacked multi-ring resonator," *Opt. Express*, vol. 13, pp. 6354–6375 (2005).

[13] O. Schwelb, and I. Chremmos, "Defect assisted resonator optical waveguide: weak perturbations," *Opt. Comm.*, vol. 283, pp. 3686–3690 (2010).

[14] A.R.M. Zain, N.P. Johnson, M. Sorel, and R.M. De La Rue, "Ultra high quality factor one dimensional photonic crystal/photonic wire micro-cavities in silicon-on-insulator (SOI)," *Opt. Express*, vol. 16, pp. 12084–12089 (2008).

[15] Y. Landobasa and M. Chin, "Defect modes in microring resonator arrays," *Opt. Express*, vol. 13, pp. 7800–7815 (2005).

[16] Y.M. Landobasa Tobing, P. Dumon, R. Baets, and M.-K. Chin, "Demonstration of defect modes in coupled microresonator arrays fabricated in silicon-on-insulator technology," *Opt. Lett.*, vol. 33, pp. 1939–1941 (2008).

[17] Z. Jiang, D.E. Leaird, and A.M. Weiner, "Line-by-line pulse shaping control for optical arbitrary waveform generation," *Opt. Express*, vol. 13, no. 25, pp. 10431–10439 (Nov. 2005).

[18] B. Xia, L.R. Chen, P. Dumais, and C.L. Callender, "Ultrafast pulse train generation with binary code patterns using planar lightwave circuits," *Electron. Lett.*, vol. 45, no. 19, pp. 1119–1120 (Sep. 2006).

[19] A. Jain, I.A. Kostko, L.R. Chen, B. Xia, P. Dumais, and C.L. Callender, "Design and analysis of a 6-stage tunable MZI PLC for BPSK generation," in *Proc. 50th IEEE Midwest Symposium on Circuits and Systems*, Montreal, QC, Canada (2007).

[20] P. Ou, Y. Zhang, and C.X. Zhang, "Optical generation of binary-phase-coded, direct-sequence ultra-wideband signals by polarization modulation and FBG-based multichannel frequency discriminator," *Opt. Express*, vol. 16, no. 7, pp. 5130–5135 (Mar. 2008).

[21] T. Nishitani, T. Konishi, and K. Itoh, "Optical coding scheme using optical interconnection for high sampling rate and high resolution photonic analog-to-digital conversion," *Opt. Express*, vol. 15, no. 24, pp. 15812–15817 (Nov. 2007).

[22] Y. Dai and J. Yao, "Optical generation of binary phase-coded direct-sequence UWB signals using a multichannel chirped fiber bragg grating," *J. Lightw. Technol.*, vol. 26, no. 15, pp. 2513–2520 (Aug. 2008).

[23] H. Huang, K. Xu, J. Li, J. Wu, X. Hong, and J. Lin, "UWB Pulse Generation and Distribution Using a NOLM Based Optical Switch," *J. Lightw. Technol.*, vol. 26, no. 15, pp. 2635–2640 (Aug. 2008).

[24] C. Teeka, P. Chaiyachet, P.P. Yupapin, N. Pornsuwancharoen, and T. Juthanggoon, "Soliton collision management in a microring resonator system," *Physics Procedia*, vol. 2, pp. 67–73 (2009).

[25] D.G. Rabus, M. Hamacher, U. Troppenz, and H. Heidrich, "Optical filters based on ring resonators with integrated semiconductor optical amplifiers in GaInAsP–InP," *IEEE J. Sel. Top. Quantum Electron.*, vol. 8, no. 6, pp. 1405–1411 (2002).

# 3

# Dark-Bright Soliton Conversion

**CHAPTER OUTLINE**

- ▶ Introduction
- ▶ Operating Principle
- ▶ Soliton Nonlinear Behaviors
- ▶ Optical Soliton
- ▶ Dark-Bright Soliton Conversion
- ▶ Dark-Bright Soliton Conversion in Add/Drop Filter
- ▶ Soliton Collision Management in a Microring Resonator
- ▶ Soliton Collision Management
- ▶ Conclusion
- ▶ References

## 3.1 Introduction

Nonlinear behaviors of light pulse traveling within a fiber optic ring resonator have been analyzed [1, 2], where the nonlinear Kerr type is the major effect within the fiber ring [3, 4]. The authors have shown that such nonlinear behaviors can be beneficial for different forms such as bistability switching, signal security and digital encoding. However, the power attenuation of the signal output becomes a problem in the long distance link. For security purpose, a dark soliton pulse is recommended to overcome such a problem so the problem of power attenuation can be solved. In addition, the low level signal detection of the dark soliton is another problematic aspect of behavior. Dark soliton is one of the soliton properties where the soliton amplitude vanishes during the propagation within the transmission line. Therefore, dark soliton detection is extremely difficult. To date, several papers have investigated the dark soliton behaviors [5–8] and one shows an interesting result in that the dark soliton can be converted into a bright soliton and finally detected. This means that the dark soliton penalty can be used as a communication carrier so that it can be retrieved by the dark-bright soliton conversion. A soliton pulse has been used to produce fast switching [9, 10] localized within a nano-waveguide and it was reported that they have designed a system which consists of micro and nanoring resonators. In this section, we present the interesting results, where a dark soliton pulse can be formed with nonlinear behaviors within the designed waveguide, which means that the usage of dark soliton is possible and if we can use such behaviors in an advantageous way we can have a reasonable detection power. The dark soliton valley signal is always low output level, which is difficult to detect, while the dark-bright conversion is required.

Dark and bright soliton behaviors have been widely investigated in different forms [11–17]. The use of soliton, i.e., bright soliton in long distance communication links has been in operation for nearly two decades. However, some questions still need to be answered in the area of communication safety, whereas the use of a dark soliton pulse within a microring resonator for communication security has been studied. In this chapter dark/bright soliton is used to enhance confidentiality in communication links. Dark soliton is one of the soliton properties, where the soliton amplitude vanishes or minimizes during the propagation in media, therefore, the detection of dark soliton is difficult. The investigation of dark soliton behaviors has been reported and one has shown the interesting result that the dark soliton can be stabilized and converted into bright soliton and finally detected. This means that we can use the dark soliton to perform the communication transmission for security, whereas the required information can be retrieved by the dark-bright soliton conversion. In fact, we are looking for a simple technique that can be employed to detect the dark soliton. Yupapin et al. [12] have reported the interesting results of light pulse propagation within a nonlinear microring device where the transfer function of the output at the resonant condition is derived and used. They found that the broad spectrum of light pulse can be transformed to the discrete pulses. In this chapter, we have shown that after the dark soliton is input and sliced for the purpose of noisy signals for security purposes within the nonlinear ring resonator system this can be transmitted into the signal via an add/drop filter by choosing to retrieve either bright or dark soliton pulses. However, the device parameters are the given keys for the end users and these can be used to form the device that can be used to retrieve the signals in the link or network. The conversion signal amplification is also a better technique and with this we found that significant amplified signals can be achieved. The novelty is that the soliton behavior within a micro and nanoring resonator has been investigated, whereas all parameters are based on the practical device parameters, which is shown the potential of applications.

## 3.2 Operating Principle

An optical soliton is recognized as a powerful laser pulse, which can be used to enlarge the optical bandwidth when propagating within the nonlinear microring resonator. Moreover, the superposition of self-phase modulation (SPM) soliton pulses can maintain the large power output. Initially, the optimum energy is coupled into the waveguide by a larger effective core area device, i.e., a ring resonator, then the smaller one is connected to form the stopping behavior. The filtering characteristic of the optical signal is presented within a ring resonator, where the suitable parameters can be controlled to obtain the required output energy. To maintain the soliton pulse propagating within the ring resonator, suitable coupling power into the device is required, whereas the interference signal has a minor effect compared to the loss associated with the direct signal passing through. We are looking for a stationary soliton pulse, which is introduced into the microring or a multi-stage microring resonator system as shown in Figure 3.1, the input optical field ($E_{in}$) of the dark soliton pulse input is given by [1]

$$E_{in} = A \tanh\left[\frac{T}{T_0}\right] \exp\left[\left(\frac{z}{2L_D}\right) - i\omega t\right] \quad (3.1)$$

where $A$ and $z$ are the optical field amplitude and propagation distance, respectively. $T$ is the soliton pulse propagation time in a frame moving at the group velocity, $T = t - \beta_1 z$, where $\beta_1$ and $\beta_2$ are the coefficients of the linear and second order terms of Taylor expansion of the propagation constant. $L_D = T_0^2/|\beta_2|$ is the dispersion length of the soliton pulse. $T_0$ in the equation is the soliton pulse propagation time at the initial input. Where $t$ is the soliton phase shift time, the frequency shift of the soliton is $\omega_0$. This solution describes a pulse that keeps its temporal width in variance as it propagates, and is thus called a temporal soliton. When a soliton peak intensity ($|\beta_2/\Gamma T_0^2|$) is given, then $T_0$ is known. For the soliton pulse in the microring device, a balance should be achieved between the dispersion length ($L_D$) and the nonlinear length ($L_{NL} = 1/g\Gamma\phi_{NL}$), where $\Gamma = n_2 k_0$, is the length scale over which dispersive or nonlinear effects make the beam wider or narrower. For a soliton pulse there is a balance between dispersion and nonlinear lengths, hence $L_D = L_{NL}$.

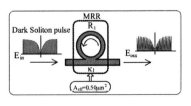
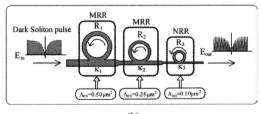

(a) (b)

**FIGURE 3.1** A schematic diagram of an all optical dark soliton pulse system, $R_s$: Ring radii, $\kappa_s$: Coupling coefficients, MRR: Microring resonator, NRR: Nanoring resonator.

When light propagates within the nonlinear material (medium), the refractive index ($n$) of light within the medium is given by:

$$n = n_0 + n_2 I = n_0 + \left(\frac{n_2}{A_{\text{eff}}}\right) P \qquad (3.2)$$

where $n_0$ and $n_2$ are the linear and nonlinear refractive indexes, respectively. $I$ and $P$ are the optical intensity and optical power, respectively. The effective mode core area of the device is given by $A_{\text{eff}}$. For the microring and nanoring resonators, the effective mode core areas range from 0.50 to 0.1 µm² [18, 19].

When a soliton pulse is input and propagated within a microring resonator as shown in Figure 3.1, which consists of a series of microring resonators. The resonant output is formed, thus, the normalized output of the light field is the ratio between the output and input fields ($E_{\text{out}}(t)$ and $E_{\text{in}}(t)$) in each roundtrip, which can be expressed as [20]

$$\left|\frac{E_{\text{out}}(t)}{E_{\text{in}}(t)}\right|^2 = (1-\gamma)\left[1 - \frac{\left(1-(1-\gamma)x^2\right)\kappa}{\left(1-x\sqrt{1-\gamma}\sqrt{1-\kappa}\right)^2 + 4x\sqrt{1-\gamma}\sqrt{1-\kappa}\sin^2\left(\frac{\phi}{2}\right)}\right] \qquad (3.3)$$

The close form of the Eq. (3.3) indicates that a ring resonator in this particular case is very similar to a Fabry-Pérot cavity, which has an input and output mirror with a field reflectivity, $(1 - \kappa)$ and a fully reflecting mirror. $\kappa$ is the coupling coefficient, $x = \exp(-\alpha L/2)$ represents a roundtrip loss coefficient, $\phi_0 = kLn_0$ and $\phi_{\text{NL}} = kLn_2|E_{\text{in}}|^2$ are the linear and nonlinear phase shifts, $k = 2\pi/\lambda$ is the wave propagation number in a vacuum, where $L$ and $\alpha$ are a waveguide length and linear absorption coefficient, respectively. In this chapter, the iterative method is introduced to obtain the results as shown in the Eq. (3.3) and similarly, when the output field is connected and input into the other ring resonators.

After the signals are multiplexed with the generated chaotic noise, the chaotic cancellation is required by the individual user. To retrieve the signals from the chaotic noise, we propose to use the add/drop device with the appropriate parameters. This is given in the details that follow. The optical circuits of ring-resonator add/drop filters for the throughput and drop port can be given by Eqs (3.4) and (3.5), respectively [20]

$$\left|\frac{E_t}{E_{\text{in}}}\right|^2 = \frac{(1-\kappa_1) - 2\sqrt{1-\kappa_1}\sqrt{1-\kappa_2}\,e^{-\frac{\alpha}{2}L}\cos(k_n L) + (1-\kappa_2)e^{-\alpha L}}{1+(1-\kappa_1)(1-\kappa_2)e^{-\alpha L} - 2\sqrt{1-\kappa_1}\sqrt{1-\kappa_2}\,e^{-\frac{\alpha}{2}L}\cos(k_n L)} \qquad (3.4)$$

and

$$\left|\frac{E_d}{E_{\text{in}}}\right|^2 = \frac{\kappa_1\kappa_2 e^{-\frac{\alpha}{2}L}}{1+(1-\kappa_1)(1-\kappa_2)e^{-\alpha L} - 2\sqrt{1-\kappa_1}\sqrt{1-\kappa_2}\,e^{-\frac{\alpha}{2}L}\cos(k_n L)} \qquad (3.5)$$

here $E_t$ and $E_d$ represent the optical fields of the throughput and drop ports respectively. $\beta = kn_{\text{eff}}$ is the propagation constant, $n_{\text{eff}}$ is the effective refractive index of the waveguide and the circumference of the ring is $L = 2\pi R$, here $R$ is the radius of the ring. In the following, new parameters will be used for simplification: $\phi = \beta L$ is the phase constant. The chaotic noise cancellation can be managed by using the specific parameters of the add/drop device so that the required signals can be retrieved by the specific users. $\kappa_1$ and $\kappa_2$ are coupling coefficient of add/drop filters, $k_n = 2\pi/\lambda$ is the wave propagation number for a vacuum and the waveguide (ring resonator) loss is $\alpha = 0{:}5$ dBmm$^{-1}$. The fractional coupler intensity loss is $\gamma = 0{:}1$. In the case of the add/drop device, the nonlinear refractive index is neglected.

## 3.3 Soliton Nonlinear Behaviors

To begin the concept, we use only one ring resonator to form the nonlinear behaviors of a dark soliton. The schematic diagram of the proposed system is as shown in Figure 3.1(a) and(b). A bright soliton pulse with 50 ns pulse

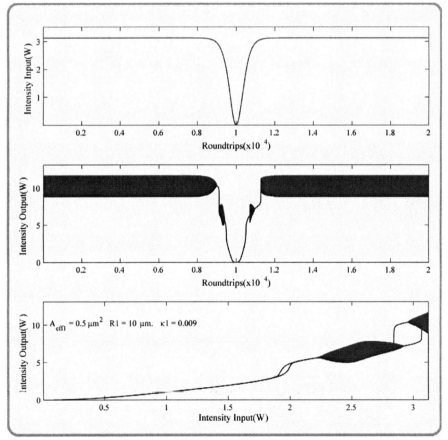

**FIGURE 3.2** Results obtained when a dark soliton pulse is input into a micro ring resonator, where $R_1 = 10$ μm, $A_{\text{eff1}} = 0.50$ μm$^2$, $\kappa_1 = 0.009$.

width, maximum power at 1.25 W is input into the system. The suitable ring parameters are used, for instance, ring radii $R_1$ = 10.0 μm. In order to make the system associate with the practical device [19], the selected parameters of the system are fixed to $\lambda_0$ = 1.55 μm, $n_0$ = 3.34 (InGaAsP/InP), $A_{eff}$ = 0.50, 0.25 μm$^2$ and 0.10 μm$^2$ for a microring and nanoring resonator, respectively, $\alpha$ = 0.5 dBmm$^{-1}$, $\gamma$ = 0.1. The coupling coefficient (kappa, $\kappa$) of the microring resonator is 0.009. The nonlinear refractive index is $n_2$ = 2.2 × 10$^{-13}$ m$^2$/W. In this case, the wave guided loss used is 0.5 dBmm$^{-1}$.

The input dark soliton pulse is chopped (sliced) into a smaller signal spreading over the spectrum as shown in Figure 3.2 (middle). This is shown that the large bandwidth signal is generated within the first ring device, however, it isthe excepted region in which the signal level is very low or vanished, about 0.85–0.15 × 10$^4$ roundtrips, which is called a stop band.

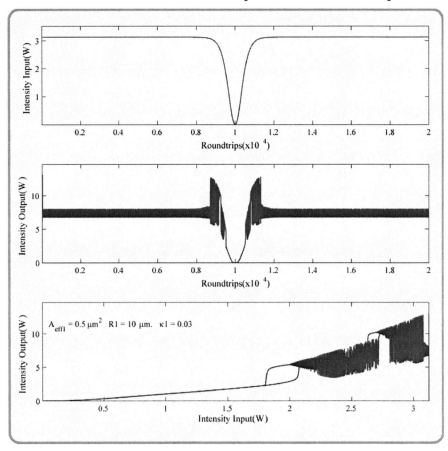

**FIGURE 3.3** Results obtained when a dark soliton pulse is input into a microring resonator, where $R_1$ = 10 μm, $A_{eff1}$ = 0.50 μm$^2$, $\kappa_1$ = 0.030.

The plot of nonlinear behaviors is shown in Figure 3.2 (bottom), where the plot between output and input power (intensity) is seen. The bistability, bifurcation and chaos occur when the input intensity is increased, for instance, the first bistability occurs when the input intensity is between 1.8–2.0 W. Similarly, results obtained when a dark soliton pulse is input into a microring resonator, where the parameters used are $R_1 = 10$ μm, $A_{\text{eff1}} = 0.50$ μm$^2$, $\kappa_1 = 0.0030$ and $\kappa_1 = 0.050$ as shown in Figures 3.3 and 3.4, respectively. The different input and output powers are used, where the different characteristics are seen. The smaller bandwidth modulation dept with smaller signal amplification occur in Figure 3.3, where the larger ones are seen in Figure 3.4.

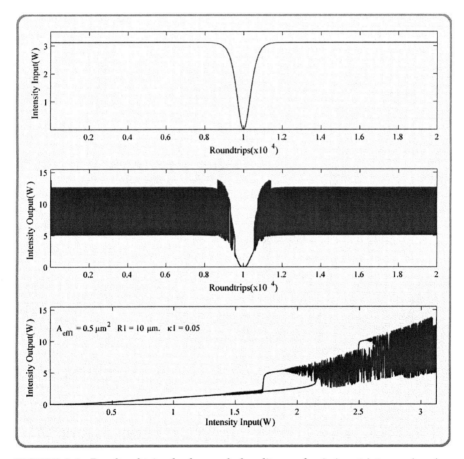

**FIGURE 3.4** Results obtained when a dark soliton pulse is input into a microring resonator, where the parameters used are $R_1 = 10$ μm, $A_{\text{eff1}} = 0.50$ μm$^2$, $\kappa_1 = 0.050$.

The attempt of using a dark soliton pulse propagating into a micro and nanoring system is shown in Figure 3.1(b). Result obtained is shown in Figure 3.5. The parameters used are (a) $R_1 = 10$ μm, $R_2 = 5$ μm, $R_3 = 2.5$ μm, center wavelength at 1,550 nm, and (b) $R_1 = 10$ μm, $R_2 = 7$ μm, $R_3 = 5$ μm, center wavelength at 1,550 nm. The coupling coefficients are the same,

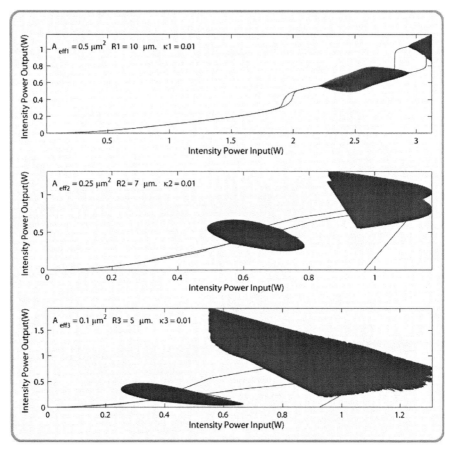

**FIGURE 3.5** Result of a dark soliton pulse propagating in a microring resonator system, where $R_1 = 10$ μm, $A_{\text{eff1}} = 0.50$ μm$^2$, $R_2 = 7$ μm, $A_{\text{eff2}} = 0.25$ μm$^2$, $R_3 = 5$ μm, $A_{\text{eff3}} = 0.10$ μm$^2$ and $\kappa_1 = 0.01$ and $\kappa_2 = \kappa_3 = 0.01$.

which is equal to 0.01. In Figure 3.5, a dark soliton pulse is input into a microring resonator system, where the parameters used are $R_1 = 10$ μm, $A_{\text{eff1}} = 0.50$ μm$^2$, $R_2 = 7$ μm, $A_{\text{eff2}} = 0.25$ μm$^2$, $R_3 = 5$ μm, $A_{\text{eff3}} = 0.10$ μm$^2$ and $\kappa_1 = 0.01$ and $\kappa_2 = \kappa_3 = 0.01$. By using the control light concept, light can be trapped within the waveguide. For example, the control of input power after a soliton propagating within the system for a certain roundtrip is plausible, where the input power can be adjusted to the values that can keep the trapping light being localized within the ring resonator.

## 3.4 Optical Soliton

Optical solitons can naturally be divided into dark and bright solitons. Dark soliton exhibits specifically, an interesting and remarkable behavior, when it is transmitted into an optical transmission system. It has the advantage of signal security, when the ambiguity of signal detection becomes a problem of the un-required users. Sarapat et al. [20] have shown that the conversion of a dark soliton into a bright one can be realized using an add/drop filter. Here the secured signals in the transmission are retrieved using a suitable add/drop filter that is connected to the transmission line. The other promising application of a dark soliton signal is confirmed by using the large guard band of two different frequencies which can be achieved by using a dark soliton generation scheme and trapping a dark soliton pulse within a nanoring resonator [21, 22]. Furthermore, the dark soliton pulse shows a more stable behavior than the bright solitons with respect to the perturbations such as amplifier noise, fiber losses and intra-pulse stimulated Raman scattering [23]. It is found that the dark soliton pulses propagation in a lossy fiber, spreads in time at approximately half the rate of bright solitons. Heidari et al. [24] have shown that the multichannel wavelength conversion uses three different types of dispersion profiles along the optical fibers. The dark solitons trapped in add/drop system is possible as optical tweezers due to scattering effects and gradient force of light propagation. Yuan et al. [25] have shown an abruptly tapered twin-core fiber optical tweezers, by using two-beam combination technique and found that a strong enough gradient forces well tapered twin-core fiber optical tweezers. Optical tweezers were also characterized in terms of the optical potential well by measuring the displacement of trapped particles experiencing a viscous drag at a fluid flow below the critical velocity [26]. We are looking for a stationary dark soliton pulse, which is introduced into the multistage microring resonators as shown in Figure 3.6. The input optical field ($E_{in}$) of the dark soliton pulse input is given by [20]

$$E_{in}(t) = A \tanh\left[\frac{T}{T_0}\right] \exp\left[\left(\frac{z}{2L_D}\right) - i\omega_0 t\right] \quad (3.6)$$

where $A$ and $z$ are the optical field amplitude and propagation distance, respectively. $T$ is a soliton pulse propagation time in a frame moving at the group velocity, $T = t - \beta_1 z$, where $\beta_1$ and $\beta_2$ are the coefficients of the linear and second-order terms of Taylor expansion of the propagation constant. $L_D = T_0^2/|\beta_2|$ is the dispersion length of the soliton pulse. $T_0$ in equation is a soliton pulse propagation time at initial input (or soliton pulse width), where $t$ is the soliton phase shift time, and the frequency shift of the soliton is $\omega_0$. This solution describes a pulse that keeps its temporal width in variance as it propagates, and is thus called a temporal soliton. When a soliton of peak intensity $(|\beta_2/\Gamma T_0^2|)$ is given then $T_0$ is known. For the soliton pulse in the microring device, a balance should be achieved between the dispersion length ($L_D$) and the nonlinear length ($L_{NL} = 1/\Gamma\phi_{NL}$), where $\Gamma = n_2 k_0$, is the length scale over which dispersive or nonlinear effects makes the beam become wider or narrower. For a soliton pulse, there is a balance between dispersion and nonlinear lengths. Hence $L_D = L_{NL}$.

When light propagates within the nonlinear medium, the refractive index ($n$) of light within the medium is given by

$$n = n_0 + n_2 I = n_0 + \frac{n_2}{A_{eff}} P \qquad (3.7)$$

where $n_0$ and $n_2$ are the linear and nonlinear refractive indexes, respectively. $I$ and $P$ are the optical intensity and optical power, respectively. The effective mode core area of the device is given by $A_{eff}$. For the microring resonator (MRR) and nanoring resonator (NRR), the effective mode core areas range from 0.50 to 0.10 μm$^2$ [19]. When a soliton pulse is input and propagated within a MRR, as shown in Figure 3.6, it consists of a series of MRRs. The resonant output is formed, thus, the normalized output of the light field is the ratio between the output and input fields [$E_{out}(t)$ and $E_{in}(t)$] in each roundtrip, which is given by [21]

$$\left|\frac{E_{out}(t)}{E_{in}(t)}\right|^2 = (1-\gamma)\left[1 - \frac{\left(1-(1-\gamma)x^2\right)\kappa}{\left(1-x\sqrt{1-\gamma}\sqrt{1-\kappa}\right)+4x\sqrt{1-\gamma}\sqrt{1-\kappa}\sin^2\left(\frac{\phi}{2}\right)}\right] \qquad (3.8)$$

The close form of Eq. (3.8) indicates that a ring resonator in this particular case is very similar to a Fabry–Pérot cavity, which has an input and output mirror with a field reflectivity, $(1-\kappa)$, and a fully reflecting mirror. $\kappa$ is the coupling coefficient, and $x = \exp(-\alpha L/2)$ represents a roundtrip loss coefficient, $\phi_0 = kLn_0$ and $\phi_{NL} = kLn_2|E_{in}|^2$ are the linear and nonlinear phase shifts, $k = 2\pi/\lambda$ is the wave propagation number in a vacuum, where $L$ and $\alpha$ are waveguide length and linear absorption coefficient, respectively. In this chapter, the iterative method is introduced to obtain the results as shown in Eq. (3.8) and similarly, when the output field is connected and input into the other ring resonators.

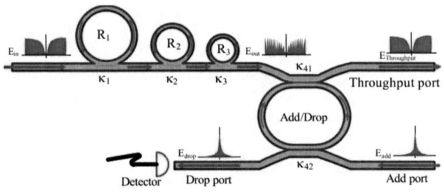

**FIGURE 3.6** A schematic diagram of a dark–bright soliton conversion system, where $R_s$ is the ring radii, $\kappa_s$ is the coupling coefficient, and $\kappa_{41}$ and $\kappa_{42}$ are the add/drop coupling coefficients.

To retrieve the signals from the chaotic noise, we propose to use the add/drop device with the appropriate parameters. This is given in the following details. The optical circuits of ring-resonator add/drop filters for the throughput and drop port can be given by Eqs (3.9) and (3.10), respectively [21].

$$\left|\frac{E_t}{E_{in}}\right|^2 = \frac{(1-\kappa_1) - 2\sqrt{1-\kappa_1}\sqrt{1-\kappa_2}e^{-\frac{\alpha}{2}L}\cos(k_n L) + (1-\kappa_2)e^{-\alpha L}}{1+(1-\kappa_1)(1-\kappa_2)e^{-\alpha L} - 2\sqrt{1-\kappa_1}\sqrt{1-\kappa_2}e^{-\frac{\alpha}{2}L}\cos(k_n L)} \quad (3.9)$$

and

$$\left|\frac{E_d}{E_{in}}\right|^2 = \frac{\kappa_1 \kappa_2 e^{-\frac{\alpha}{2}L}}{1+(1-\kappa_1)(1-\kappa_2)e^{-\alpha L} - 2\sqrt{1-\kappa_1}\sqrt{1-\kappa_2}e^{-\frac{\alpha}{2}L}\cos(k_n L)} \quad (3.10)$$

where $E_t$ and $E_d$ represent the optical fields of the throughput and drop ports, respectively. $\beta = k n_{eff}$ is the propagation constant, $n_{eff}$ is the effective refractive index of the waveguide, and the circumference of the ring is $L = 2\pi R$ with $R$ as the radius of the ring. In the following, new parameter is used for simplification with $\phi = \beta L$ as the phase constant. The chaotic noise cancellation can be managed by using the specific parameters of the add/drop device, and the required signals can be retrieved by the specific users. $\kappa_1$ and $\kappa_2$ are the coupling coefficients of the add/drop filters, $k_n = 2\pi/\lambda$ is the wave propagation number for in a vacuum, and where the waveguide (ring resonator) loss is $\alpha = 0.5$ dBmm$^{-1}$. The fractional coupler intensity loss is $\gamma = 0.1$. In the case of the add/drop device, the nonlinear refractive index is neglected.

## 3.5 Dark-Bright Soliton Conversion

Dark-bright soliton control has been investigated clearly by the authors in reference [20], where one of the advantages is that the dark soliton peak signal is always low level, which can be useful for secured signal communication in the transmission link. The other is formed when the high optical field is configured as an optical tweezer or potential well, which is available for atom/molecule trapping. Optical tweezers technique has become a powerful tool for manipulation of micrometer-sized particles in three spatial dimensions [27]. Initially, the useful static tweezer is recognized, and the dynamic tweezer is now realized in practical work. Typically, by using the continuous-wave (cw) lasers, the spatial control of atoms, beyond their trapping in stationary potentials, has been continuously gaining importance in investigations of ultra cold gases and in the application of atomic ensembles and single atoms for cavity quantum electrodynamics (QED) and quantum information studies. Recent progress includes the trapping and control of single atoms in dynamic potentials [28, 29], the submicron positioning of individual atoms with standing-wave potentials [30, 31], micro-structured and dynamic traps for Bose-Einstein condensates [32–34] and, as another example, the realization of chaotic dynamics in atom-optics "billiards" [35–37]. Schulz et al. [38] have shown that the transfer of trapped atoms between two optical potentials could be performed. In this chapter, we present a novel system of the optical tweezers storage using a dark-bright soliton pulse propagating within an add/drop optical filter. The multiplexing signals with different wavelengths of the dark solitions are controlled and amplified within the system. The dynamic behaviors of dark bright soliton interaction are analyzed and described. The storage signals are controlled and tuned to be an optical probe which is known as the optical tweezers. The optical tweezers storages are obtained by using the embedded nanoring resonators within the add/drop optical filter system. The controlled light pulses are added into the add port of the add/drop filter. By using the bright soliton input, the different in time of the first two dynamic wells of 1 ns is noted, while the potential well stability is seen when the Gaussian pulse is input into the add port. In application, the optical tweezers can be stored and trapped light/atom, which can be formed the dynamic tweezers and tweezers memory.

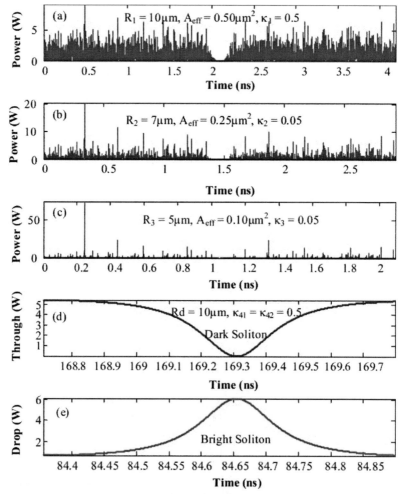

**FIGURE 3.7** Results of the soliton signals within the ring resonator system, where (a) $R_1$, (b) $R_2$, (c) $R_3$, and (d) – (e) dark-bright solitons conversion at the add/drop filter. The input dark soliton power is 2 W.

Experimentally, the generated dark soliton pulse, for instance, with 50 ns pulse width, and a maximum power of 0.65 W is input into the dark-bright soliton conversion system, as shown in Figure 3.6. The suitable ring parameters are used, such as ring radii where $R_1$ = 10.0 μm, $R_2$ = 7.0 μm, and $R_3$ = 5.0 μm. In order to make the system associate with the practical device [19], whereas the selected parameters of the system are fixed to $\lambda_0$ = 1.50 μm, $n_0$ = 3.34 (InGaAsP/InP). The effective core areas are $A_{eff}$ = 0.50, 0.25 and 0.10 μm² for a MRR and NRR, respectively. The waveguide and coupling loses are $\alpha$ = 0.5 dBmm$^{-1}$ and $\gamma$ = 0.1, respectively, and the coupling coefficients $\kappa$s of the MRR are ranged from 0.05 to 0.90. However, more parameters are used as shown in Figure 6. The nonlinear refractive index is $n_2$ = 2.2 × 10$^{-13}$ m²/W. In this case, the waveguide loss used is 0.5 dBmm$^{-1}$. The input dark soliton pulse is chopped (sliced) into the smaller signals, where the filtering signals within the rings $R_2$ and $R_3$ are seen. We find that the output signals from $R_3$ are smaller than from $R_1$, which is more difficult to detect when it is used in the link. In fact, the multistage ring system is proposed due to the different core effective areas of the rings in the system, where the effective areas can be transferred from 0.50 to 0.10 μm² with some losses. The soliton signals in $R_3$ is entered in the add/drop filter, where the dark-bright soliton conversion can be performed by using Eqs (3.9) and (3.10). Results obtained when a dark soliton pulse is input into a MRR and NRR system as shown in Figure 3.6. The add/drop filter is formed by using two couplers and a ring with radius ($R_d$) of 10 μm, the coupling constants ($\kappa_{41}$ and $\kappa_{42}$) are the same values (0.50). When the add/drop filter is connected to the third ring ($R_3$), the dark-bright soliton conversion can be seen. The bright and dark solitons are detected by the through (throughput) and drop ports as shown in Figure 3.7(a)–(e), respectively.

## 3.6 Dark-Bright Soliton Conversion in Add/Drop Filter

In operation, dark-bright soliton conversion using a ring resonator optical channel dropping filter (OCDF) is composed of two sets of coupled waveguides, as shown in Figure 3.8. The relative phase of the two output light signals after coupling into the optical coupler is $\pi/2$ before coupling into the ring and the input bus, respectively. This means that the signals coupled into the drop and through ports are acquired a phase of $\pi$ with respect to the input port signal. In application, if we engineer the coupling coefficients appropriately, the field coupled into the through port on resonance would completely extinguish the resonant wavelength, and all power would be coupled into the drop port. We will show that this is possible later in this section.

$$E_{ra} = -j\kappa_1 E_i + \tau_1 E_{rd} \qquad (3.11)$$

$$E_{rb} = \exp(j\omega T/2)\exp(-\alpha L/4)E_{ra} \qquad (3.12)$$

$$E_{rc} = \tau_2 E_{rb} - j\kappa_2 E_a \qquad (3.13)$$

$$E_{rd} = \exp(j\omega T/2)\exp(-\alpha L/4)E_{rc} \qquad (3.14)$$

$$E_t = \tau_1 E_i - j\kappa_1 E_{rd} \qquad (3.15)$$

$$E_d = \tau_2 E_a - j\kappa_2 E_{rb} \qquad (3.16)$$

where $E_i$ is the input field, $E_a$ is the add (control) field, $E_t$ is the through field, $E_d$ is the drop field, $E_{ra} \ldots E_{rd}$ are the fields in the ring at points $a \ldots d$, $\kappa_1$ is the field coupling coefficient between the input bus and ring, $\kappa_2$ is the field coupling coefficient between the ring and output bus, $L$ is the circumference of the ring, $T$ is the time taken for one round-trip (round-trip time), and $\alpha$ is the power loss in the ring per unit length. We assume that this is the lossless coupling, i.e., $\tau_{1,2} = \sqrt{1 - \kappa_{1,2}^2}$ $T = Ln_{\text{eff}}/c$.

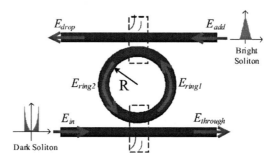

**FIGURE 3.8** A schematic diagram of dark-bright soliton conversion using a ring resonator optical channel dropping filter (OCDF).

The output power/intensities at the drop and through ports are given by

$$|E_d|^2 = \left| \frac{-\kappa_1 \kappa_2 A_{1/2} \Phi_{1/2}}{1 - \tau_1 \tau_2 A \Phi} E_i + \frac{\tau_2 - \tau_1 A \Phi}{1 - \tau_1 \tau_2 A \Phi} E_a \right|^2 \qquad (3.17)$$

$$|E_t|^2 = \left| \frac{\tau_2 - \tau_1 A \Phi}{1 - \tau_1 \tau_2 A \Phi} E_i + \frac{-\kappa_1 \kappa_2 A_{1/2} \Phi_{1/2}}{1 - \tau_1 \tau_2 A \Phi} E_a \right|^2 \qquad (3.18)$$

where $A_{1/2} = \exp(-\alpha L/4)$ (the half-round-trip amplitude), $A = A_{1/2}^2$, $\Phi_{1/2} = \exp(j\omega T/2)$ (the half-round-trip phase contribution), and $\Phi = \Phi_{1/2}^2$.

The input and control fields at the input and add ports are formed by the dark-bright optical soliton as shown in Eqs (3.19)–(3.20)

$$E_{\text{in}}(t) = E_0 \tanh\left[\frac{T}{T_0}\right] \exp\left[\left(\frac{z}{2L_D}\right) - i\omega_0 t\right] \qquad (3.19)$$

$$E_{\text{in}}(t) = E_0 \operatorname{sech}\left[\frac{T}{T_0}\right] \exp\left[\left(\frac{z}{2L_D}\right) - i\omega_0 t\right] \qquad (3.20)$$

where $E_0$ and $z$ are the optical field amplitude and propagation distance, respectively. $T = t-\beta_1 z$, where $\beta_1$ and $\beta_2$ are the coefficients of the linear and second-order terms of Taylor expansion of the propagation constant. $L_D = T_0^2/|\beta_2|$ is the dispersion length of the soliton pulse. $T_0$ in equation is a soliton pulse propagation time at initial input (or soliton pulse width), where $t$ is the soliton phase shift time, and the frequency shift of the soliton is $\omega_0$. When the optical field is entered into the nanoring resonator as shown in Figure 3.9, where the coupling coefficient ratio $\kappa_1:\kappa_2$ are 50:50, 90:10, 10:90. By using (a) dark soliton is input into input and control ports, (b) dark and bright soliton are used for input and control signals, (c) bright and dark soliton are used for input and control signals, and (d) bright soliton is used for input and control signals. The ring radii $R_{ad} = 5$ μm, $A_{eff} = 0.25$ μm², $n_{eff} = 3.14$ (for InGaAsP/InP), $\alpha = 5$ dB/mm, $\gamma = 0.1$, $\lambda_0 = 1.51$ μm.

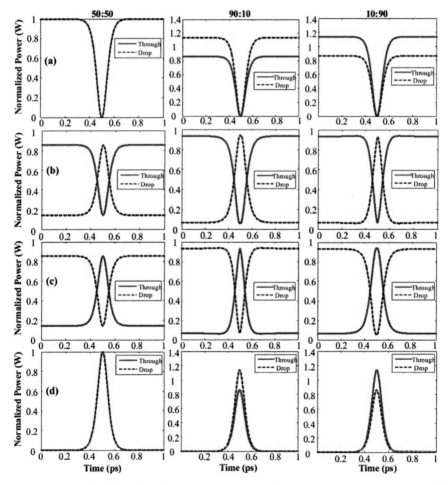

**FIGURE 3.9** Dark-bright soliton conversion results using a ring resonator optical channel dropping filter (OCDF).

In application, the dynamic optical tweezers occurs, when we added bright soliton input at the add port with shown in Figure 3.8, the parameters of system are set the same as the previous section. The bright soliton was generated at the central wavelength $\lambda_0 = 1.5$ μm, when the bright soliton propagating into the add/drop system, the occurrence of dark-bright soliton collision in add/drop system is shown in Figure 3.10(a)–(d) and Figure 3.11(a)–(d).

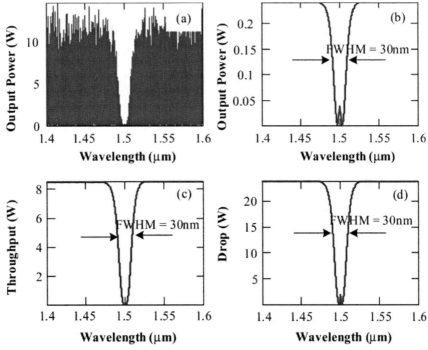

**FIGURE 3.10** The dynamic optical tweezers output within the add/drop filter, when the bright soliton input with the central wavelength $\lambda_0 = 1.5$ μm, where (a) add/drop signal, (b) dark-bright soliton collision, (c)optical tweezers at throughput port, and (d) optical tweezers at drop port.

The optical tweezers probe can be trapped/confined atom/light by using the appropriate probe, which can be tuned to meet the specific requirement. The stability of the dual Brillouin is shown in Figure 3.7. The dark soliton valley dept, i.e. potential well, is changed when it was modulated by the trapping energy (dark-bright solitons interaction) as shown in Figure 3.11(a)–(d). The trapping of photon within the dark well occurs and is seen, the recovery photon can be obtained by using the dark-bright soliton conversion, which is well analyzed by Sarapat et al. [20], where the trapped photon or molecule can be released and seen separately from the dark soliton pulse, in practice, in this case the bright soliton is become alive and seen. The results of the dynamic dark-bright solitons conversion are shown in Figures 3.12 and 3.13.

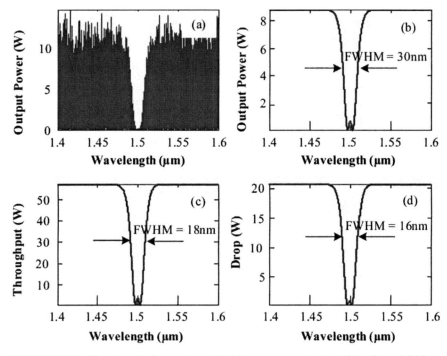

**FIGURE 3.11** The tuned dynamic optical tweezers output within the add/drop filter, when the bright soliton input with the central wavelength $\lambda_0 = 1.5$ μm, where (a) the add/drop signal, (b) dark-bright soliton collision, (c) optical tweezers at throughput port, and (d) optical tweezers at drop port.

## 3.7 Soliton Collision Management in a Microring Resonator

One of the techniques that can be employed well in long distance communication is a soliton communication system. Many earlier works of soliton applications in either theory and experimental works are found in a soliton application book by Hasegawa [39]. Many of the soliton related concepts in fiber optic are discussed by Agrawal [23]. The problems of soliton–soliton interactions [40], collision [41], rectification [42] and dispersion management [43] are required to solve and address. Therefore, in this chapter we are looking for a powerful laser source with broad spectrum that can be used for redundancy incorporating with the transmission link. One of the interesting results have been reported, where the use of a chaotic soliton to form a fast light generation within a tiny device known as a microring resonator (waveguide) has been reported by Yupapin et al. [12]. They have shown that the large bandwidth of a soliton input pulse can be broadly generated. However, we have the evidence of the practical applications, whereas such devices have been fabricated and

# Nanoscale Nonlinear PANDA Ring Resonator

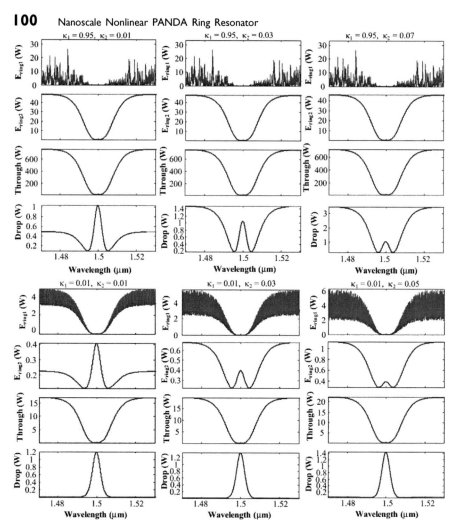

**FIGURE 3.12** Results of the dark-bright soliton conversion, where the coupling coefficients are varied within Figure 3.8.

used [19] in various forms. In this chapter, the broad wavelength of the soliton pulse is generated by using a microring resonator incorporating and add/drop filter. There are more wavelengths of the soliton pulse that can be used incorporating with the soliton wavelength. Therefore, more channels, i.e. wavelengths via DWDM within the soliton regime are also available. By using the nano-waveguide, the selected channel, i.e. a soliton pulse with a specific wavelength can be amplified which is available for link redundancy.

In operation, the large bandwidth signal within the microring device can be generated by using a common soliton pulse input into the nonlinear microring resonator. This means that the broad spectrum of light can be generated after the soliton pulse is input into the ring resonator system. The schematic diagram of the proposed system is as shown in Figure 1.4.

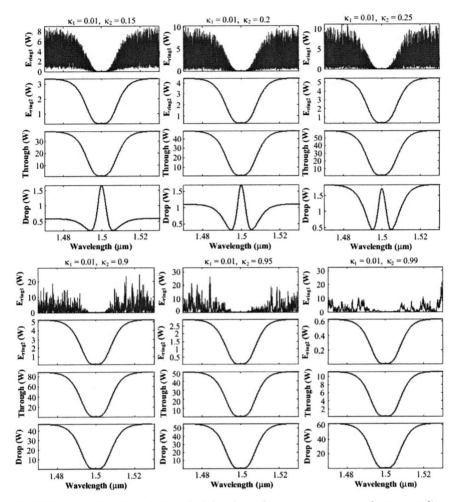

**FIGURE 3.13** Results of the dark-bright soliton conversion, where coupling coefficients are varied within Figure 3.8.

A soliton pulse with 50 ns pulse width, peak power at 2 W is input into the system. The suitable ring parameters are used, for instance, ring radii $R_1 = 15.0$ μm, $R_2 = 10.0$ μm, $R_3 = R_s = 5.0$ μm and $R_5 = R_d = 20.0$ μm. In order to make the system associate with the practical device [19], the selected parameters of the system are fixed to $\lambda_0 = 1.55$ μm, $n_0 = 3.34$ (InGaAsP/InP), $A_{\text{eff}} = 0.50, 0.25$ μm$^2$ and $0.10$ μm$^2$ for a microring and nanoring resonator, respectively, $\alpha = 0.5$ dBmm$^{-1}$, $\gamma = 0.1$. The coupling coefficient (kappa, $\kappa$) of the microring resonator ranged from 0.1 to 0.95. The nonlinear refractive index is $n_2 = 2.2 \times 10^{-13}$ m$^2$/W. In this case, the wave guided loss used is 0.5 dBmm$^{-1}$. The input soliton pulse is chopped (sliced) into the smaller signals spreading over the spectrum (i.e. broad wavelength) as shown in Figure 3.15(b), which is shown that the large bandwidth signal is generated within the first ring device. The biggest output amplification is obtained within the nano-waveguides (rings $R_3$ and $R_4$) as shown in Figure 3.15(d) and (e), whereas the maximum power of 10 W is obtained at the center wavelength of 1.50 µm. The coupling coefficients are given as shown in the figures. The coupling loss is included due to the different core effective areas between micro and nanoring devices, which is given by 0.1 dB.

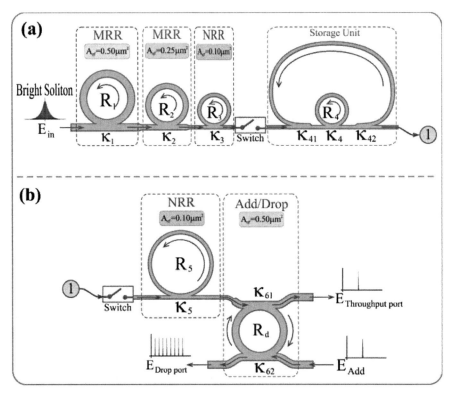

**FIGURE 3.14** A broadband source generation system, (a) a storage unit, (b) a soliton band selector, where $R_s$: ring radii, $\kappa_s$: coupling coefficients, $\kappa_{41}$, $\kappa_{42}$: coupling losses, $\kappa_{61}$ and $\kappa_{61}$ are the add/drop coupling coefficients.

## 3.8 Soliton Collision Management

The obtained results have shown that a large bandwidth of the optical signals with the specific wavelength can be generated within the microring resonator system as shown in Figure 3.14. The amplified signals with broad spectrum can be generated, stored and regenerated within the nano-waveguide. In Figures 3.15 and 3.16 are the results when the output soliton bands at the center wavelengths at 1.99 and 2.48 µm, respectively. The obtained free spectrum range (FSR) and spectral width (Full Width at Half Maximum, FWHM) of the soliton bands is 10 nm and 100 pm, with the center wavelength at 2.48 µm. The maximum stored power of

10 W is obtained as shown in Figure 3.17(d) and (e), where the average regenerated optical output power of 4 W is achieved via and a drop port of an add/drop filter as shown in Figure 3.17(h)-(k), which is a broad spectra of light cover the large bandwidth as shown in Figure 3.17(g). However, to make the system being realistic, the waveguide and connection losses are required to address in the practical device, which may be affected the signal amplification. The storage light pulse within a storage ring ($R_s$ or $R_4$) is achieved, which has also been reported by Ref. [21].

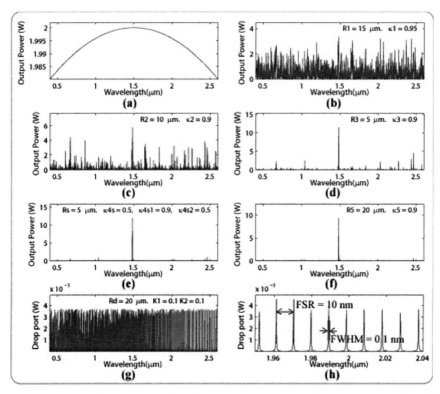

**FIGURE 3.15** Solitons output band with center wavelength at 1.99 μm, where (a) input soliton, (b) ring $R_1$, (c) ring $R_2$, (d) ring $R_3$, (e) storage ring ($R_s$), (f) ring $R_5$, (g) and (h) drop port signals.

In applications, the increasing in communication channel and network capacity can be formed by using the different soliton bands (center wavelength) as shown in Figure 3.17, where Figure 3.17(h) 0.51 μm, Figure 3.17(i) 0.98 μm, Figure 3.17(j) 1.48 μm and Figure 3.17(k) 2.46 μm are the generated center wavelengths of the soliton bands. The selected wavelength center can be performed by using the designed add/drop filter, whereas the required FWHM and FSR are obtained, the channel spacing and bandwidth are represented by FSR and FWHM, respectively, for instance, the FSR and FWHM of 2.3 nm and 100 pm are obtained, respective as shown in Figure 3.17(i). The collision problem of the solitons can be controlled by using the suitable add/drop filter parameters such as radius, coupling input power and ring coupling coefficients as shown in the results in Figures 3.15–3.17. From the results, the sign of the good collision management is the gap between the soliton pulses which is called a FSR, for instance, the FSR in Figures 3.15 and 3.16 is 10 and 14 nm respectively, which is large comparing to the spectral width, i.e. 1 nm, which is enough space to avoid the collision. Moreover, the quality of the signal is presented by the signal cross-talk, which is extremely good because of the soliton behaviors, where the constant gain and phase are the properties that can always be seen. Apart from the soliton behaviors, the different center wavelengths of the selected solitons are also supported the collision management, whereas the transparence of the different wavelength is key beneficial of the collision control management.

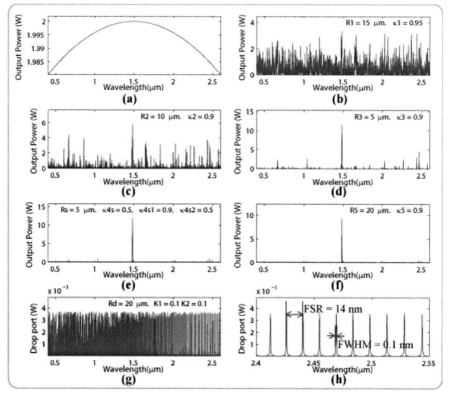

**FIGURE 3.16** Solitons output with center wavelength at 2.48 µm, where (a) input soliton, (b) ring $R_1$, (c) ring $R_2$, (d) ring $R_3$, (e) storage ring ($R_s$), (f) ring $R_5$, (g) and (h) drop port signals.

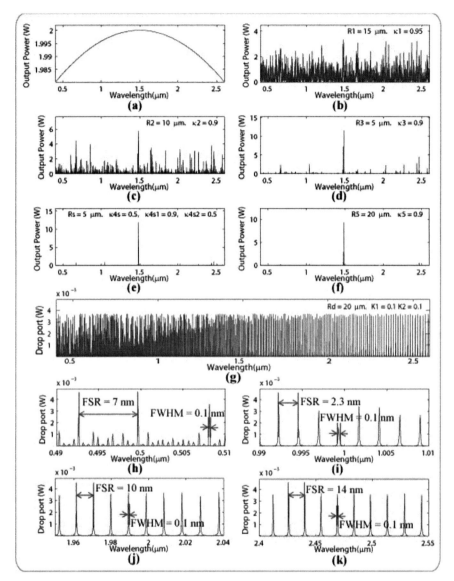

**FIGURE 3.17** Soliton output with different center wavelengths at 1.5 mm, where (a) input soliton, (b) ring $R_1$, (c) ring $R_2$, (d) ring $R_3$, (e) storage ring ($R_s$), (f) ring $R_5$, (g) drop port signals. The different soliton bands (center wavelength) are as shown, where (h) 0.51 μm, (i) 0.98 μm, (j) 1.99 μm and (k) 2.48 μm.

## 3.9 Conclusion

We have demonstrated that the dark soliton pulse was generated by using the pumped laser system. We have shown that the propagating dark soliton within the MRR and NRR system can be converted to be a bright soliton by using the ring resonator system, incorporating the add/drop multiplexer, moreover, the amplification and tenability of the dark soliton pulse can be obtained. By using the reasonable dark-bright soliton input power, the tunable optical tweezers can be controlled, which can be used as the dynamic optical tweezers probe. In application, such a behavior can be used to confine the suitable size of light pulse or molecule, which can be employed in the same way of the optical tweezers. But in this case the terms dynamic probing comes to be a realistic function. Moreover, the transportation of the trapped pulse or molecule is plausible.

The very interesting results when the multi-soliton bands can be generated by using a common soliton wavelength propagating within the nonlinear waveguide system. The device parameters have been designed closely to the practical values, which can be fabricated within a single system. In applications, the large bandwidth of the soliton bands can provide the large users in the future demand, however, in practice, the problem of soliton collision within the device is required to solve, which is well analyzed by Ref. [5]. The best results obtained of FSR and FWHM are 14 nm and 100 pm, respectively, which allows the increasing in channel capacity of at least 10 times in only one center wavelength. This means the idea of personnel wavelength (network) being realistic for the large demand user due to un-limit wavelength discrepancy, whereas the specific soliton band can be generated using the proposed system. The large bandwidth of the arbitrary wavelength of a soliton pulse can be enlarged and stored within a nano-waveguide, which is available for trapping within the waveguide. The selected light pulse can be trapped and controlled by light. Furthermore, the large signal amplification due to the effects of a dark soliton pulse in the nonlinear waveguide may introduce the unexpected applications, where the use for signal security in long distance communication and network can be performed. Moreover, the idea of trapping a dark soliton may introduce the concept of storing a dark soliton, which is a new surprise in this subject.

A novel system of the tunable dynamic optical tweezers using a generated dark soliton in the fiber optic loop is proposed. A dark soliton known as an optical tweezers is amplified and tuned within the ring resonator system. The required tunable tweezers with different width and power can be controlled. The analysis of dark-bright soliton conversion using a dark soliton pulse propagating within a microring resonator system is analyzed. The dynamic behaviors of soliton conversion in add/drop filter is also analyzed. The control dark soliton is input into the system via the add port of the add/drop filter, where the dynamic behavior of the dark-bright soliton conversion is seen. The required stable signal is obtained via a drop and throughput ports of the add/drop filter with some suitable parameters. In application, the dynamic optical tweezers can be configured by the dark-bright soliton conversion system. Therefore, the use of trapped light/atom can be covered by using the proposed system.

# REFERENCES

[1] S. Mitatha, "Dark soliton behaviors within the nonlinear micro and nanoring resonators and applications," *Progress In Electromagnetics Research*, PIER, vol. 99, pp. 383–404 (2009).

[2] M.F.S. Ferreira, "Nonlinear effects in optical fibers: Limitations and benefits," *Proc. SPIE*, vol. 6793, 679302 (2008).

[3] S.A. Akhmanov and S.Y. Nikitin, *Physical Optics*, 657, Oxford: Oxford University Press, Clarendon (1997).

[4] P.P. Yupapin, W. Suwanchareon, and S. Suchat, "Nonlinearity penalties and benefits of light traveling in a fiber optic ring resonator," *Int. J. of Light and Electron. Opt.*, vol. 121(2), pp. 159–167 (2010).

[5] P.P. Yupapin and N. Pornsuwancharoen, *Guided Wave Optics and Photonics: Microring Resonator Design for Telephone Network Security*, New York: Nova Science Publishers (2008).

[6] P.P. Yupapin and P. Saeung, *Photonics and Nanotechnology*, Singapore: World Scientific (2008).

[7] Y. Su, F. Liu, and Q. Li, "System performance of slow-light buffering and storage in silicon nano-waveguide," *Proc. SPIE*, vol. 6783, 679302P (2007).

[8] F.G. Gharakhili, M. Shahabadi, and M. Hakkak, "Bright and dark soliton generation in a left-handed nonlinear transmission line with series nonlinear capacitors," *Progress in Electromagnetics Research*, PIER, vol. 96, pp. 237–249 (2009).

[9] M. Ballav and A.R. Chowdhury, "On a study of diffraction and dispersion managed soliton in a cylindrical media," *Progress in Electromagnetics Research*, PIER, vol. 63, pp. 33–50 (2006).

[10] S. Konar and A. Biswas, "Soliton–soliton interaction with power law nonlinearity," *Progress in Electromagnetics Research*, PIER , vol. 54, pp. 95–108 (2005).

[11] R. Gangwar, S.P. Singh, and N. Singh, "Soliton based optical communication," *Progress in Electromagnetics Research*, PIER, vol. 74, pp. 157–166 (2007).

[12] P.P. Yupapin, N. Pornsuwanchroen, and S. Chaiyasoonthorn, "Attosecond pulse generation using nonlinear microring resonators," *Microw. Opt. Technol. Lett.*, vol. 50, pp. 3108–3111 (2008).

[13] N. Sangwara, K. Sarapat, K. Srinuanjan, and P.P. Yupapin, "A novel dark-bright optical solitons conversion system and power amplification," *Opt. Eng.*, vol. 48, no. 4, p. 045004 (2009).

[14] W. Zhao and E. Bourkoff, "Propagation properties of dark solitons," *Opt. Lett.*, vol. 14, pp. 703–705 (1989).

[15] I.V. Barashenkov, "Stability criterion for dark soliton," *Phys. Rev. Lett.*, vol. 77, pp. 1193–1195 (1996).

[16] D.N. Christodoulides, T.H. Coskun, M. Mitchell, Z. Chen, and M. Segev, "Theory of incoherent dark solitons," *Phys. Rev. Lett.*, vol. 80, pp. 5113–5115 (1998).

[17] A.D. Kim, W.L. Kath, and C.G. Goedde, "Stabilizing dark solitons by periodic phase-sensitive amplification," *Opt. Lett.*, vol. 21, pp. 465–467 (1996).
[18] C. Fietz and G. Shvets, "Nonlinear polarization conversion using microring resonators," *Opt. Lett.*, vol. 32, pp. 1683–1685 (2007).
[19] Y. Kokubun, Y. Hatakeyama, M. Ogata, S. Suzuki, and N. Zaizen, "Fabrication technologies for vertically coupled microring resonator with multilevel crossing bus line and ultracompact-ring radius," *IEEE J. of Sel. Topics in Quantum Electron.*, vol. 11, pp. 4–10 (2005).
[20] K. Sarapat, N. Sangwara, K. Srinuanjan, P.P. Yupapin, and N. Pornsuwancharoen. "Novel dark-bright optical solitons conversion system and power amplification," *Optical Engineering*, vol. 48, p. 045004-1 (2009).
[21] S. Mitatha, N. Pornsuwancharoen, and P.P. Yupapin, "A simultaneous short-wave and millimeter-wave generation using a soliton pulse within a nano-waveguide," *IEEE Photon. Technol. Lett.*, vol. 21, pp. 932–934 (2009).
[22] A. Charoenmee, N. Pornsuwancharoen, and P.P. Yupapin, "Trapping a dark soliton pulse within a nanoring resonator," *International Journal of Light and Electron Optics*, vol. 121(18), pp. 1670–1673 (2010).
[23] G.P. Agrawal, *Nonlinear Fiber Optics*, New York: Academic Press, 4th edition (2007).
[24] M.E. Heidari, M.K. Moravvej-Farshi, and A. Zariffkar, "Multichannel wavelength conversion using fourth-order soliton decay," *J. Lightwave Technol.*, vol. 25, pp. 2571–2578 (2007).
[25] L. Yuan, Z. Liu, J. Yang, and C. Guan, "Twin-core fiber optical tweezers," *Optics Express*, vol. 16, pp. 4559–4566 (2008).
[26] N. Malagninoa, G. Pescea, A. Sassoa, and E. Arimondo, "Measurements of trapping efficiency and stiffness in optical tweezers," *Opt. Commun.*, vol. 214, pp. 15–24 (2002).
[27] A. Ashkin, J.M. Dziedzic, J.E. Bjorkholm, and S. Chu, "Observation of a single-beam gradient force optical trap for dielectric particles," *Opt. Lett.*, vol. 11, pp. 288–290 (1986).
[28] S. Bergamini, B. Darqui, M. Jones, L. Jacubowiez, A. Browaeys, and P. Grangier, "Holographic generation of microtrap arrays for single atoms by use of a programmable phase modulator", *J. Opt. Soc. Am. B*, vol. 21, pp. 1889–1894 (2004).
[29] D.D. Yavuz, P.B. Kulatunga, E. Urban, T.A. Johnson, N. Proite, T. Henage, T.G. Walker, and M. Saffman, "Fast ground state manipulation of neutral atoms in microscopic optical traps", *Phys. Rev. Lett.*, vol. 96, p. 063001 (2006).
[30] D. Schrader, I. Dotsenko, M. Khudaverdyan, Y. Miroshnychenko, A. Rauschenbeutel, and D. Meschede, "Neutral atom quantum register", *Phys. Rev. Lett.*, vol. 93, p. 150501 (2004).

[31] J.A. Sauer, K.M. Fortier, M.S. Chang, C.D. Hamley, and M.S. Chapman, "Submicrometer position control of single trapped neutral atoms", *Phys. Rev.*, A 69, 051804(R) (2004).

[32] T.P. Meyrath, F. Schreck, J.L. Hanssen, C.-S. Chuu, and M.G. Raizen, "Bose-Einstein condensate in a box," *Phys. Rev. A*, vol. 71, p. 041604(R) (2005).

[33] V. Boyer, R.M. Godun, G. Smirne, D. Cassettari, C.M. Chandrashekar, A.B. Deb, Z.J. Laczik, and C.J. Foot, "Dynamic manipulation of Bose-Einstein condensates with a spatial light modulator", *Phys. Rev. A*, vol. 73, p. 031402(R) (2006).

[34] A.V. Carpentier, J. Belmonte-Beitia, H. Michinel, and V.M. Perez-Garcia, "Laser tweezers for atomic solitons," *J. of Mod. Opt.*, vol. 55, no. 17, pp. 2819–2829 (2008).

[35] V. Milner, J.L. Hanssen, W.C. Campbell, and M.G. Raizen, "Optical billiards for atoms," *Phys. Rev. Lett.*, vol. 86, pp. 1514–1516 (2001).

[36] N. Friedman, A. Kaplan, D. Carasso, and N. Davidson, "Observation of chaotic and regular dynamics in atom-optics billiards," *Phys. Rev. Lett.*, vol. 86, pp. 1518–1520 (2001).

[37] M. Li and J. Arlt, "Trapping multiple particles in single optical tweezers", *Opt. Commun.*, vol. 281, pp. 135–140 (2008).

[38] M. Schulz, H. Crepaz, F. Schmidt-Kaler, J. Eschner, and R. Blatt, "Transfer of trapped atoms between two optical tweezer potentials" *J. of Mod. Opt.*, vol. 54, no. 11, pp. 1619–1626 (2007).

[39] A. Hasegawa, *Massive WDM and TDM Soliton Transmission Systems*, Boston: Kluwer Academic Publishers, (2000).

[40] Yu. A. Simonov and J.A. Tjon, "Soliton-soliton interaction in confining models," *Phys. Lett. B*, vol. 85, pp. 380–384 (1979).

[41] J.K. Drohm, L.P. Kok, Yu. A. Simonov, J.A. Tjon, and A.I. Veselov, "Collision and rotation of soliton in three space time dimensions", *Phys. Lett. B*, vol. 101, pp. 204–208 (1981).

[42] Takeshi Iizuka and Yuri S. Kivshar, "Optical gap solitons in non-resonant quadratic media", *Phys. Rev. E*, vol. 59, pp. 7148–7151 (1999).

[43] R. Ganapathy, K. Porsezian, A. Hasegawa, V.N. Serkin, "Soliton interaction under soliton dispersion management", *IEEE Quantum Electron*, vol. 44, no. 4, pp. 383–390 (2008).

# 4
# Dynamic Optical Tweezers

## CHAPTER OUTLINE

- Introduction
- The Add/Drop Optical Filter
- Storage Trapping Tool
- Dynamic Potential Well Generation
- Dynamic Optical Tweezers via a Wavelength Router
- Trapping Forces
- Trapping Stability
- Trapping and Transportation Mechanism
- Atom/Molecule Transmission and Transportation via Wavelength Router
- Conclusion
- References

## 4.1 Introduction

Optical tweezers technique has become a powerful tool for manipulation of micrometer-sized particles in three spatial dimensions [1]. Initially, the useful static tweezer is recognized, and the dynamic tweezer is now realized in practical work. Typically by using the continuous-wave (cw) lasers, the spatial control of atoms, beyond their trapping in stationary potentials, has been continuously gaining importance in investigations of ultra cold gases, application of atomic ensembles and single atoms for cavity quantum electrodynamics (QED) and quantum information studies. Recent progress includes the trapping and control of single atoms in dynamic potentials [2, 3], the sub-micron positioning of individual atoms with standing-wave potentials [4, 5], micro-structured and dynamic traps for Bose-Einstein condensates [6, 7, 8] and the realization of chaotic dynamics in atom-optics "billiards" [9, 10, 11]. Recently, Matthias et al. [12] have shown that the transfer of trapped atoms between two optical potentials could be performed. In this chapter, we present a novel system of the optical tweezers storage using a dark-bright soliton pulses propagating within an add/drop optical filter. The multiplexing signals with different wavelengths of the dark solitions are controlled and amplified within the system. Dynamic behaviors of dark bright soliton interaction is analyzed and described. The storage signals are controlled and tuned from optical probe which is known as the optical tweezers. The optical tweezers storage is obtained by using the embedded nanoring resonator within the add/drop optical filter system. In application, the optical tweezers can be used to store and trap light/atom, which can form the tweezers memory, which is available for long distance link of high density atoms transmission.

We are looking for a stationary dark soliton pulse, which is introduced into the multistage microring resonators as shown in Figure 4.1 [13]. The input optical field ($E_{in}$) of the dark soliton pulse input and the add optical field ($E_{add}$) of the bright soliton pulse at add port are given by [14]

$$E_{in}(t) = A\tanh\left[\frac{T}{T_0}\right]\exp\left[\left(\frac{z}{2L_D}\right) - i\omega_0 t\right] \quad (4.1a)$$

$$E_{add}(t) = A\operatorname{sech}\left[\frac{T}{T_0}\right]\exp\left[\left(\frac{z}{2L_D}\right) - i\omega_0 t\right] \quad (4.1b)$$

$$E_{add}(t) = E_0 \exp\left[\left(\frac{z}{2L_D}\right) - i\omega_0 t\right] \quad (4.1c)$$

where $A$ and $z$ are the optical field amplitude and propagation distance, respectively. $T$ is a soliton pulse propagation time in a frame moving at the group velocity, $T = t - \beta_1 z$, where $\beta_1$ and $\beta_2$ are the coefficients of the linear and second-order terms of Taylor expansion of the propagation constant. $L_D = T_0^2/|\beta_2|$ is the dispersion length of the soliton pulse. $T_0$ in equation is a soliton pulse propagation time at initial input (or soliton pulse width), where $t$ is the soliton phase shift time, and the frequency shift of the soliton is $\omega_0$. This solution describes a pulse that keeps its temporal width in variance as it propagates, and thus is called a temporal soliton. When a soliton peak intensity ($|\beta_2/\Gamma T_0^2|$) is given then $T_0$ is known. For the soliton pulse in the microring device, a balance should be achieved between the dispersion length ($L_D$) and the nonlinear length ($L_{NL} = 1/\Gamma\phi_{NL}$). $\Gamma = n_2 k_0$, is the length scale over which dispersive or nonlinear effects makes the beam become wider or narrower. For a soliton pulse, there is a balance between dispersion and nonlinear lengths, hence $L_D = L_{NL}$.

When light propagates within the nonlinear material (medium), the refractive index ($n$) of light within the medium is given by

$$n = n_0 + n_2 I = n_0 + \frac{n_2}{A_{\text{eff}}} P \qquad (4.2)$$

where $n_0$ and $n_2$ are the linear and nonlinear refractive indexes, respectively. $I$ and $P$ are the optical intensity and optical power, respectively. The effective mode core area of the device is given by $A_{\text{eff}}$. For the microring resonator (MRR) and nanoring resonator (NRR), the effective mode core areas range from 0.50 to 0.10 μm².

## 4.2 The Add/Drop Optical Filter

When a soliton pulse is input and propagated within a MRR, as shown in Figure 4.1, which consists of a series MRRs. The resonant output is formed, thus, the normalized output of the light field is the ratio between the output and input fields [$E_{\text{out}}(t)$ and $E_{\text{in}}(t)$] in each roundtrip, which is given by

$$\left|\frac{E_{\text{out}}(t)}{E_{\text{in}}(t)}\right|^2 = (1-\gamma)\left[1 - \frac{(1-(1-\gamma)x^2)\kappa}{(1-x\sqrt{1-\gamma}\sqrt{1-\kappa})^2 + 4x\sqrt{1-\gamma}\sqrt{1-\kappa}\sin^2\left(\frac{\varphi}{2}\right)}\right] \qquad (4.3)$$

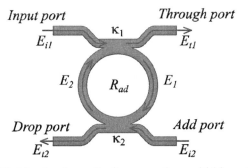

**FIGURE 4.1** A schematic diagram of an add/drop filter.

The close form of Eq. (4.3) indicates that a ring resonator in this particular case is very similar to a Fabry–Pérot cavity, which has an input and output mirror with a field reflectivity, $(1 - \kappa)$, and a fully reflecting mirror. $\kappa$ is the coupling coefficient, and $x = \exp(-\alpha L/2)$ represents a roundtrip loss coefficient. $\phi_0 = kLn_0$ and $\phi_{\text{NL}} = kLn_2|E_{\text{in}}|^2$ are the linear and nonlinear phase shifts. $k = 2\pi/\lambda$ is the wave propagation number in a vacuum, where $L$ and $\alpha$ are waveguide length and linear absorption coefficient, respectively. In this chapter, the iterative method is introduced to obtain the results as shown in Eq. (4.3), similarly, when the output field is connected and input into the other ring resonators.

To retrieve the signals from the chaotic noise, we propose to use the add/drop device with the appropriate parameters. This is given in the following details. The optical circuits of ring-resonator add/drop filters for the throughput and drop port can be given by Eqs (4.4) and (4.5), respectively [13, 14]

$$\left|\frac{E_t}{E_{in}}\right|^2 = \frac{(1-\kappa_1) - 2\sqrt{1-\kappa_1} \cdot \sqrt{1-\kappa_2}e^{-\frac{\alpha}{2}L}\cos(k_n L) + (1-\kappa_2)e^{-\alpha L}}{1+(1-\kappa_1)(1-\kappa_2)e^{-\alpha L} - 2\sqrt{1-\kappa_1} \cdot \sqrt{1-\kappa_2}e^{-\frac{\alpha}{2}L}\cos(k_n L)} \quad (4.4)$$

$$\left|\frac{E_d}{E_{in}}\right|^2 = \frac{\kappa_1 \kappa_2 e^{-\frac{\alpha}{2}L}}{1+(1-\kappa_1)(1-\kappa_2)e^{-\alpha L} - 2\sqrt{1-\kappa_1} \cdot \sqrt{1-\kappa_2}e^{-\frac{\alpha}{2}L}\cos(k_n L)} \quad (4.5)$$

where $E_t$ and $E_d$ represent the optical fields of the throughput and drop ports, respectively. $\beta = kn_{eff}$ is the propagation constant, $n_{eff}$ is the effective refractive index of the waveguide. The circumference of the ring is $L = 2\pi R$, with $R$ as the radius of the ring. In the following, new parameters will be used for simplification with $\phi = \beta L$ as the phase constant. The chaotic noise cancellation can be managed by using the specific parameters of the add/drop device, and the required signals can be retrieved by the specific users. $\kappa_1$ and $\kappa_2$ are the coupling coefficient of the add/drop filters, $k_n = 2\pi/\lambda$ is the wave propagation number in vacuum. The waveguide (ring resonator) loss is $\alpha = 0.5$ dBmm$^{-1}$. The fractional coupler intensity loss is $\gamma = 0.1$. In the case of the add/drop device, the nonlinear refractive index is neglected.

By controlling the bright soliton which is input into the add port, as shown in Figure 4.1, the dynamic behavior of optical tweezers occurs and is seen. The used parameters are the add/drop optical filter radius $R_{ad} = 15$ µm, the coupling coefficients $\kappa_1 = 0.35$ and $\kappa_2 = 0.7$. The dark and bright solitons are generated at the central wavelength $\lambda_0 = 1.5$ µm. The parameters were obtained by using related practical material parameters (InGaAsP/InP) [15]. When the bright soliton propagating into the add/drop system, the dark-bright soliton collision within the add/drop system is occurred as shown in Figures 4.2(a)–(d). The smallest tweezer width (full width at half maximum, FWHM) of 11 nm is obtained. The maximum power is 50 W, which means that the optical tweezers probe in the form of intense field, i.e. potential well is formed. It can be used to trap/confine atom/light. The dark soliton valley depth, i.e. potential well, is changed when it is modulated by the trapping energy (dark-bright solitons interaction) as shown in Figure 4.2(b). The trapping of photon within the dark well is occurred and seen. The recovery of trapped atom/photon can be obtained by using the dark-bright soliton conversion behavior, which is well analyzed by Sarapat et al. [13]. The trapped photon or molecule can be released and seen separately from the dark soliton pulse. In practice, the bright soliton becomes alive and is seen.

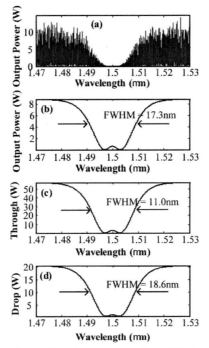

**FIGURE 4.2** Result of the optical tweezers in the add/drop filter system, where $R_{ad} = 15$ μm, $\kappa_1 = 0.35$ and $\kappa_2 = 0.7$: (a) dark soliton input power, (b) dynamic tweezer, (c) throughput port signal and (d) drop port signal.

## 4.3 Storage Trapping Tool

The schematic diagram of the optical tweezers storage is designed and shown in Figure 4.3. In operation, to form the memory unit, a nanoring resonator is embedded within the add/drop optical filter system. The nanoring resonator radius ($R_{ring}$) and the coupling coefficient ($\kappa$) are 100 nm and 0.15, respectively.

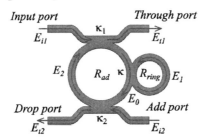

**FIGURE 4.3** A schematic diagram of storage optical tweezers.

The parameters of the add/drop optical filter system are the same as in the previous section. In the system design, the dark soliton pulse is input into the input port through the coupler with the coupling coefficient is $\kappa_1 = 0.35$. It is partially input into the nanoring resonator with 20,000 roundtrips, where the storage signal is observed, the memory time is noted.

The output field ($E_{t1}$) at throughput port is expressed by

$$E_{t1} = -x_1 x_2 y_2 \sqrt{\kappa_1} E_{i2} e^{-\frac{\alpha L}{2} - jk_n \frac{L}{2}} \tag{4.6}$$
$$+ \left[ \frac{x_2 x_3 \kappa_1 \sqrt{\kappa_2} E_0 E_{i1} e^{-\alpha L - jk_n L} + x_3 x_4 y_1 y_2 \sqrt{\kappa_1} \sqrt{\kappa_2} E_0 E_{i2} e^{-\frac{3\alpha L}{2} - jk_n \frac{3L}{2}}}{1 - x_1 x_2 y_1 y_2 E_0 e^{-\alpha L - jk_n L}} \right]$$

where the $x_1 = \sqrt{1-\gamma_1}$, $x_2 = \sqrt{1-\gamma_2}$, $x_3 = 1-\gamma_1$, $x_4 = 1-\gamma_2$, $y_1 = \sqrt{1-\kappa_1}$ and $y_2 = \sqrt{1-\kappa_2}$.

The power output ($P_{t1}$) at throughput port is given by

$$P_{t1} = |E_{t1}|^2 \tag{4.7}$$

The output field ($E_{t2}$) at drop port is

$$E_{t2} = x_2 y_2 E_{i2} + \left[ \frac{x_1 x_2 \sqrt{\kappa_1} \sqrt{\kappa_2} E_0 E_{i1} e^{-\frac{\alpha L}{2} - jk_n \frac{L}{2}} + x_1 x_3 y_1 y_2 \sqrt{\kappa_2} E_0 E_{i2} e^{-\alpha L - jk_n L}}{1 - x_1 x_2 y_1 y_2 E_0 e^{-\alpha L - jk_n L}} \right] \tag{4.8}$$

The power output ($P_{t2}$) at drop port is

$$P_{t2} = \left| x_2 y_2 E_{i2} + \left[ \frac{x_1 x_2 \sqrt{\kappa_1} \sqrt{\kappa_2} E_0 E_{i1} e^{-\frac{\alpha L}{2} - jk_n \frac{L}{2}} + x_1 x_3 y_1 y_2 \sqrt{\kappa_2} E_0 E_{i2} e^{-\alpha L - jk_n L}}{1 - x_1 x_2 y_1 y_2 E_0 e^{-\alpha L - jk_n L}} \right] \right|^2 \tag{4.9}$$

The optical tweezers storage signals within the add/drop system is as shown in Figure 4.4. We found that the storage time is 1.2 ns, the tweezer widths of the storage tweezers in add/drop at the throughput and drop ports are 19.2, 17.6 and 18.6 ns, respectively. In Figures 4.4 (a)–(b), the tweezers in the form of potential wells are seen. It can be used for atom/molecule trapping. The potential well depth (peak valley) can be controlled by adjusting the system parameters, for instance, the bright soliton input power at the add port and the coupling coefficients. The potential well of the tweezers is tuned to be the single well and seen at the drop port, as shown in Figure 4.4(c). In application, the optical tweezers in the design system can be tuned and amplified as shown in Figures 4.2 and 4.4. Therefore, the tunable optical tweezers can be controlled by the dark-bright soliton collision within the add/drop optical system by adjusting the parameters of the input power at the input and add ports, respectively. The output power at the throughput port is shown in Figure 4.2(c), where the single potential well with the optical power of 15 W is seen.

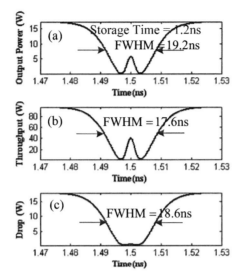

**FIGURE 4.4** Result of the optical tweezers storage signals in the add/drop system, where $R_{ad} = 15$ μm and $R_{ring} = 100$ nm, $\kappa = 0.15$, $\kappa_1 = 0.35$ and $\kappa_2 = 0.1$: (a) storage tweezer, (b) throughput port signal, and (c) drop port signal.

## 4.4 Dynamic Potential Well Generation

Figure 4.5, consists of add/drop optical multiplexing used for generated random binary coded light pulse and add/drop optical filter device for decoded binary code signal. The resonator output field, $E_{t1}$ and $E_1$ consists of the transmitted and circulated components within the add/drop optical multiplexing system, which can perform the driven force to photon/molecule/atom.

When the input light pulse passes through the first coupling region of the add/drop optical multiplexing system, the transmitted and circulated components can be written as

$$E_{t1} = \sqrt{1-\gamma_1}\left[\sqrt{1-\kappa_1}E_{i1} + j\sqrt{\kappa_1}E_4\right] \quad (4.10)$$

$$E_1 = \sqrt{1-\gamma_1}\left[\sqrt{1-\kappa_1}E_4 + j\sqrt{\kappa_1}E_{i1}\right] \quad (4.11)$$

$$E_2 = E_0 e^{-\frac{\alpha L}{2 2} - jk_n\frac{L}{2}} \quad (4.12)$$

where $\kappa_1$ is the intensity coupling coefficient, $\gamma_1$ is the fractional coupler intensity loss, $\alpha$ is the attenuation coefficient, $k_n = 2\pi/\lambda$ is the wave propagation number, $\lambda$ is the input wavelength light field and $L = 2\pi R_{ad}$, $R_{ad}$ is the radius of add/drop device.

For the second coupler of the add/drop optical multiplexing system,

$$E_{t2} = \sqrt{1-\gamma_2}\left[\sqrt{1-\kappa_2}\,E_{i2} + j\sqrt{\kappa_2}\,E_2\right] \quad (4.13)$$

$$E_3 = \sqrt{1-\gamma_2}\left[\sqrt{1-\kappa_2}\,E_2 + j\sqrt{\kappa_2}\,E_{i2}\right] \quad (4.14)$$

$$E_4 = E_{0L}e^{-\frac{\alpha L}{2 2}-jk_n\frac{L}{2}} \quad (4.15)$$

where $\kappa_2$ is the intensity coupling coefficient, $\gamma_2$ is the fractional coupler intensity loss. The circulated light fields, $E_0$ and $E_{0L}$ are the light field circulated components of the nanoring radii, $R_r$ and $R_L$ which coupled into the right and left sides of the add/drop optical multiplexing system, respectively. The light field transmitted and circulated components in the right nanoring, $R_r$, are given by

$$E_2 = \sqrt{1-\gamma_0}\left[\sqrt{1-\kappa_0}\,E_1 + j\sqrt{\kappa_0}\,E_{r2}\right] \quad (4.16)$$

$$E_{r1} = \sqrt{1-\gamma_0}\left[\sqrt{1-\kappa_0}\,E_{r2} + j\sqrt{\kappa_0}\,E_1\right] \quad (4.17)$$

$$E_{r2} = E_{r1}e^{-\frac{\alpha}{2}L_1 - jk_n L_1} \quad (4.18)$$

where $k_0$ is the intensity coupling coefficient, $\lambda$ is the fractional coupler intensity loss, $\alpha$ is the attenuation coefficient, $k_n = 2\pi/\lambda$ is the wave propagation number, $\lambda$ is the input wavelength light field and $L_1 = 2\pi R_r$, $R_r$, is the radius of right nanoring.

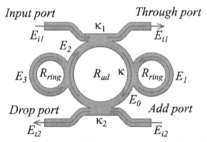

**FIGURE 4.5** A schematic diagram of dynamic potential well within two nanoring resonators coupled to add/drop optical filter.

From Eqs (4.16)–(4.18), the circulated roundtrip light fields of the right nanoring radii, $R_r$, are given in Eqs (4.19) and (4.20), respectively.

$$E_{r1} = \frac{j\sqrt{1-\gamma_0}\sqrt{\kappa_0}\,E_1}{1-\sqrt{1-\gamma_0}\sqrt{1-\kappa_0}\,e^{-\frac{\alpha}{2}L_r - jk_n L_r}} \quad (4.19)$$

$$E_{r2} = \frac{j\sqrt{1-\gamma_0}\sqrt{\kappa_0}\,E_1 e^{-\frac{\alpha}{2}L_r - jk_n L_r}}{1-\sqrt{1-\gamma_0}\sqrt{1-\kappa_0}\,e^{-\frac{\alpha}{2}L_r - jk_n L_r}} \quad (4.20)$$

Thus, the output circulated light field, $E_0$, for the right nanoring is given by

$$E_0 = E_1 \left\{ \frac{\sqrt{(1-\gamma_0)(1-\kappa_0)} - (1-\gamma_0)e^{-\frac{\alpha}{2}L_r - jk_n L_r}}{1 - \sqrt{(1-\gamma_0)(1-\kappa_0)}e^{-\frac{\alpha}{2}L_r - jk_n L_r}} \right\} \quad (4.21)$$

Similarly, the output circulated light field, $E_{0L}$, for the left nanoring at the left side of the add/drop optical multiplexing system is given by

$$E_{0L} = E_3 \left\{ \frac{\sqrt{(1-\gamma_3)(1-\kappa_3)} - (1-\gamma_3)e^{-\frac{\alpha}{2}L_3 - jk_n L_3}}{1 - \sqrt{(1-\gamma_3)(1-\kappa_3)}e^{-\frac{\alpha}{2}L_3 - jk_n L_3}} \right\} \quad (4.22)$$

where $\kappa_3$ is the intensity coupling coefficient, $\gamma_3$ is the fractional coupler intensity loss, $\alpha$ is the attenuation coefficient, $k_n = 2\pi/\lambda$ is the wave propagation number, $\lambda$ is the input wavelength light field and $L_2 = 2\pi R_L$, $R_L$ is the radius of left nanoring.

From Eqs (4.10)–(4.22), the circulated light fields, $E_1$, $E_3$ and $E_4$ are defined by

$$E_1 = \frac{j\sqrt{1-\gamma_1}\sqrt{\kappa_1}E_{i1} + j\sqrt{1-\gamma_1}\sqrt{1-\gamma_2}\sqrt{1-\kappa_1}\sqrt{\kappa_2}E_{0L}E_{i2}e^{-\frac{\alpha L}{2\,2} - jk_n \frac{L}{2}}}{1 - \sqrt{1-\gamma_1}\sqrt{1-\gamma_2}\sqrt{1-\kappa_1}\sqrt{1-\kappa_2}E_0 E_{0L} e^{-\frac{\alpha L}{4\,2} - jk_n \frac{L}{2}}} \quad (4.23)$$

$$E_3 = \sqrt{1-\gamma_2}\left[\sqrt{1-\kappa_2}E_1\left\{\frac{\sqrt{1-\gamma_0}\sqrt{1-\kappa_0} + (1-\gamma_0)e^{-\frac{\alpha}{2}L_r - jk_n L_r}}{1 - \sqrt{1-\gamma_0}\sqrt{1-\kappa_0}e^{-\frac{\alpha}{2}L_r - jk_n L_r}}\right\} \times \right.$$
$$\left. e^{-\frac{\alpha L}{4\,4} - jk_n \frac{L}{4}} + j\sqrt{\kappa_2}E_{i2}\right] \quad (4.24)$$

$$E_4 = E_3 \left\{ \frac{\sqrt{1-\gamma_3}\sqrt{1-\kappa_3} + (1-\gamma_3)e^{-\frac{\alpha}{2}L_3 - jk_n L_3}}{1 - \sqrt{1-\gamma_3}\sqrt{1-\kappa_3}e^{-\frac{\alpha}{2}L_3 - jk_n L_3}} \right\} e^{-\frac{\alpha L}{4\,4} - jk_n \frac{L}{4}} \quad (4.25)$$

where

$$E_0 E_{0L} = \left\{\frac{\sqrt{1-\gamma_0}\sqrt{1-\kappa_0} - (1-\gamma_0)e^{-\frac{\alpha}{2}L_r - jk_n L_r}}{1 - \sqrt{1-\gamma_0}\sqrt{1-\kappa_0}e^{-\frac{\alpha}{2}L_r - jk_n L_r}}\right\} \times$$
$$\left\{\frac{\sqrt{1-\gamma_3}\sqrt{1-\kappa_3} - (1-\gamma_3)e^{-\frac{\alpha}{2}L_3 - jk_n L_3}}{1 - \sqrt{1-\gamma_3}\sqrt{1-\kappa_3}e^{-\frac{\alpha}{2}L_3 - jk_n L_3}}\right\} \quad (4.26)$$

Thus, from Eqs (4.10), (4.12), (4.23)–(4.26), the output optical field of the through port ($E_{t1}$) is expressed by

$$E_{t1} = \sqrt{1-\gamma_1}\sqrt{1-\kappa_1}E_{i1} +$$

$$j\sqrt{1-\gamma_1}\sqrt{1-\gamma_2}\sqrt{\kappa_1}\sqrt{1-\kappa_2}E_1 e^{-\frac{\alpha L}{4 2}-jk_n\frac{L}{2}} \times$$

$$\left\{\frac{\sqrt{1-\gamma_0}\sqrt{1-\kappa_0}-(1-\gamma_0)e^{-\frac{\alpha}{2}L_r-jk_nL_r}}{1-\sqrt{1-\gamma_0}\sqrt{1-\kappa_0}e^{-\frac{\alpha}{2}L_r-jk_nL_r}}\right\} \times$$

$$\left\{\frac{\sqrt{1-\gamma_3}\sqrt{1-\kappa_3}-(1-\gamma_3)e^{-\frac{\alpha}{2}L_3-jk_nL_3}}{1-\sqrt{1-\gamma_3}\sqrt{1-\kappa_3}e^{-\frac{\alpha}{2}L_3-jk_nL_3}}\right\} -$$

$$\sqrt{1-\gamma_1}\sqrt{1-\gamma_2}\sqrt{\kappa_1}\sqrt{\kappa_2}E_{i2}e^{-\frac{\alpha L}{4 2}-jk_n\frac{L}{2}} \times$$

$$\left\{\frac{\sqrt{1-\gamma_3}\sqrt{1-\kappa_3}-(1-\gamma_3)e^{-\frac{\alpha}{2}L_3-jk_nL_3}}{1-\sqrt{1-\gamma_3}\sqrt{1-\kappa_3}e^{-\frac{\alpha}{2}L_3-jk_nL_3}}\right\} \quad (4.27)$$

The power output of the through port ($P_{t1}$) is written by

$$P_{t1} = (E_{t1})\cdot(E_{t1})^* = |E_{t1}|^2 \quad (4.28)$$

Similarly, from Eqs (4.12), (4.15), (4.23)–(4.26), the output optical field of the drop port ($E_{t2}$) is given by

$$E_{t2} = \sqrt{1-\gamma_2}\sqrt{1-\kappa_2}E_{i2} +$$

$$j\sqrt{1-\gamma_2}\sqrt{\kappa_2}\left\{\frac{\sqrt{1-\gamma_0}\sqrt{1-\kappa_0}-(1-\gamma_3)e^{-\frac{\alpha}{2}L_0-jk_nL_0}}{1-\sqrt{1-\gamma_0}\sqrt{1-\kappa_0}e^{-\frac{\alpha}{2}L_0-jk_nL_0}}\right\}e^{-\frac{\alpha L}{4 4}-jk_n\frac{L}{4}} \times$$

$$\frac{j\sqrt{1-\gamma_1}\sqrt{\kappa_1}E_{i1}+j\sqrt{1-\gamma_1}\sqrt{1-\gamma_2}\sqrt{1-\kappa_1}\sqrt{\kappa_2}E_{i2}E_{0L}e^{-\frac{\alpha L}{4 4}-jk_n\frac{L}{4}}}{1-\sqrt{1-\gamma_1}\sqrt{1-\gamma_2}\sqrt{1-\kappa_1}\sqrt{1-\kappa_2}E_0 E_{0L}e^{-\frac{\alpha L}{4 4}-jk_n\frac{L}{4}}} \quad (4.29)$$

The power output of the drop port ($P_{t2}$) is expressed by

$$P_{t2} = (E_{t2})\cdot(E_{t2})^* = |E_{t2}|^2 \quad (4.30)$$

In order to retrieve the required signals, we propose to use the add/drop optical multiplexing device with the appropriate parameters. This is given in the following details. The optical circuits of a PANDA ring resonator for the through port and drop port can be given by Eqs (4.28) and (4.30), respectively. The chaotic noise cancellation can be managed by using the specific parameters of the add/drop multiplexing device, and the required signals can be retrieved by the specific users. $\kappa_1$ and $\kappa_2$ are the coupling coefficients of the add/drop filters, $k_n = 2\pi/\lambda$ is the wave propagation number in a vacuum, and the waveguide (ring resonator) loss is $\alpha = 5 \times 10^{-5}$ dBmm$^{-1}$. The fractional coupler intensity loss is $\gamma = 0.01$. In the case of the add/drop multiplexing device, the nonlinear refractive index is neglected.

More results are as shown in Figures 4.6 and 4.7. In Figure 4.6, the coupling coefficients are $\kappa_0 = 0.7$, $\kappa_1 = 0.35$, $\kappa_2 = 0.1$ and $\kappa_3 = 0.9$, respectively. The ring radii are $R_{add} = 15$ μm, $R_{right} = 10$ μm and $R_{left} = 6$ μm. The $A_{eff}$ are 50, 25 and 5 μm$^2$ for add/drop, right and left ring resonators, respectively, where (a) $E_1$, (b) $E_2$, (c) $E_3$, (d) $E_4$, (e) and (f) are the through port and drop port signals, respectively. The input signals are dark and bright solitons. In Figure 4.7, the coupling coefficients are $\kappa_0 = 0.1$, $\kappa_1 = 0.35$, $\kappa_2 = 0.1$ and $\kappa_3 = 0.2$, respectively. The ring radii are $R_{add} = 15$ μm, $R_{right} = 10$ μm and $R_{left} = 6$ μm. $A_{eff}$ are 50, 25 and 5 μm$^2$ for add/drop, right and left ring resonators respectively. (a) $E_1$, (b) $E_2$, (c) $E_3$, (d) $E_4$, (e) and (f) are the through port and drop port signals respectively. The input signals are a dark soliton and a Gaussian pulse. In application, the term dynamics can be realized and available for dynamic wells/tweezers control, which means the movement of tweezers/wells can be formed within the system. From the results, they have shown that the forms of tweezers/wells can be configured whether they are stable/unstable configurations. For instance, the optical tweezers in the forms of peaks/valleys are kept in the stable forms within the add/drop filter, which can be seen at the through port in Figures 4.6 and 4.7. One special case is seen in Figure 4.6 when the double well is occurred at $E_3$.

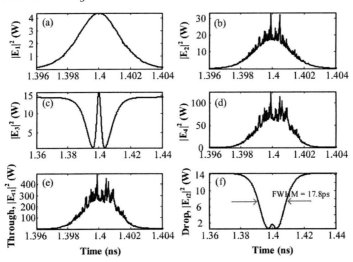

**FIGURE 4.6** Results of the dynamic tweezers (wells), where (a) $E_1$, (b) $E_2$, (c) $E_3$, (d) $E_4$, (e) are the through port and (f) drop port signals, the input signals are dark and bright solitons.

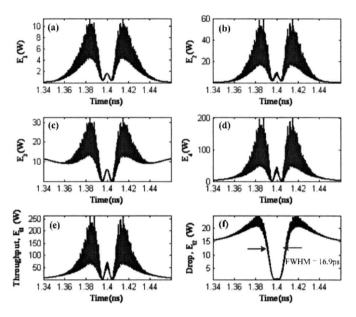

**FIGURE 4.7** Results of the dynamic tweezers, where (a) $E_1$, (b) $E_2$, (c) $E_3$, (d) $E_4$, (e) are the through port and (f) drop port signals, the input signals are dark soliton and a Gaussian pulse.

## 4.5 Dynamic Optical Tweezers via a Wavelength Router

Optical tweezers are a powerful tool used in the three-dimensional rotation and translation (location manipulation) of nano-structures such as micro- and nano-particles as well as living micro-organisms [17]. Many research works have been concentrated on the static tweezers [18–22], which it cannot move. The benefit offered by optical tweezers is the ability to interact with nano-scaled objects in a non-invasive manner, i.e. there is no physical contact with the sample, thus preserving many important characteristics of the sample, such as the manipulation of a cell with no harm to the cell. Optical tweezers are now widely used and they are particularly powerful in the field of microbiology [23–25] to study cell–cell interactions, manipulate organelles without breaking the cell membrane and to measure adhesion forces between cells. In this chapter we describe a new concept of developing an optical tweezers source using a dark soliton pulse. The developed tweezers has many potential applications in electron, ion, atom and molecule probing and manipulation as well as DNA probing and transportation. Furthermore, the soliton pulse generator is a simple and compact design, making it more commercially viable. In this chapter, we present the theoretical background in the physical model concept, where potential well can be formed by the barrier of optical field. The change in potential value, i.e. gradient of potential can produce force that can be used to confine/trap atoms/molecule. Furthermore, the change in potential well is still stable in some conditions, which means that the dynamic optical tweezers is plausible, therefore, the transportation of atoms/molecules in the optical network via a dark soliton being realized in the near future.

## 4.6 Trapping Forces

The optical forces generated by the mW of visible light are more than enough to overwhelm the random thermal forces which drive the dynamics of microparticles. The goal in creating an optical tweezers is to direct the optical forces from a single laser beam to trap a particle in all three dimensions. While quite general formulations of this problem have been developed [26–29], a simplified discussion suffices to motivate the design of optical tweezers arrays. We consider the forces exerted by monochromatic light of wave-number $k$ on a dielectric sphere of radius $a$ in the Rayleigh limit, where $a*2p/k$. The total optical force, F, is the sum of two contributions is given by [30]

$$F = F_\nabla + F_s \tag{4.31}$$

where the first of which arises from gradients in the light's intensity and the second of which is due to scattering of light by the particle. The gradient force on a particle of dielectric constant $\varepsilon$ immersed in a medium of dielectric constant $\varepsilon_0$ and subjected to an optical field with Poynting vector $S$, where

$$F_\nabla = 2\pi a^3 \frac{\sqrt{\varepsilon_0}}{c} \left( \frac{\varepsilon - \varepsilon_0}{\varepsilon + 2\varepsilon_0} \right) \nabla |S| \tag{4.32}$$

It tends to draw the particle toward the region of highest intensity. The scattering force is expressed by

$$F_s = \frac{8}{3} \pi (ka)^4 a^2 \frac{\sqrt{\varepsilon_0}}{c} \left( \frac{\varepsilon - \varepsilon_0}{\varepsilon + 2\varepsilon_0} \right)^2 S \tag{4.33}$$

It drives the particle along the direction of propagation of the light. When dark soliton pulse is generated in the fiber optic system, the optical field ($E_{in}$) of the dark soliton pulse is given by an Eq. (4.34) as [31].

$$E_{in}(t) = A \tanh\left[\frac{T}{T_0}\right] \exp\left[\left(\frac{z}{2L_D}\right) - i\omega_0 t\right] \tag{4.34}$$

where $A$ and $z$ are the optical field amplitude and propagation distance, respectively. $T$ is a soliton pulse propagation time in a frame moving at the group velocity, $T = t - \beta_1 * z$, where $\beta_1$ and $\beta_2$ are the coefficients of the linear and second order terms of Taylor expansion of the propagation constant. $L_D = T_0^2/|\beta_2|$ is the dispersion length of the soliton pulse. $T_0$ in equation is a soliton pulse propagation time at initial input. Where $t$ is the soliton phase shift time, and the frequency shift of the soliton is $\omega_0$. This solution describes a pulse that keeps its temporal width in variance as it propagates, and thus is called a temporal soliton. When a soliton peak intensity ($|\beta_2/\Gamma T_0^2|$) is given, then $T_0$ is known. For the soliton pulse in the microring device, a balance should be achieved between the dispersion length ($L_D$) and the nonlinear length ($L_{NL} = (1/\Gamma\phi_{NL})$, where $\Gamma = n_2^* k_0$, is the length scale over which dispersive or nonlinear effect makes the beam becomes wider or narrower. For a soliton pulse, there is a balance between dispersion and nonlinear lengths, hence $L_D = L_{NL}$.

## 4.7 Trapping Stability

We propose the trapping stability of an atom/molecule by using the trapping photon within a potential well that generates within the fiber grating, where two photon components are trapped within the well. Firstly, wave propagation in optical fibers is analyzed by solving Maxwell's equation with appropriate boundary conditions. In the presence of Kerr nonlinearity, using the coupled-mode theory, the nonlinear coupled mode equation is defined under the absence of material and waveguide dispersive effects. The dispersion arising from the periodic structure dominates near Bragg resonance conditions and it is valid only for wavelengths near the Bragg wavelength. By substituting the stationary solution into the coupled mode equation and by assuming $E_\pm(z,t) = e_\pm(z)e^{-i\tilde{\delta}ct/\bar{n}}$, we obtain

$$i\frac{de_f}{dz} + \hat{\delta}e_f + \kappa e_b + \left(\Gamma_S|e_f|^2 + 2\Gamma_X|e_b|^2\right)e_f = 0$$

and

$$-i\frac{de_b}{dz} + \hat{\delta}e_b + \kappa e_f + \left(\Gamma_S|e_b|^2 + 2\Gamma_X|e_f|^2\right)e_b = 0 \quad (4.35)$$

Equation (4.35) represents the time-independent light transmission through the gratings structure where $e_f$ and $e_b$ are the forward and backward propagating modes [32]. In order to explain the formation of Bragg soliton, consider the Stokes parameter since it provides useful information about the total energy and energy difference between the forward and backward propagating modes. In this study, we consider the following Stokes parameter [33].

and

$$A_0 = |e_f|^2 + |e_b|^2$$
$$A_1 = e_f e_b^* + e_f^* e_b$$
$$A_2 = i\left(e_f e_b^* - e_f^* e_b\right)$$
$$A_3 = |e_f|^2 - |e_b|^2$$

(4.36)

with the constraint $A_0^2 = A_1^2 + A_2^2 + A_3^2 + A_4^2$ where we introduce $A_4 = g(e)$ defined as the unknown function. In the FBG theory, the nonlinear coupled-mode (NLCM) equation requires that the total power $P_0 = A_3 = |e_f|^2 - |e_b|^2$ inside the grating is constant along the grating structures. Rewriting the NLCM equations in terms of Stokes parameter gives

$$\frac{dA_0}{dz} = -2\kappa A_2, \frac{dA_1}{dz} = 2\hat{\delta} A_2 + 3\Gamma A_0 A_2$$

$$\frac{dA_2}{dz} = -2\hat{\delta} A_1 - 2\kappa A_0 - 3\Gamma A_0 A_1, \frac{dA_3}{dz} = 0, \frac{dA_4}{dz} = g'(e) \quad (4.37)$$

In Eq. (4.37), we drop the distinction between the self-phase modulation (SPM) and cross phase modulation effects and hence it becomes $3\Gamma = 2\Gamma_x + \Gamma_s$. It can be clearly shown that the total power, $P_0 (=A_3)$ inside the grating is found to be conserved along the grating structure [18]. In the construction of the anharmonic oscillator type equation, it is necessary to use the conserved quantity, and it is obtained in the form $\hat{\delta} A_0 + (3/4)\Gamma A_0^2 + \kappa A_1 = C$, where $C$ is the constant of integration and $\hat{\delta}$ is the detuning parameter. Using Eq. (4.37), we obtain

$$\frac{d^2 A_0}{dz^2} - \alpha A_0 + \beta A_0^2 + \gamma A_0^3 + \theta A_0^4 = 4\hat{\delta} C \quad (4.38)$$

where

$$\alpha = 2\left[2\hat{\delta}^2 - 2\kappa^2 - 3\Gamma C\right]$$

$$\beta = 9\Gamma\hat{\delta}$$

$$\gamma = \frac{9}{4}\Gamma^2 \text{ and } \theta = f(\theta)$$

To simplify Eq. (4.38), it is assumed the parameters of $\alpha$, $\beta$ and $\gamma$ is independent with respect to parameter $\theta$. Equation (4.38) contains all the physical parameter of the NLCM equation.

In order to describe the motion of a particle moving with the classic anharmonic potential, we have the solution as follows:

$$V(A_0) = -\alpha \frac{A_0^2}{2} + \beta \frac{A_0^3}{3} + \gamma \frac{A_0^4}{4} + \theta \frac{A_0^5}{5} \quad (4.39)$$

It represents the potential energy distribution in Fiber Bragg Grating structures. where $\hat{\delta}$ is the potential energy distribution in the FBG structure [35].

Figure 4.8 depicts the motion of photon in a dynamic potential well when nonlinear parameters are taken into account as shown by Eq. (4.39) for a potential well with $\alpha = 0.9$, $\beta = 0.3$, $\theta = 0.09$ and $\gamma$ is varied from 0.3 to 0.9. The photon is trapped by the $\alpha$ parameter which is depicted by $V$. When $\alpha$ is too large, the potential well produces an increase in $A_0$ and have a wider double well. The $\gamma$ parameter is represented by $X$. When $\gamma$ is large, the potential well produces an increase in $A_0$. Suppose that the power source is imposed on the FBG and the initial power is used to generate the particles. It shows that the double well potential well is not symmetrical and the potential energy decreases within in a certain region indicated by $Y$. The other effect is the disturbance of the potential energy shown by $Z$ where the photon cannot be trapped symmetrically, and it is not equilibrium. Ultimately because of instability, it will lead to losses.

In terms of parametric function, we can describe it as follows. The change in $\alpha$ will affect the dip of the potential well. If $\alpha$ is approximately too small, the shape of the potential well turns into a single potential well. The occurrence of $\beta$ effect on the motion of photon affects the negative region which means $A_0 < 0$. The effect of $\gamma$ shows that the width of the potential well decreases if we increased the value of $\gamma$. Therefore if we increased the value of gamma, we can be assumed that the photon is localized and can be trapped. The addition of another nonlinear factor $\theta$, affects the profile of potential well rapidly. We could say that if we include the existence of $\theta$, the shape of potential well becomes chaotic. The photon does not only move in certain region that is known as potential well.

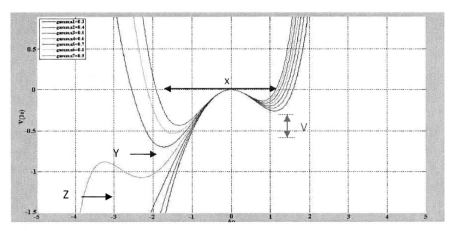

**FIGURE 4.8** The motion of photon in potential well for $\alpha = 0.9$, $\beta = 0.3$, $\theta = 0.09$ and $\gamma$ varies from 0.3 to 0.9.

## 4.8 Trapping and Transportation Mechanism

The experimental setup is shown in Figure 4.9. The dark soliton generator consists of a fibre laser based on a non-linear gain medium that is placed in a linear cavity. The non-linear gain medium is a 7.7 km Dispersion Compensating Fibre (DCF) which is pumped by a 1500 nm Brillouin Pump (BP) at 1.96 dBm. An Optical Circulator (OC) is used at one end of the setup to act as a fibre based mirror, with Port 3 connected to Port 1 while

Port 2 is connected to the rest of the experimental setup. Another OC is also used in the experimental setup to guide the incoming and outgoing signals. An Optical Spectrum Analyzer (OSA) with a resolution of 0.07 nm is used to analyze the output of the proposed setup. The operation of the experimental setup also follows: the BP generates a 1500 nm signal at 1.96 dBm, where it enters Port 1 of the first OC. The signal then travels onward to the DCF, where the non-linear interactions will provide the first Stokes wavelength. The BP and Stokes then travels onwards to the second OC where it is reflected back to the DCF and again to Port 2 of the first OC, where it will now exit via Port 3 which is connected to the OSA.

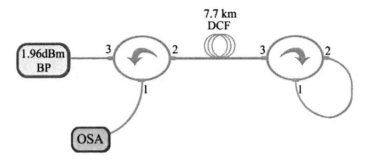

**FIGURE 4.9** Experimental setup.

Figure 4.10 shows the generated dark soliton pulse. As can be seen in the figure, the proposed setup is able to generate two wavelength peaks at the BP at approximately 1500 nm and also at the 1st Stokes at approximately 1500.09 nm. Both the BP peak and the 1st Stokes peak have a power of approximately −12 dBm. This represents the darks soliton and the trap pulse which is formed by gaps of the two intensities. The trap pulse can be employed for trapping an atom or molecule, much akin to a pair of

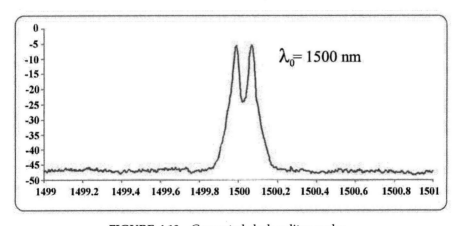

**FIGURE 4.10** Generated dark soliton pulse.

tweezers (the atom/molecule size would ideally be the same as the gap, which is approximately 0.09 nm). The dark soliton can also be converted into a bright soliton by adding an add/drop filter. The same mechanism also allows for the recovery of the transported medium such as the atom or molecule being transported as well as the light pulse. Figure 4.11 shows the pulse train of the trap pulse over a time of 10 minutes. From Figure 4.10, the soliton propagation over time can be obtained. As can be seen in the figure, the soliton pulse maintains its shape through the time of testing with no observable fluctuation in the power or wavelength. This is critical as any slight fluctuation will cause the beam to lose its hold over the transported atom or molecule, effectively dropping it.

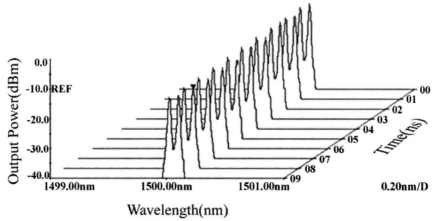

FIGURE 4.11  Soliton propagation over time (similar to a potential well, which can be seen clearly in a 3-D).

## 4.9 Atom/Molecule Transmission and Transportation via Wavelength Router

From the above reasons, the transmission of atoms/molecules from dark soliton pulses via a wavelength router is plausible, which can be described by the following reasons: (i) a dark soliton pulse can propagate into the optical device/media, (ii) atom/molecule being trapped by tweezers force during the movement, the atom/molecule recovery can be realized by using the optical detection scheme, where the dark-bright conversion technique is also available [20]. From Figure 4.12, the transmission atoms/molecules can be formed by the dark soliton pulse. The atoms/molecules recovery can be taken by using the add/drop filter. However, the separation of atoms/molecules from light pulse is required to have the specific environment, which becomes the interesting research area. Light with the specific wavelength ($\lambda_i$) is detected by a detector, while the required molecule is absorbed by the specific environment.

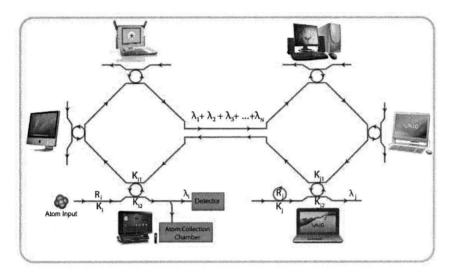

**FIGURE 4.12** A schematic diagram of atomic/molecular router and network system, where $R_i$, $R_j$: ring radii and $\kappa_{is}$, $\kappa_{js}$ are the coupling coefficients.

## 4.10 Conclusion

In this chapter, we have shown that the dynamic optical tweezers can be generated by using dark and bright solitons collision control within the add/drop filter. The tweezers storage can be performed by adding a nanoring resonator within design system, incorporating the add/drop multiplexer. By using the reasonable dark-bright soliton input power, the dynamic optical tweezers can be controlled and stored. The smallest tweezer width (full width at half maximum, FWHM) of 11 nm is obtained. The maximum power and memory time of 50 W and 1.2 ns are obtained respectively. In application, such a behavior can be used to confine the suitable size of light pulse or molecule, which can be employed in the same way of the optical tweezers. But in this case the terms dynamic probing is become the realistic function. Moreover, the trapped pulse, atom or molecule within the period of time (memory) is plausible, which is available for high density long distance transmission. Moreover, the multiple tweezers/wells storage configuration is allowed. The storage array of tweezers/wells which is available for high density tweezers/wells storage and high capacity molecular transportation which will be our continuous work.

In addition, we have demonstrated a dark soliton pulse beam that has potential applications in the probing and transport of atoms or molecules as an optical tweezers. The pulse beam consists of a BP signal at 1500.00 nm and Stokes signal at 1500.09 nm, in which the gap formed between the two intensities of −12 dBm forms the trap pulse for the optical tweezers. The generated beam is also highly stable and shows no fluctuations over a test period of 10 minutes, thus showing that the beam can be used as a probe or transporter without the risk of losing the medium being probed or transported. The theoretical background of the trapped atom/molecule is analyzed. It is shown that the dynamic optical tweezers for atoms/molecules transportation may be realized. The atomic/molecular network is also described.

# REFERENCES

[1] A. Ashkin, J.M. Dziedzic, J.E. Bjorkholm, and S. Chu, "Observation of a single-beam gradient force optical trap for dielectric particles", *Opt. Lett.*, 11, 288–290 (1986).

[2] S. Bergamini, B. Darqui, M. Jones, L. Jacubowiez, A. Browaeys, and P. Grangier, "Holographic generation of microtrap arrays for single atoms by use of a programmable phase modulator", *J. Opt. Soc. Am.*, B 21, 1889–1894 (2004).

[3] D.D. Yavuz, P.B. Kulatunga, E. Urban, T.A. Johnson, N. Proite, T. Henage, T.G. Walker, and M. Saffman, "Fast ground state manipulation of neutral atoms in microscopic optical traps", *Phys. Rev. Lett.*, 96, 063001 (2006).

[4] D. Schrader, I. Dotsenko, M. Khudaverdyan, Y. Miroshnychenko, A. Rauschenbeutel, and D. Meschede, "Neutral atom quantum register", *Phys. Rev. Lett.*, 93, 150501 (2004).

[5] J.A. Sauer, K.M. Fortier, M.S. Chang, C.D. Hamley, and M.S. Chapman, "Submicrometer position control of single trapped neutral atoms", *Phys. Rev., A* 69, 051804(R) (2004).

[6] T.P. Meyrath, F. Schreck, J. L. Hanssen, C.-S. Chuu, and M.G. Raizen, "Bose–Einstein condensate in a box", *Phys. Rev. A* 71, 041604(R) (2005).

[7] V. Boyer, R.M. Godun, G. Smirne, D. Cassettari, C.M. Chandrashekar, A.B. Deb, Z.J. Laczik, and C.J. Foot, "Dynamic manipulation of Bose–Einstein condensates with a spatial light modulator", *Phys. Rev., A* 73, 031402(R) (2006).

[8] A.V. Carpentier, J. Belmonte-Beitia, H. Michinel, and V.M. Perez-Garcia, "Laser tweezers for atomic solitons", *J. Mod. Opt.*, 55(17), 2819–2829 (2008).

[9] V. Milner, J.L. Hanssen, W.C. Campbell and M.G. Raizen, "Optical billiards for atoms", *Phys. Rev. Lett.*, 86, 1514–1516 (2001).

[10] N. Friedman, A. Kaplan, D. Carasso, and N. Davidson, "Observation of chaotic and regular dynamics in atom-optics billiards", *Phys. Rev. Lett.*, 86, 1518–1520 (2001).

[11] M. Li and J. Arlt, "Trapping multiple particles in single optical tweezers", *Opt. Commun.*, 281, 135–140 (2008).

[12] M. Schulz, H. Crepaz, F. Schmidt-Kaler, J. Eschner, and R. Blatt, "Transfer of trapped atoms between two optical tweezer potentials", *J. Mod. Opt.*, 54(11), 1619–1626 (2007).

[13] K. Sarapat, N. Sangwara, K. Srinuanjan, P.P. Yupapin, and N. Pornsuwancharoen, "Novel dark-bright optical solitons conversion system and power amplification", *Opt. Eng.*, 48, 045004 (2009).

[14] S. Mitatha, N. Pornsuwancharoen, and P.P. Yupapin, "A simultaneous short-wave and millimeter-wave generation using a soliton pulse within a nano-waveguide", *IEEE Photon. Technol. Lett.*, 21, 932–934 (2009).

[15] Y. Kokubun, Y. Hatakeyama, M. Ogata, S. Suzuki, and N. Zaizen, "Fabrication technologies for vertically coupled microring resonator with multilevel crossing busline and ultracompact-ring radius", *IEEE J. Sel. Top. Quantum Electron.*, 11, 4–10 (2005).

[16] T. Threepak, X. Luangvilay, S. Mitatha, and P.P. Yupapin, "Novel quantum-molecular transporter and networking via a wavelength router", *Microwave and Optical Technology Letters*, vol. 52(12), pp. 2703–2706 (2010).

[17] K. Svoboda and S.M. Block, "Biological applications of optical forces", *Annu. Rev. Biophys. Biomol.Struct.*, 23, 247–283 (1994).

[18] S.M. Block, "Making light work with optical tweezers". *Nature*, 360, 493–495 (1992).

[19] R.M. Simmons, J.T. Finer, S. Chu, and J.A. Spudich, "Quantitative measurements of force and displacement using an optical trap". *J. Biophys*, 70(4), 1813–1822 (1996).

[20] I.M. Peters, B.G. de Grooth, J.M. Schins, C.G. Figdor, and J. Greve, "Three dimensional single-particle tracking with nanometer resolution". *Review of Scientific Instruments* 69(7), 2762-2766 (1998).

[21] M.J. Lang, C.L. Asbury, J.W. Shaevitz, and S.M. Block, "An automated two-dimensional optical force clamp for single molecule studies", *J. Biophys* 83(1), 491–501 (2002).

[22] J. Pine and G. Chow, "Moving live dissociated neurons with an optical tweezer", *IEEE Transaction on Biomedical Engineering*, 56(4), 1184–1188 (2009).

[23] A. Ashkin, J.M. Dziedzic, and T. Yamane, "Optical trapping and manipulation of single cells using infrared laser beams", *Nature*, 330, 769–771 (1987).

[24] K. Svoboda and S.M. Block, "Biological applications of optical forces", *Annu. Rev. Biophys. Biomolec. Struct.*, 23, 247–285 (1994).

[25] J. Conia, B.S. Edwards, and S. Voelkel, "The micro-robotic laboratory: optical trapping and scissing for the biologist", *J. Clin. Lab. Anal.*, 11, 28–38 (1997).

[26] G. Gouesbet, B. Maheu, and G. Gréhan, "Light scattering from a sphere arbitrarily located in a Gaussian beam, using a Bromwich formulation ", *J. Opt. Soc. Am. A* 5, 1427–1443 (1988).

[27] A. Ashkin, "Forces of a single-beam gradient laser trap on a dielectric sphere in the ray optics regime", *J. Biophys.* 61, 569–582 (1992).

[28] K.F. Ren, G. Gréhan, and G. Gouesbet, "Prediction of reverse radiation pressure by generalized Lorenz-Mie theory", *Appl. Opt.*, 35, 2702–2710 (1996).

[29] O. Farsund and B.U. Felderhof, "Force, torque, and absorbed energy for a body of arbitrary shape and constitution in an an electromagnetic", *Physica A* 227, 108–130 (1996).

[30] Y. Harada and T. Asakura, "Radiation forces on a dielectric sphere in the Rayleigh scattering regime", *Opt. Commun.* 124, 529–541 (1996).

[31] G.P. Agrawal, *"Nonlinear Fiber Optics"*, 4th edition, New York: Academic Press (2007).

[32] Y.S. Kivshar and G.P. Agrawal, *Optical Soliton : From Fibers to Photonics Crystal*, New York: Academic Press (2003).

[33] E. Collet, *Polarized Light - Concept and Applications*, New York: Marcel-Decker (1989).

[34] K. Senthilnathan and K. Porsezian, "Symmetry-breaking instability in gap soliton", *Opt. Commun.*, 227, 295–299 (2003).

[35] D.L. Mills and S.E. Trullinger, "Gap soliitons in nonlinear periodic structures", *Phys. Rev., B.* 36, 947–952 (1987).

[36] N. Sangwara, K. Sarapat, K. Srinuanjan, and P.P. Yupapin, "A Novel Dark-Bright Optical Solitons Conversion System and Power Amplification", *Optical Engineering*, 48(4), 045004 (2009).

# 5

# Hybrid Interferometer

**CHAPTER OUTLINE**
- Introduction
- Theoretical Background
- Hybrid Interferometer
- Conclusions
- References

## 5.1 Introduction

The use of soliton, i.e. bright soliton in long distance communication link has been implemented for nearly two decades. However, the interesting work of using bright soliton in communications still remain, whereas the use of a soliton pulse within a micro ring resonator for communication security has been studied [1]. Dark soliton is one of the soliton properties in which the soliton amplitude is eliminated or minimized during the propagation in media. Therefore, the dark soliton detection is difficult. The investigation of dark soliton behaviors has been reported [2]. With a promising and interesting results, the required users can retrieve the original signal via an add/drop filter. They can choose to retrieve in either bright or dark soliton pulses depending on the device parameter for signal processing or networking.

Dark-bright soliton control within a semiconductor add/drop multiplexer has shown promising applications. It has been investigated clearly by the authors in references [3, 4]. One of the advantages is that the dark soliton peak signal is always at a low level, which is useful for secured signal communication in the transmission link. The other is formed when the high optical field is configured as an optical tweezer or potential well [5, 6]. This is available for atom/molecule trapping. An optical tweezer uses forces exerted by intensity gradients in a strongly focused beam of light to trap and move a microscopic volume of matter. Optical tweezers technique has become a powerful tool for manipulation of micrometer-sized particles in three spatial dimensions. It has the unique ability to manipulate matter at mesoscopic scales that has led to widespread applications in biology [7, 8], and the physical sciences [9, 10]. Initially, the useful static tweezer is recognized, and the dynamic tweezer is now realized in practical work [11–13]. Recently, Schulz et al. [14] have shown that the transfer of trapped atoms between two optical potentials could be performed.

In this chapter, we present a novel system of the dynamic optical tweezers generation using a dark-bright soliton pulse propagating within an add/drop optical multiplexer. The multiplexing signals with slightly different wavelengths of the dark solitons are controlled and amplified within the system. The dynamic behaviors of dark bright soliton interaction are analyzed and described. The tweezer stability is seen when the Gaussian pulse is used to control via the add port. In application, the optical tweezers can be used to store and trap light, atom, molecule or particle within the proposed system, which can be used to form the dynamic tweezers. However, many research works in optical tweezers were reported in various techniques and applications, which can be found for further reading [15–19]. Finally, the use of optical multiplexer to form a hybrid interferometer using the generated tweezers is discussed in detail.

## 5.2 Theoretical Background

We are looking for a stationary dark soliton pulse, which is introduced into the add/drop optical filter system as shown in Figure 5.1. The input optical field ($E_{in}$) and the add port optical field ($E_{add}$) of the dark, bright soliton or Gaussian pulse are given by [20].

$$E_{in}(t) = A \tanh\left[\frac{T}{T_0}\right] \exp\left[\left(\frac{x}{2L_D}\right) - i\omega_0 t\right] \quad (5.1a)$$

$$E_{add}(t) = A \operatorname{sech}\left[\frac{T}{T_0}\right] \exp\left[\left(\frac{x}{2L_D}\right) - i\omega_0 t\right] \quad (5.1b)$$

$$E_{add}(t) = E_0 \exp\left[\left(\frac{x}{2L_D}\right) - i\omega_0 t\right] \quad (5.1c)$$

where $A$ and $x$ are the optical field amplitude and propagation distance, respectively. $T$ is a soliton pulse propagation time in a frame moving at the group velocity, $T = t - \beta_1^* x$, where $\beta_1$ and $\beta_2$ are the coefficients of the linear and second-order terms of Taylor expansion of the propagation constant. $L_D = T_0^2 / |\beta_2|$ is the dispersion length of the soliton pulse. $T_0$ in the equation is a soliton pulse propagation time at initial input (or soliton pulse width), where $t$ is the soliton phase shift time, and the frequency shift of the soliton is $\omega_0$. This solution describes a pulse that keeps its temporal width invariance as it propagates, and thus is called a temporal soliton. When a soliton of peak intensity ($|\beta / \Gamma T_0^2|$) is given then $T_0$ is known. For the soliton pulse in the microring device, a balance should be achieved between the dispersion length ($L_D$) and the nonlinear length ($L_{NL} = 1/\Gamma\phi_{NL}$). Here $\Gamma = n_2 * k_0$ is the length scale over which dispersive or nonlinear effects makes the beam become wider or narrower. For a soliton pulse, there is a balance between dispersion and nonlinear lengths. Hence $L_D = L_{NL}$. For Gaussian pulse in Eq. (5.1c), $E_0$ is the amplitude of optical field.

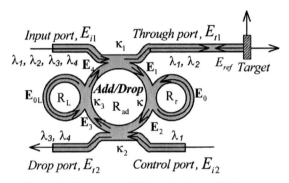

**FIGURE 5.1** A schematic diagram of light signal multiplexer controlled by light.

When light propagates within the nonlinear medium, the refractive index ($n$) of light within the medium is given by

$$n = n_0 + n_2 I = n_0 + \frac{n_2}{A_{eff}} P \qquad (5.2)$$

with $n_0$ and $n_2$ as the linear and nonlinear refractive indexes, respectively. $I$ and $P$ are the optical intensity and the power, respectively. The effective mode core area of the device is given by $A_{eff}$. For the add/drop optical filter design, the effective mode core areas range from 0.50 to 0.10 µm$^2$, in which the parameters were obtained by using the related practical material parameters [(InGaAsP/InP)] [21]. When a dark soliton pulse is input and propagated within a add/drop optical filter as shown in Figure 5.1, the resonant output is formed.

The resonator output field, $E_{t1}$ and $E_1$ consists of the transmitted and circulated components within the add/drop optical filter system, which can perform the driven force to photon/molecule/atom.

For the first coupling device of the add/drop optical filter system, the transmitted and circulated components can be written as

$$E_{t1} = \sqrt{1-\gamma_1}\left[\sqrt{1-\kappa_1}E_{i1} + j\sqrt{\kappa_1}E_4\right] \tag{5.3}$$

$$E_1 = \sqrt{1-\gamma_1}\left[\sqrt{1-\kappa_1}E_4 + j\sqrt{\kappa_1}E_{i1}\right] \tag{5.4}$$

$$E_2 = E_0 E_1 e^{-\frac{\alpha L}{2 2}-jk_n\frac{L}{2}} \tag{5.5}$$

where $\kappa_1$ is the intensity coupling coefficient, $\gamma_1$ is the fractional coupler intensity loss, $\alpha$ is the attenuation coefficient, $k_n = 2\pi/\lambda$ is the wave propagation number, $\lambda$ is the input wavelength light field and $L = 2\pi R_{ad}$, $R_{ad}$ is the radius of add/drop device.

For the second coupler of the add/drop system,

$$E_{t2} = \sqrt{1-\gamma_2}\left[\sqrt{1-\kappa_2}E_{i2} + j\sqrt{\kappa_2}E_2\right] \tag{5.6}$$

$$E_3 = \sqrt{1-\gamma_2}\left[\sqrt{1-\kappa_2}E_2 + j\sqrt{\kappa_2}E_{i2}\right] \tag{5.7}$$

$$E_4 = E_{0L} E_3 e^{-\frac{\alpha L}{2 2}-jk_n\frac{L}{2}} \tag{5.8}$$

where $\kappa_2$ is the intensity coupling coefficient, $\gamma_2$ is the fractional coupler intensity loss. The circulated light fields, $E_0$ and $E_{0L}$ are the light field circulated components of the nanoring radii, $R_r$ and $R_L$ which coupled into the right and left sides of the add/drop optical filter system, respectively. The light field transmitted and circulated components in the right nanoring, $R_r$, as shown in Figure 5.2, are given by

$$E_2 = \sqrt{1-\gamma}\left[\sqrt{1-\kappa_0}E_1 + j\sqrt{\kappa_0}E_{r2}\right] \tag{5.9}$$

$$E_{r1} = \sqrt{1-\gamma}\left[\sqrt{1-\kappa_0}E_{r2} + j\sqrt{\kappa_0}E_1\right] \tag{5.10}$$

$$E_{r2} = E_{r1} e^{-\frac{\alpha}{2}L_1-jk_n L_1} \tag{5.11}$$

where $\kappa_0$ is the intensity coupling coefficient, $\gamma$ is the fractional coupler intensity loss, $\alpha$ is the attenuation coefficient, $k_n = 2\pi/\lambda$ is the wave propagation number, $\lambda$ is the input wavelength light field and $L_1 = 2\pi R_r$, $R_r$ is the radius of right nanoring.

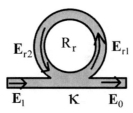

**FIGURE 5.2** A schematic diagram of nanoring resonator.

From Eqs (5.9)–(5.11), the circulated roundtrip light fields of the right nanoring radii, $R_r$, are given in Eqs (5.12) and (5.13), respectively.

$$E_{r1} = \frac{j\sqrt{1-\gamma}\sqrt{\kappa_0}E_1}{1-\sqrt{1-\gamma}\sqrt{1-\kappa_0}e^{-\frac{\alpha}{2}L_1 - jk_nL_1}} \tag{5.12}$$

$$E_{r2} = \frac{j\sqrt{1-\gamma}\sqrt{\kappa_0}E_1 e^{-\frac{\alpha}{2}L_1 - jk_nL_1}}{1-\sqrt{1-\gamma}\sqrt{1-\kappa_0}e^{-\frac{\alpha}{2}L_1 - jk_nL_1}} \tag{5.13}$$

Thus, the output circulated light field, $E_0$, for the right nanoring is given by

$$E_0 = E_1 \left\{ \frac{\sqrt{(1-\gamma)(1-\kappa_0)} - (1-\gamma)e^{-\frac{\alpha}{2}L_1 - jk_nL_1}}{1-\sqrt{(1-\gamma)(1-\kappa_0)}e^{-\frac{\alpha}{2}L_1 - jk_nL_1}} \right\} \tag{5.14}$$

Similarly, the output circulated light field, $E_{0L}$, for the left nanoring at the left side of the add/drop system is given by

$$E_{0L} = E_3 \left\{ \frac{\sqrt{(1-\gamma_3)(1-\kappa_3)} - (1-\gamma_3)e^{-\frac{\alpha}{2}L_2 - jk_nL_2}}{1-\sqrt{(1-\gamma_3)(1-\kappa_3)}e^{-\frac{\alpha}{2}L_2 - jk_nL_2}} \right\} \tag{5.15}$$

where $\kappa_3$ is the intensity coupling coefficient, $\gamma_3$ is the fractional coupler intensity loss, $\alpha$ is the attenuation coefficient, $k_n = 2\pi/\lambda$ is the wave propagation number, $\lambda$ is the input wavelength light field and $L_2 = 2\pi R_L$, $R_L$ is the radius of left nanoring.

From Eqs (5.3)–(5.15), the circulated light fields, $E_1$, $E_3$ and $E_4$ are defined by given $x_1 = (1-\gamma_1)^{1/2}$, $x_2 = (1-\gamma_2)^{1/2}$, $y_1 = (1-\kappa_1)^{1/2}$, and $y_2 = (1-\kappa_2)^{1/2}$.

$$E_1 = \frac{jx_1\sqrt{\kappa_1}E_{i1} + jx_1x_2y_1\sqrt{\kappa_2}E_{0L}E_{i2}e^{-\frac{\alpha}{2}\frac{L}{2} - jk_n\frac{L}{2}}}{1 - x_1x_2y_1y_2E_0E_{0L}e^{-\frac{\alpha}{2}L - jk_nL}} \tag{5.16}$$

$$E_3 = x_2y_2E_0E_1e^{-\frac{\alpha}{2}\frac{L}{2} - jk_n\frac{L}{2}} + jx_2\sqrt{\kappa_2}E_{i2} \tag{5.17}$$

$$E_4 = x_2y_2E_0E_{0L}E_1e^{-\frac{\alpha}{2}L - jk_nL} + jx_2\sqrt{\kappa_2}E_{0L}E_{i2}e^{-\frac{\alpha}{2}\frac{L}{2} - jk_n\frac{L}{2}} \tag{5.18}$$

Thus, from Eqs (5.3), (5.5), (5.16)–(5.18), the output optical field of the through port ($E_{t1}$) is expressed by

$$E_{t1} = AE_{i1} - BE_{i2}e^{-\frac{\alpha}{2}\frac{L}{2} - jk_n\frac{L}{2}} \left[ \frac{CE_{i1}e^{-\frac{\alpha}{2}L - jk_nL} + DE_{i2}e^{-\frac{3\alpha}{2}\frac{L}{2} - jk_n\frac{3L}{2}}}{1 - Fe^{-\frac{\alpha}{2}L - jk_nL}} \right] \tag{5.19}$$

where

$$A = x_1 x_2$$
$$B = x_1 x_2 y_2 \sqrt{\kappa_1} E_{0L}$$
$$C = x_1^2 x_2 \kappa_1 \sqrt{\kappa_2} E_0 E_{0L}$$
$$D = (x_1 x_2)^2 y_1 y_2 \sqrt{\kappa_1 \kappa_2} E_0 E_{0L}^2 \text{ and}$$
$$F = x_1 x_2 y_1 y_2 E_0 E_{0L}.$$

The power output of the through port ($P_{t1}$) is written by

$$P_{t1} = (E_{t1}) \cdot (E_{t1})^* = |E_{t1}|^2 \quad (5.20)$$

Similarly, from Eqs (5.5), (5.6), (5.16)–(5.18), the output optical field of the drop port ($E_{t2}$) is given by

$$E_{t2} = x_2 y_2 E_{i2} - \left[ \frac{x_1 x_2 \sqrt{\kappa_1 \kappa_2} E_0 E_{i1} e^{-\frac{\alpha}{2}\frac{l}{2} - jk_n \frac{L}{2}} + x_1 x_2^2 y_1 y_2 \sqrt{\kappa_2} E_0 E_{0L} E_{i2} e^{-\frac{\alpha}{2}l - jk_n L}}{1 - x_1 x_2 y_1 y_2 E_0 E_{0L} e^{-\frac{\alpha}{2}l - jk_n L}} \right] \quad (5.21)$$

The power output of the drop port ($P_{t2}$) is expressed by

$$P_{t2} = (E_{t2}) \cdot (E_{t2})^* = |E_{t2}|^2 \quad (5.22)$$

In order to retrieve the required signals, we propose to use the add/drop device with the appropriate parameters. This is given in the following details. The optical circuits of ring resonator add/drop filters for the through port and drop port can be given by Eqs (5.20) and (5.22), respectively. The chaotic noise cancellation can be managed by using the specific parameters of the add/drop device, and the required signals can be retrieved by the specific users. $\kappa_1$ and $\kappa_2$ are the coupling coefficients of the add/drop filters, $k_n = 2\pi/\lambda$ is the wave propagation number for in a vacuum, and the waveguide (ring resonator) loss is $\alpha = 5 \times 10^{-5}$ dBmm$^{-1}$. The fractional coupler intensity loss is $\gamma = 0.01$. In the case of the add/drop device, the nonlinear refractive index is neglected.

## 5.3 Hybrid Interferometer

Simulation results of the dynamic optical tweezers signals within the light signal multiplexer are as shown in Figure 5.3. In this case the bright soliton is input into the add port, and the dynamic tweezers are shown in Figure 5.3(A)–(D). In Figure 5.4(A)–(F), the tweezers in the form of potential wells are shown. These tweezers can be used for atom/molecule trapping and transportation. The potential well depth (peak valley) can be controlled by adjusting the system parameters, for instance, the bright soliton input power at the add port and the coupling coefficients. The potential well of the tweezers is tuned to be the multi wells and seen at the add port, as shown in Figure 5.4(F). In application, the optical tweezers in the design system can be tuned and amplified as shown in Figures 5.3 and 5.4. Therefore, the tunable optical tweezers can be controlled by the dark-bright soliton collision within the light signal multiplexer. This can be done by adjusting the parameters of the input power at the input and add ports, respectively.

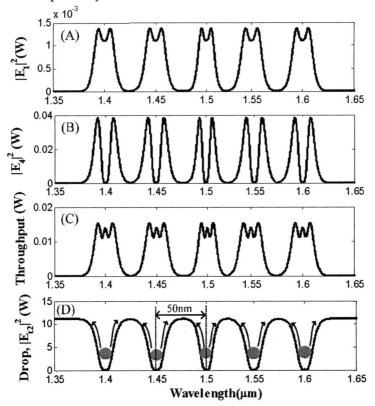

**FIGURE 5.3** Simulation result of the dynamic tweezers with five different center wavelengths, where (A) $|E_1|^2$, (B) $|E_4|^2$ and (C) are the through port and (D) drop port signals. The input signals are dark and bright soliton pulses at the input and control ports, respectively.

The output power at the through port is shown in Figure 5.4(F). The potential well with the optical power of 15 W is observed. More results are as shown in Figure 5.5. The coupling coefficients are given as $\kappa_0 = 0.1$, $\kappa_1 = 0.35$, $\kappa_2 = 0.1$ and $\kappa_3 = 0.2$, respectively. The ring radii are $R_{add} = 300$ μm, $R_{right} = 30$ μm and $R_{left} = 15$ μm. $A_{eff}$ are 0.50, 0.25 and 0.25 mm$^2$ [21] for add/drop, right and left ring resonators respectively. Simulation results of the ring resonator interferometer for electron spectroscopy with center wavelengths are at $\lambda_1 = 1400$ nm, $\lambda_2 = 1450$ nm, $\lambda_3 = 1,500$ nm, where (A) $|E_1|^2$, (B) $|E_2|^2$, (C) $|E_3|^2$, (D) $|E_4|^2$, (E) are the reflected outputs from the throughput port, and (F) is the output

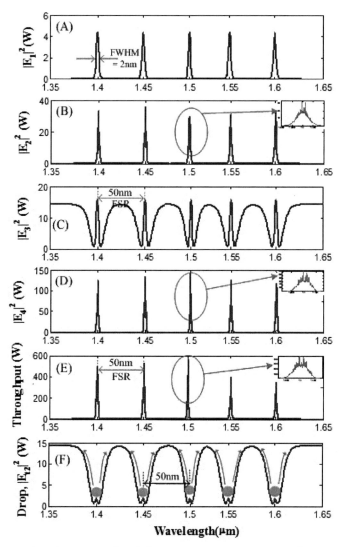

**FIGURE 5.4** Simulation result of dynamic tweezers array with trapped atoms for five different center wavelengths, where (A) $|E_1|^2$, (B) $|E_2|^2$, (C) $|E_3|^2$, (D) $|E_4|^2$, (E) through port and (F) drop port signals.

at the drop port. Here (A) $E_1$, (B) $E_2$, (C) $E_3$, (D) $E_4$, (E) are the through port and (F) drop port signals.

Over the last three decades, interferometric measurement methods have been applied in research and industry for the investigation of deformation and vibration behavior of mechanical components [22, 23]. In applications, the term dynamics can be realized and is suitable for dynamic wells/tweezers control. This means that the movement of tweezers/wells can be formed within the system. From the results, they have shown that the types of tweezers/wells can be configured whether they are in the stable/

or unstable configurations. For instance, the optical tweezers in the forms of peaks/valleys are kept in the stable forms within the add/drop filter. This can be seen at the through port in Figures 5.4 and 5.5. One special case is observed in Figure 5.5 when the double well is formed at $E_3$ as shown in Figure 5.5(F). In Figures 5.6 and 5.7, the used parameters are $R_r = R_L = 30$ µm, $R_{ad} = 300$ µm and $\alpha = 0.05$ dBmm$^{-1}$ for hybrid interferometer with the center wavelength at $\lambda_0 = 1,400$ nm.

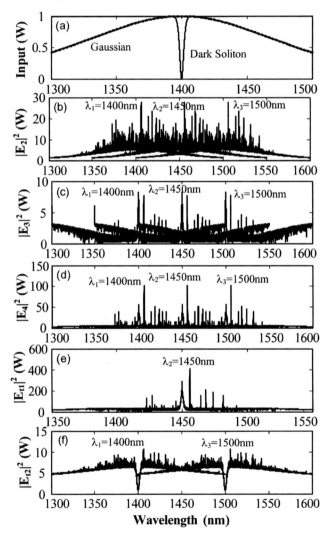

FIGURE 5.5 Simulation result of hybrid interferometer with two different center wavelengths $\lambda_1 = 1,400$ nm and $\lambda_2 = 1,500$ nm for hybrid interferometer applications using optical tweezers, where (A) $|E_1|^2$, (B) $|E_2|^2$, (C) $|E_3|^2$, (D) $|E_4|^2$, (E) are through port and (F) drop port signals, where dark solitons and bright solitons pulse are used as the input and control ports respectively.

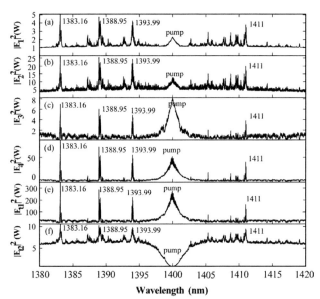

**FIGURE 5.6.** Simulation result of hybrid interferometer with center wavelengths $\lambda_0 = 1{,}400$ nm for hybrid interferometer applications using optical tweezers, where (a) $|E_1|^2$, (b) $|E_2|^2$, (c) $|E_3|^2$, (d) $|E_4|^2$, (e) are through port and (f) drop port signals, where $R_r = R_L = 30$ nm, $R_{ad} = 300$ nm and $\alpha = 0.05$ dBmm$^{-1}$.

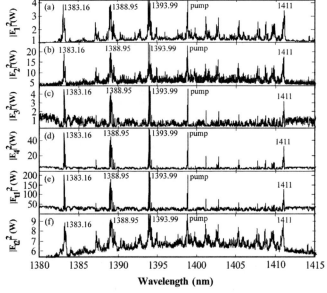

**FIGURE 5.7** Simulation result of hybrid interferometer with center wavelengths $\lambda_0 = 1{,}400$ nm for hybrid interferometer applications using optical tweezers, where (a) $|E_1|^2$, (b) $|E_2|^2$, (c) $|E_3|^2$, (d) $|E_4|^2$, (e) are through port and (f) drop port signals, where $R_r = 15$ nm, $R_L = 30$ nm, $R_{ad} = 300$ nm and $\alpha = 5 \times 10^{-5}$ dBmm$^{-1}$.

In Figure 5.8(A)–(F), the coupling coefficients are given as $\kappa_0 = 0.2$, $\kappa_1 = 0.35$, $\kappa_2 = 0.1$ and $\kappa_3 = 0.1$, respectively. The simulation results are shown, where (A) $|E_1|^2$, (B) $|E_2|^2$, (C) $|E_3|^2$, (D) $|E_4|^2$, (E) are the through port and (F) drop port signals. We found that the optical potential wells are stable and seen within the system. Moreover, the optical potential wells can be amplified and tuned as shown in Figure 5.8(B)–(E), whereas the multi-wells are seen. For instance, the four wells are generated, whereas the well width of 16 nm with the peak power of 20 W is noted, as shown in Figure 5.8(F). The photon/molecular/atomic free spectrum range of 50 nm is obtained, which becomes the trapping length limit of two adjacent molecule/atoms in the optical tweezers in the system (optical multiplexer/hybrid interferometer).

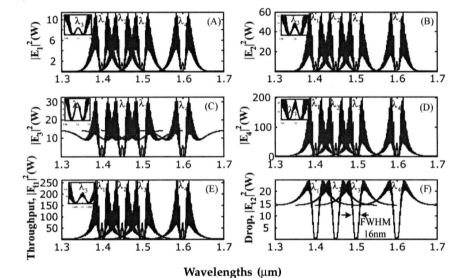

**FIGURE 5.8** Simulation result of hybrid interferometer with four different center wavelengths $\lambda_1 = 1{,}400$ nm, $\lambda_2 = 1{,}450$ nm, $\lambda_3 = 1{,}500$ nm and $\lambda_4 = 1{,}600$ nm for hybrid interferometer applications using the dynamic optical tweezers, which are controlled by the Gaussian signal at the control port, where (A) $|E_1|^2$, (B) $|E_2|^2$, (C) $|E_3|^2$, (D) $|E_4|^2$, (E) are through port and (F) drop port signals, where dark solitons and Gaussian pulse are used as the input and control ports respectively.

In operation, the optical signal multiplexer can be used to form a hybrid interferometer, photons/molecules/atoms can be fed into the light signal multiplexer by dark soliton (optical tweezer). The output photon/atoms/molecules from the through port pass through the target and are reflected back to the light signal multiplexer. The induced change of the collision (coupling effects) can be controlled by the add port input signal. Finally the interference signal is seen at the drop and through ports. The measurement can be made by balancing and adjusting the controlled parameters via the add port(control port). For instance, the drop port signals in Figures 5.6 and 5.7 show the imbalanced case of the hybrid interferometer, therefore, the use of controlled port parameter is required to keep the balancing position, which means that the measurement is in operation in this case. Finally, the through and drop port signals are recovered to the balanced position. The changes in controlled parameters are the measurement quantity respecting to the interested physical parameters.

## 5.4 Conclusions

In this chapter, we have shown that the hybrid interferometer can be formed by using the dynamic tweezers. This can be formed by using two nanoring resonators incorporating into the proposed light signal multiplexer. In this case, the dynamic behavior can be controlled and used to form the hybrid devices. In application, such a behavior can be used to confine the suitable size of light pulse, atom or molecule, which is then employed in the same way as the optical tweezers. But in this case the terms dynamic probing is becoming the realistic. The trapped pulses or molecules within the period of time (memory) are plausible. In applications, the use of the proposed concept and system can be used to form a new nanoscale interpretation of hybrid interferometer, whereas the trapped photons/molecules/atoms can be used to process the nanoscale signal processing interpretation based on an interferometric technique, where the balancing parameters can be found (measured). Thus, by using the dynamic tweezers, the hybrid interferometer using photons/atoms/molecules trapping and transportation within the system can be realized.

## REFERENCES

[1] D.N. Christodoulides, T.H. Coskun, M. Mitchell, Z. Chen, and M. Segev, "Theory of incoherent dark solitons", *Phys. Rev. Lett.*, 80, 5113–5115 (1998).

[2] A.D. Kim, W.L. Kath, and C.G. Goedde, "Stabilizing dark solitons by periodic phase-sensitive amplification", *Opt. Lett.*, 21, 465–467 (1996).

[3] K. Sarapat, N. Sangwara, K. Srinuanjan, P.P. Yupapin, and N. Pornsuwancharoen, "Novel dark-bright optical solitons conversion system and power amplification", *Opt. Eng.*, 48, 045004-1–5 (2009).

[4] S. Mitatha, N. Chaiyasoonthorn, and P.P. Yupapin: "Dark-bright optical solitons conversion via an optical add/drop filter", *Microw. Opt. Technol. Lett.*, 51, 2104–2107 (2009).

[5] T. Threepak, X. Luangvilay, S. Mitatha, and P.P. Yupapin, "Novel quantum-molecular transporter and networking via a wavelength router", *Microw. Opt. Technol. Lett.*, 52(6), 1353–1357 (2010).

[6] K. Kulsirirat, W. Techithdeera, and P.P. Yupapin, "Dynamic potential well generation and control using double resonators incorporating in an add/drop filter", *Mod. Phys. Lett. B* (2010), in press.

[7] J. Younse, "Projection Display Systems Based on the Digital Micromirror Device", Proceedings of Micromechanical structures and Microelectromechanical Devices for Optical Processing and Multimedia Applications, 64 (1995).

[8] B.C. Kress and P. Meyrueis, "Applied Digital Optics—From Micro-optics to Nanophotonics", John Wiley and Sons Ltd., UK (2009).

[9] A. Ashkin, J.M. Dziedzic, J.E. Bjorkholm, and S. Chu, "Observation of a single-beam gradient force optical trap for dielectric particles", *Opt. Lett.*, 11, 288–290 (1986).

[10] A. Ashkin, "Optical trapping and manipulation of neutral particles using lasers", Proceedings of theNational Academy of Sciences USA, 4853 (1997).

[11] S.F. Hanim, J. Ali, and P.P. Yupapin, "Dark soliton generation using dual Brillouin fiber laser in a fiber optic ring resonator", *Microw. Opt. Technol. Lett.*, 52, 881–883 (2010).

[12] T. Saktio, J. Ali, and P.P. Yupapin, "Novel design of multiplexed sensors using a dual FBGs scheme", *Microw. Opt. Technol. Lett.*, 52, 1218–1221 (2010).

[13] P.P. Yupapin, T. Saktioto, and J. Ali, "Photon trapping model within a fiber Bragg grating for dynamic optical tweezers use", *Microw. Opt. Technol. Lett.*, 52, 959–961 (2010).

[14] M. Schulz, H. Crepaz, F. Schmidt-Kaler, J. Eschner, and R. Blatt, "Transfer of trapped atoms between two optical tweezer potentials", *J. Mod. Opt.*, 54(11), 1619–1626 (2007).

[15] P.J. Reece, E.M. Wright, and K. Dholakia, "Experimental observation of modulation instability and optical spatial soliton arrays in soft condensed matter", *Phys. Rev. Lett.*, 98, 203–902 (2007).

[16] G. Brambilla, G.S. Murugan, J.S. Wilkinson, and D.J. Richardson, "Optical manipulation of microspheres along a subwavelength optical wire", *Opt. Lett.*, 32, 3041–3043 (2007).

[17] D. McGloin, V. Garces-Chavez, and K. Dholakia, "Interfering Bessel beams for optical micromanipulation", *Opt. Lett.*, 28, 657–659 (2003).

[18] L.E. Helseth "Mesoscopic orbitals in strongly focused light", *Opt. Comm.*, 224, 255–261 (2003).

[19] V.G. Shvedov, A.V. Rode, Y.V. Izdebskaya, A.S. Desyatnikov, W. Krolokowski, and Y.S. Kivshar, "Selective trapping of multiple particles by volume speckle field", *Opt. Express*, 18, 3137–3142 (2010).

[20] S. Mitatha, N. Pornsuwancharoen, and P.P. Yupapin, "A simultaneous short-wave and millimeter-wave generation using a soliton pulse within a nano-waveguide", *IEEE Photon. Technol. Lett.*, 21, 932–934 (2009).

[21] Y. Kokubun, Y. Hatakeyama, M. Ogata, S. Suzuki, and N. Zaizen, "Fabrication technologies for vertically coupled microring resonator with multilevel crossing busline and ultracompact-ring radius", *IEEE J. Sel. Top. Quantum Electron.*, 11, 4–10 (2005).

[22] J. Engelsberger, E-H. Nösekabel, and M. Steinbichler, "Proc. *Application of Interferometry and Electronic Speckle Pattern Interferometry (ESPI) for Measurements on MEMS*, Session 4, FRINGE 2005, 488–493 (2006).

[23] A.B. Matsko, *Practical Application of Micro Resonators in Optics and Photonics*, CRC Press, Taylor & Francis Group, New York, USA (2009).

# 6

# Hybrid Transceiver

**CHAPTER OUTLINE**
- Introduction
- Theory
- Hybrid Transceiver and Repeater
- Hybrid Transceiver
- Hybrid Repeater
- Conclusion
- References

## 6.1 Introduction

Nanocommunication and networking has become the challenge and interesting aspect of research and investigation recently. Most of the time, the microscale and nanoscale devices are the basic components for such systems. Microring/nanoring resonator has also become an interesting device, which has been widely studied and investigated in many research areas [1–4]. One of them has shown that ring resonator device can be used to trap/retrieve and bring photons/atoms for long distance transmission. Several researchers have shown that the dynamic optical trapping tools can be formed by controlling the dark-bright soliton behaviors within a semiconductor add/drop multiplexer (ring resonator), which have been clearly investigated [5-8]. In those investigations, the high optical field is configured as the optical tweezers or potential wells [9, 10], which is available for photon/atom trapping and transportation. However, the searching of new suitable technique for photons/atoms trapping and transportation remains, in which there are plenty of rooms required to investigate and accommodate.

Optical tweezers technique has become a powerful tool for manipulation of micrometer-sized particles in three spatial dimensions. Initially, the useful static tweezers are recognized, and the dynamic tweezers now realized in practical works [11–13]. Recently, Schulz et al. [14] have shown that the transfer of trapped atoms between two optical potentials could be performed. In principle, an optical tweezers use forces exerted by intensity gradients in the strongly focused beams of light to trap and move the microscopic volumes of matters. Moreover, the other combination of force is induced by the interaction between photons, which is caused by the photon scattering effects. In application, the field intensity can be adjusted and tuned to form the suitable trapping potential. The desired gradient field and scattering force can be formed the suitable trapping force. Hence, the appropriated force can be configured for the transmitter/receiver part, which can be performed the long distance transportation.

In this chapter, the dynamic optical vortices are generated using a dark soliton, bright soliton and Gaussian pulse propagating within an add/drop optical multiplexer incorporating two nanoring resonators (PANDA ring resonator). The dynamic behaviors of solitons and Gaussian pulses are analyzed and described. To increase the channel multiplexing, the dark solitons with slightly different wavelengths are controlled and amplified within the tiny system. The trapping force stability is simulated and seen when the Gaussian pulse is used to control via the add(control) port. In application, the optical vortices (dynamic tweezers) can be used to store (trap) photon, atom, molecule, DNA, ion, or particle, which can perform the dynamic tweezers. By using the hybrid devices, the hybrid transceiver and repeater can be integrated and performed the required functioned by using a single system. The use of the transceiver and repeater to form the hybrid communication of those microscopic volumes of matters in the nanoscale regime can be realized.

## 6.2 Theory

In operation, the optical tweezers use forces that are exerted by the intensity gradients in the strongly focused beams of light to trap and move the microscopic volumes of matters, in which the optical forces are customarily defined by the relationship [15].

$$F = \frac{Q n_m P}{c} \tag{6.1}$$

where $Q$ is a dimensionless efficiency, $n_m$ is the index of refraction of the suspending medium, $c$ is the speed of light, and $P$ is the incident laser power, measured at the specimen. $Q$ represents the fraction of power utilized to exert force. For plane waves incident on a perfectly absorbing particle, $Q$ is equal to 1. To achieve stable trapping, the radiation pressure must create a stable, three-dimensional equilibrium. Because biological specimens are usually contained in aqueous medium, the dependence of $F$ on $n_m$ can be rarely exploited to achieve higher trapping forces. Increasing the laser power is possible, but only over a limited range due to the possibility of optical damage. $Q$ itself is therefore the main determinant of trapping force. It depends upon the NA (numerical aperture), laser wavelength, light polarization state, laser mode structure, relative index of refraction and geometry of the particle.

Furthermore, in the Rayleigh regime, trapping forces decompose naturally into two components. Since, in this limit, the electromagnetic field is uniform across the dielectric, particles can be treated as induced point dipoles. The scattering force, $F_\text{scatt}$ is given by

$$F_\text{scatt} = n_m \frac{\langle S \rangle \sigma}{c} \qquad (6.2)$$

where

$$\sigma = \frac{8}{3}\pi(kr)^4 r^2 \left(\frac{m^2 - 1}{m^2 + 2}\right)^2 \qquad (6.3)$$

is the scattering cross section of a Rayleigh sphere with radius $r$. $S$ is the time-averaged Poynting vector, n is the index of refraction of the particle, $m = n/n_m$ is the relative index, and $k = 2\pi n_m/\lambda$ is the wave number of the light. Scattering force is proportional to the energy flux and points along the direction of propagation of the incident light. The gradient force, $F_\text{grad}$ is the Lorentz force acting on the dipole induced by the light field. It is given by

$$F_\text{grad} = \frac{\alpha}{2}\nabla\langle E^2 \rangle \qquad (6.4)$$

where

$$\alpha = n_m^2 r^3 \left(\frac{m^2 - 1}{m^2 + 2}\right) \qquad (6.5)$$

is the polarizability of the particle. The gradient force is proportional and parallel to the gradient in energy density (for $m > 1$). The large gradient force is formed by the large depth of the laser beam, in which the stable trapping requires that the gradient force in the $-\hat{z}$ direction, where it is a dark soliton valley in this proposal, which is against the direction of incident light, be greater than the scattering force. Increasing the NA decreases the focal spot size and increases the gradient strength [16], which can be formed within the tiny system, for instance, nanoscale device.

We are looking for the system that can generate the dynamic tweezers (optical vortices), in which the microscopic volume can be trapped and transmission via the communication link. Firstly, the stationary and strong pulse that can propagate within the dielectric material (waveguide) for period of time is required. Moreover, the gradient field is an important property required in this case. Therefore, a dark soliton is satisfied and recommended to perform those requirements. Secondly, we are looking for the device that optical tweezers can propagate and form the long distance link, in which the gradient field (force) can be transmitted and received by using the same device. Here, the add/drop multiplexer in the form of a PANDA ring resonator which is well known and is introduced for this proposal, as shown in Figures 6.1 and 6.2. The multifunction operation can be formed by using bright soliton and Gaussian pulse as input pulse into the system. The input optical field ($E_{in}$) and the add port optical field ($E_{add}$) of the dark soliton, bright soliton and Gaussian pulses are given by [17], respectively.

$$E_{in}(t) = A_0 \tanh\left[\frac{T}{T_0}\right] \exp\left[\left(\frac{z}{2L_D}\right) - i\omega_0 t\right] \quad (6.6a)$$

$$E_{control}(t) = A \operatorname{sech}\left[\frac{T}{T_0}\right] \exp\left[\left(\frac{z}{2L_D}\right) - i\omega_0 t\right] \quad (6.6b)$$

$$E_{control}(t) = E_0 \exp\left[\left(\frac{z}{2L_D}\right) - i\omega_0 t\right] \quad (6.6c)$$

where $A$ and $z$ are the optical field amplitude and propagation distance, respectively. $T$ is a soliton pulse propagation time in a frame moving at the group velocity, $T = t - \beta_1 z$, where $\beta_1$ and $\beta_2$ are the coefficients of the linear and second-order terms of Taylor expansion of the propagation constant. $L_D = T_0^2/|\beta_2|$ is the dispersion length of the soliton pulse. $T_0$ in equation is a soliton pulse propagation time at initial input (or soliton pulse width), where $t$ is the soliton phase shift time, and the frequency shift of the soliton is $\omega_0$. This solution describes a pulse that keeps its temporal width invariance as it propagates, and is thus called a temporal soliton.

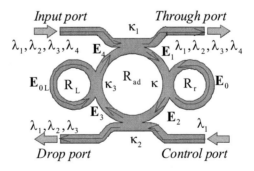

**FIGURE 6.1** A schematic diagram of a proposed hybrid transceiver.

When a solito of peak intensity ($|\beta_2/\Gamma T_0^2|$) is given, then $T_0$ is known. For the soliton pulse in the microring device, a balance should be achieved between the dispersion length ($L_D$) and the nonlinear length ($L_{NL}=1/\Gamma\phi_{NL}$).

Here $\Gamma = n_2 k_0$ is the length scale over which dispersive or nonlinear effects makes the beam become wider or narrower. For a soliton pulse, there is a balance between dispersion and nonlinear lengths. Hence $L_D = L_{NL}$. For Gaussian pulse in Eq. (6.6c), $E_0$ is the amplitude of optical field.

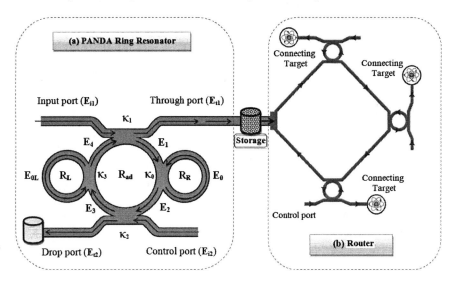

**FIGURE 6.2** A schematic diagram of a hybrid transceiver and a router.

When light propagates within the nonlinear medium, the refractive index ($n$) of light within the medium is given by

$$n = n_0 + n_2 I = n_0 + \frac{n_2}{A_{\text{eff}}} P \qquad (6.7)$$

with $n_0$ and $n_2$ as the linear and nonlinear refractive indexes, respectively. $I$ and $P$ are the optical intensity and the power, respectively. The effective mode core area of the device is given by $A_{\text{eff}}$. For the add/drop optical filter design, the effective mode core areas range from 0.50 to 0.10 µm², in which the parameters were obtained by using the related practical material parameters (InGaAsP/InP) [18]. When a dark soliton pulse is input and propagated within a add/drop optical filter as shown in Figure 6.1, the resonant output is formed.

In order to retrieve the required signals, we propose to use the add/drop device with the appropriate parameters. This is given in the following details. The optical circuits of ring resonator add/drop filters for the through port and drop port can be given by Eqs (6.8) and (6.9), respectively [20].

$$\left|\frac{E_t}{E_{in}}\right|^2 = \frac{\left[(1-\kappa_1)+(1-\kappa_2)e^{-\alpha L}-2\sqrt{1-\kappa_1}\cdot\sqrt{1-\kappa_2}e^{-\frac{\alpha}{2}L}\cos(k_n L)\right]}{\left[1+(1-\kappa_1)(1-\kappa_2)e^{-\alpha L}-2\sqrt{1-\kappa_1}\cdot\sqrt{1-\kappa_2}e^{-\frac{\alpha}{2}L}\cos(k_n L)\right]} \qquad (6.8)$$

$$\left|\frac{E_d}{E_{in}}\right|^2 = \frac{\kappa_1\kappa_2 e^{-\frac{\alpha}{2}L}}{1+(1-\kappa_1)(1-\kappa_2)e^{-\alpha L} - 2\sqrt{1-\kappa_1}\cdot\sqrt{1-\kappa_2}e^{-\frac{\alpha}{2}L}\cos(k_n L)} \quad (6.9)$$

Here $E_t$ and $E_d$ represent the optical fields of the through port and drop ports, respectively. $\beta = kn_{eff}$ is the propagation constant, $n_{eff}$ is the effective refractive index of the waveguide, and the circumference of the ring is $L = 2\pi R$, with $R$ as the radius of the ring. The filtering signal can be managed by using the specific parameters of the add/drop device, and the required signals can be retrieved via the drop port output. $\kappa_1$ and $\kappa_2$ are the coupling coefficients of the add/drop filters, $k_n = 2\pi/\lambda$ is the wave propagation number for in a vacuum, and the waveguide (ring resonator) loss is $\alpha = 0.5$ dBmm$^{-1}$. The fractional coupler intensity loss is $\gamma = 0.1$. In the case of the add/drop device, the nonlinear refractive index is not effect to the system, therefore, it is neglected.

From Eq. (6.8), the output field ($E_{t1}$) at the through port is given by

$$E_{t2} = AE_{i1} - BE_{i2}e^{-\frac{\alpha L}{2}-jk_n\frac{L}{2}} - \left[\frac{CE_{i1}e^{-\frac{\alpha L}{2}-jk_n\frac{L}{2}} + DE_{i2}e^{-\frac{3\alpha L}{2}-jk_n\frac{3L}{2}}}{1-YE_0E_{0L}e^{-\alpha L - jk_n L}}\right] \quad (6.10)$$

where

$A = \sqrt{(1-\gamma_1)(1-\gamma_2)}$

$B = \sqrt{(1-\gamma_1)(1-\gamma_2)\kappa_1(1-\kappa_2)}E_{0L}$

$C = \kappa_1(1-\gamma_1)\sqrt{(1-\gamma_2)\kappa_2}E_0 E_{0L}$

$D = (1-\gamma_1)(1-\gamma_2)\sqrt{\kappa_1(1-\kappa_1)\kappa_2(1-\kappa_2)}E_0 E_{0L}^2$ and

$Y = \sqrt{(1-\gamma_1)(1-\gamma_2)(1-\kappa_1)(1-\kappa_2)}$.

The electric fields $E_0$ and $E_{0L}$ are the field circulated within the nanoring at the right and left side of add/drop optical filter.

The power output ($P_{t1}$) at through port is written as

$$P_{t1} = |E_{t1}|^2 \quad (6.11)$$

The output field ($E_{t2}$) at drop port is expressed as

$$E_{t2} = \sqrt{(1-\gamma_2)(1-\kappa_2)}E_{i2} - \left[\frac{\sqrt{(1-\gamma_1)(1-\gamma_2)\kappa_1\kappa_2}E_0 E_{i1}e^{-\frac{\alpha L}{2}-jk_n\frac{L}{2}} + XE_0 E_{0L}E_{i2}e^{-\alpha L - jk_n L}}{1-YE_0E_{0L}e^{-\alpha L - jk_n L}}\right] \quad (6.12)$$

where

$X = (1-\gamma_2)\sqrt{(1-\gamma_1)(1-\kappa_1)\kappa_2(1-\kappa_2)}$

$Y = \sqrt{(1-\gamma_1)(1-\gamma_2)(1-\kappa_1)(1-\kappa_2)}$.

The power output ($P_{t2}$) at drop port is

$$P_{t2} = |E_{t2}|^2 \qquad (6.13)$$

## 6.3 Hybrid Transceiver and Repeater

Simulation results of trapped multi photons within the optical vortices are as shown in Figure 6.3, the coupling coefficients are given as $\kappa_0 = 0.1$, $\kappa_1 = 0.35$, $\kappa_2 = 0.1$ and $\kappa_3 = 0.2$, respectively. The ring radii are $R_{add} = 10$ μm, $R_{right} = 1.5$ μm and $R_{left} = 1.5$ μm, in which the evidence of the practical device was reported by the authors in reference [21, 22]. $A_{eff}$ is 0.50, 0.25 and 0.25 μm². Figure 6.3 (a)–(e), shows the results of four potential

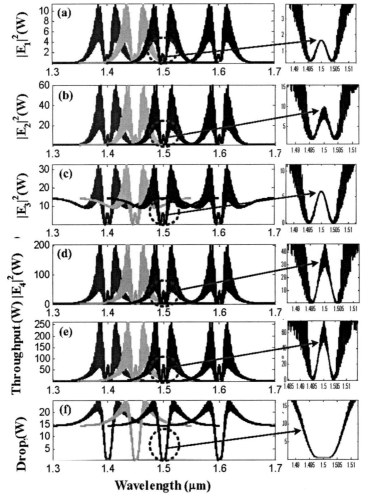

**FIGURE 6.3** Simulation result of four optical vortices/tweezers with four different center wavelengths.

wells (tweezers) with four different center wavelengths $\lambda_1 = 1.4$ μm, $\lambda_2 = 1.45$ μm, $\lambda_3 = 1.5$ μm and $\lambda_4 = 1.6$ μm for multi photons trapping using dynamic potential wells, in which they are controlled by the Gaussian signal with 1 W peak power at the control port, where the signals are (a) $|E_1|^2$, (b) $|E_2|^2$, (c) $|E_3|^2$, (d) $|E_4|^2$, (e) through port and (f) drop port signals, the dark soliton and Gaussian pulse are used as the input and control ports respectively. The important aspect is that the tunable receiver can be operated by tuning (controlling) the add (control) port input signal, in which the required number of microscopic volume (atom/photon/molecule) can be obtained via the drop port, otherwise, they are propagated within a PANDA device for a period of time before collapsing in the device. However, the use of amplification concept can keep the system being active in the same way as a hybrid memory, whereas the microscopic volume can be stored.

More results of optical tweezers are shown in Figure 6.4, where in this figure the multi tweezers can be generated with slightly different wavelengths, and the tunable gradient fields are obtained by varying the coupling coefficients, which can be formed in the fabrication process. The trapped multi photons (a) and tweezers (b) at the drop port are seen, in which the tunable coupling coefficients ($\kappa_2$) are (1) 0.1, (2) 0.35, (3) 0.6 and (4) 0.75, respectively, the other coupling coefficients are given as $\kappa_0 = 0.5$, $\kappa_1 = 0.35$, and $\kappa_3 = 0.1$. The multi photons are trapped with center wavelengths $\lambda_1 = 1.4$ μm, $\lambda_2 = 1.45$ μm, $\lambda_3 = 1.5$ μm and $\lambda_4 = 1.6$ μm, where (a) $|E_1|^2$, (b) $|E_2|^2$, (c) $|E_3|^2$, (d) $|E_4|^2$, (e) are outputs from the through port, and (f) are the output at the drop port, respectively. We found that the multi photons can be trapped and seen by controlling the second coupler parameter (coupling coefficient) of the add/drop device, moreover, the trapped multi photons can be amplified and tuned as shown in Figure 6.4 (a)–(f), in which the multi trapped photons with different coupling coefficients ($\kappa_2$) are increased and seen, whereas the trapping depth is increased. More simulation results of the dynamic optical vortices within the PANDA device are as shown in Figure 6.5. In this case the bright soliton is input into the control port, and the received atoms/molecules are seen as shown in Figure 6.6 (a)–(f). One more interesting aspect is the spiral phase structure or optical vortices in a PANDA ring, which is circulated in the spiral rotation. In this chapter, we found that the right and left microring resonators within a PANDA ring are moved faster than the center of PANDA ring (add/drop filter), which are shown in Figure 6.7(a) and (b).

### 6.3.1 Hybrid Transceiver

In application, the transmitter and receiver are formed by using through port and drop port outputs respectively, whereas such the device can be operated both functions by the single device, which is called a hybrid transceiver. The required amount of microscopic volume can be controlled

Hybrid Transceiver **155**

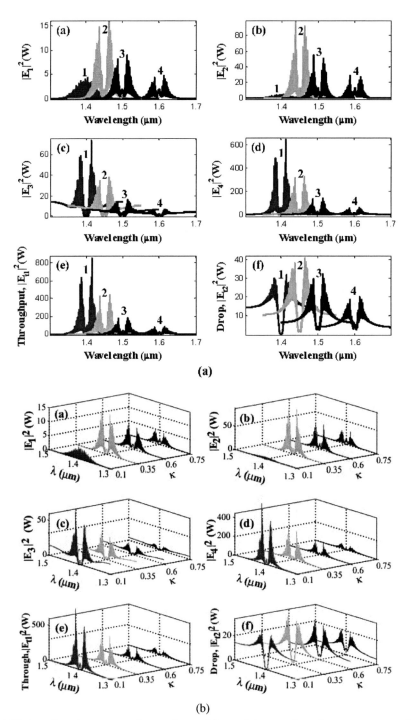

**FIGURE 6.4** Simulation results, where (a) multi wavelength tweezers and (b) tunable tweezers.

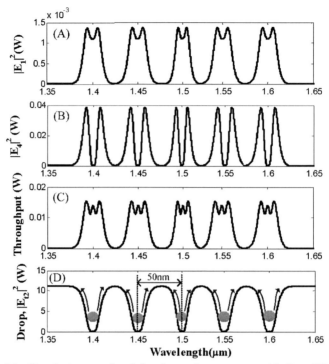

**FIGURE 6.5** Simulation result of the dynamic tweezers with five different center wavelengths.

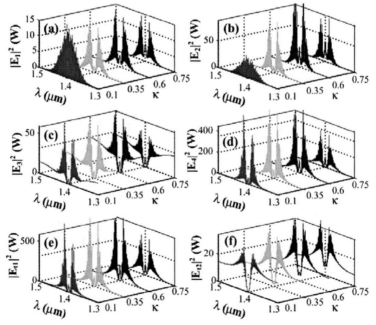

**FIGURE 6.6** Simulation result of the repeater manipulation for four different coupling coefficient.

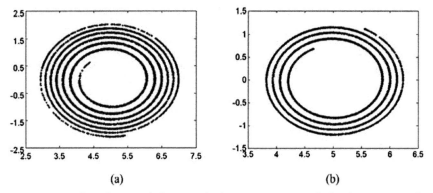

**FIGURE 6.7** Simulation of the spiral phase structure in hybrid repeater, where (a) ring $R_r$ and $R_L$ and (b) ring $R_{ad}$.

by using the hybrid transceiver and forming the transmission link via the router as shown in Figure 6.2, in which the long distance link for nano communication can be realized, where finally each end user can perform in the same way of the hybrid transceiver. The storage tank is introduced by the sender (transmitter) to store the specimen that is required to use in the communication, whereas the required specimen is received and stored by the receiver end.

### 6.3.2 Hybrid Repeater

In operation, the repeater is operated in the way that there are two functions required for the repeater concept, firstly, the repeater is performed when the number of microscopic volume is repeated as the initial signal is transmitted by the transmitter. The second one is that the amplified gradient field is also obtained. In Figure 6.6, the repeater feature can be obtained by controlling the amount of the microscopic volume via the add port signal, in which the through and drop port signals in the form of gradient fields or trapping microscopic volume can be obtained.

## 6.4 Conclusion

We have shown that the multi photons (microscopic volumes) can be trapped by optical vortices (tweezers), which generates a PANDA ring resonator. By utilizing the reasonable dark soliton input power, the dynamic multi photons can be controlled and stored (trapped) within the system. In this case, the dynamic optical vortices can be controlled and used to form the multi photons trapping tools. In application, such behaviors can be used to confine the suitable size of light pulse, atom or photon, which is then employed in the same way as the optical tweezers. But in this case the dynamic trapping photon/atom is realized, in which the trapped pulses or photons within the period of time (memory) within the system (PANDA ring) is plausible. Finally, we have shown that the use of a hybrid transceiver and repeater to form the long distance atom/molecule transportation will make the nano communication being realized in the near future. However, the problems of large atom/molecule and neutral atom/molecule may occur, in which the searching for new atom/molecule guide pipe and medium, for instance, nano tube and new medium will be the issue of investigation.

# REFERENCES

[1] K. Sarapat, N. Sangwara, K. Srinuanjan, P.P Yupapin, and N. Pornsuwancharoen, "Novel dark-bright optical solitons conversion system and power amplification", *Opt. Eng.*, 48(4), 045004 (2009).

[2] T. Phatharaworamet, C. Teeka, R. Jomtarak, S. Mitatha, and P.P. Yupapin, "Random binary code generation using dark-bright soliton conversion control within a PANDA ring resonator", *IEEE J. Lightw. Techn.*, 28(19), 2804–2809 (2010).

[3] S. Mitatha, "Dark soliton behaviors within the nonlinear micro and nanoring resonators and applications", *Progress in Electromagnetic Research (PIER)*, 99, 383–404 (2009).

[4] K. Kulsirirat, W. Techithdeera, and P.P. Yupapin, "Dynamic potential well generation and control using double resonators incorporating in an add/drop filter", *Mod. Phys. Lett. B*, 24, 1–9 (2010).

[5] T. Threepak, X. Louangvilay, S. Mitatha, and P.P. Yupapin, "Novel quantum-molecular transporter and networking via a wavelength router", *Microw. and Opt. Techn. Lett.*, 52, 1353–1357 (2010).

[6] M. Tasakorn, C. Teeka, R. Jomtarak, and P.P. Yupapin, "Multitweezers generation control within a nanoring resonator system", *Opt. Eng.*, 49, 075002 (2010).

[7] B. Piyatamrong, C. Teeka, R. Jomtarak, S. Mitatha, and P.P. Yupapin, "Multi photons trapping within optical vortices in an add/drop multiplexer", *Submitted to Opt. Lett.*

[8] M.A. Jalil, B. Piyatamrong, S. Mitatha, J. Ali, and P.P. Yupapin, "Molecular Transporter Generation for quantum-molecular transmission via an optical transmission line", *Nano Communication Network*, 1(2), 96–101 (2010).

[9] S.F. Hanim, J. Ali, and P.P. Yupapin, "Dark soliton generation using dual Brillouin fiber laser in a fiber optic ring resonator", *Microwave and Opt. Techn. Lett.*, 52, 881–883 (2010).

[10] P.P., Yupapin, T. Saktioto, and J. Ali, "Photon trapping model within a fiber Bragg grating for dynamic optical tweezers use", *Microwave and Opt. Techn. Lett.*, 52, 959–961 (2010).

[11] A. Ashkin, J.M. Dziedzic, J.E. Bjorkholm, and S. Chu, "Observation of a single-beam gradient force optical trap for dielectric particles", *Opt. Lett.*, 11, 288–290 (1986).

[12] K. Egashira, A. Terasaki, and T. Kondow, "Photon-trap spectroscopy applied to molecules adsorbed on a solid surface: probing with a standing wave versus a propagating wave", *Appl. Opt.*, 80, 5113–5115 (1998).

[13] A.V. Kachynski, A.N. Kuzmin, H.E. Pudavar, D.S. Kaputa, A.N. Cartwright, and P.N. Prasad, "Measurement of optical trapping forces by use of the two-photon-excited fluorescence of microspheres", *Opt. Lett.*, 28, 2288–2290 (2003).

[14] M. Schulz, H. Crepaz, F. Schmidt-Kaler, J. Eschner, and R. Blatt, "Transfer of trapped atoms between two optical tweezer potentials", *J. Mod. Opt.*, 54, 1619–1626 (2007).

[15] A. Ashkin, "Forces of a single-beam gradient laser trap on a dielectric sphere in the ray optics regime", *J. Biophysics*, 61, 569–582 (1992).

[16] K. Svoboda and S.M. Block, "Biological applications of optical forces", *Annual Review Biophysics and Bio-molecule Structure*, 23, 247–282 (1994).

[17] S. Mithata, N. Pornsuwancharoen, and P.P. Yupapin, "A simultaneous short wave and millimeter wave generation using a soliton pulse within a nano-waveguide", *IEEE J. Photon. Techn. Lett.*, 21(13), 932–934 (2009).

[18] Y. Kokubun, Y. Hatakeyama, M. Ogata, S. Suzuki, and N. Zaizen, "Fabrication technologies for vertically coupled microring resonator with multilevel crossing busline and ultracompact-ring radius", *IEEE J. Selec. Top. in Quan. Electron.*, 11, 4–10 (2005).

[19] P.P. Yupapin and W. Suwancharoen, "Chaotic signal generation and cancellation using a microring resonator incorporating an optical add/drop multiplexer", *Opt. Comm.*, 280, 343–350 (2007).

[20] P.P. Yupapin, P. Saeung, and C. Li, "Characteristics of complementary ring-resonator add/drop filters modeling by using graphical approach", *Opt. Comm.*, 272, 81–86 (2007).

[21] J. Zhu, S.K. Ozdemir, Y.F. Xiao, L. Li, L. He, D.R. Chen, and L. Yang, "On-chip single nanoparticle detection and sizing by mode splitting in an ultrahigh-Q microresonator", *Nat. Photon.*, 4, 46–49 (2010).

[22] Q. Xu, D. Fattal, and R.G. Beausoleil, "Silicon microring resonators with 1.5 μm radius", *Opt. Express*, 16(6), 4309–4315 (2008).

# 7
# Nanocommunication

**CHAPTER OUTLINE**
- Introduction
- Multi Variable Quantum Tweezers Generation and Modulation
- Molecular Transporter Generation for Quantum-Molecular Transmission
- Conclusion
- References

## 7.1 Introduction

By using the dark-bright soliton control within an add/drop filter, the dynamic potential well can be formed [1], which is available for molecule/atom trapping and transportation. The use of optical tweezers for molecule transportation within a wavelength router was also reported [2], where in this case the transport states are identified by using the correlated photons, where the entangled states of the molecular transporters were established. In general, optical tweezers are a powerful tool for use in the three-dimensional rotation of and translation (location manipulation) of nano-structures such as micro- and nano-particles as well as living micro-organisms [3]. Many research works have been concentrated on the static tweezers [4–8], which it cannot move. The benefit offered by optical tweezers is the ability to interact with nano-scaled objects in a non-invasive manner, i.e. there is no physical contact with the sample, thus preserving many important characteristics of the sample, such as the manipulation of a cell with no harm to the cell. Optical tweezers are now widely used and they are particularly powerful in the field of microbiology [9–11] to study cell–cell interactions, manipulate organelles without breaking the cell membrane and to measure adhesion forces between cells. In this chapter we describe a new concept of developing an optical tweezers source using a dark soliton pulse. The developed tweezers have shown many potential applications in electron, ion, atom and molecule probing and manipulation as well as DNA probing and transportation. Furthermore, the soliton pulse generator is a simple and compact design, making it more commercially viable.

In this chapter, we present the theoretical background in the physical model concept, where a potential well can be formed by the barrier of optical field. The change in potential value, i.e. gradient of potential can produce force that can be used to confine/trap atoms/molecule. Furthermore, the change in potential well is still stable in some conditions, which means that the dynamic optical tweezers is plausible, therefore, the transportation of atoms/molecules in the optical network via a dark soliton is being realized in the near future. In application, the high capacity tweezers can be formed by using the tweezers array (multi tweezers) [12], which is available for high capacity transportation via optical wireless link [13]. The photon entanglement using a quantum processor is also reviewed.

## 7.2 Multi Variable Quantum Tweezers Generation and Modulation

In principle, the optical forces are customarily defined by the relationship [14]

$$F = \frac{Q n_m P}{c} \tag{7.1}$$

where $Q$ is a dimensionless efficiency, $n_m$ is the index of refraction of the suspending medium, $c$ is the speed of light and P is the incident laser power, measured at the specimen. $Q$ represents the fraction of power utilized to exert force. For plane waves incident on a perfectly absorbing particle, $Q = 1$. To achieve stable trapping, the radiation pressure must create a stable, three-dimensional equilibrium. Because biological specimens are usually contained in aqueous medium, the dependence of $F$ on $n_m$ can rarely be exploited to achieve higher trapping forces. Increasing the laser power is possible, but only over a limited range due to the possibility of optical damage. $Q$ itself is therefore the main determinant of trapping force. It depends upon the numerical aperture (NA), laser wavelength, light polarization state, laser mode structure, relative index of refraction, and geometry of the particle.

In the Rayleigh regime, trapping forces decompose naturally into two components. Since, in this limit, the electromagnetic field is uniform across the dielectric, particles can be treated as induced point dipoles. The scattering force is given by

$$F_{scatt} = n_m \frac{\langle S \rangle \sigma}{c} \tag{7.2}$$

where

$$\sigma = \frac{8}{3}\pi(kr)^4 r^2 \left(\frac{m^2-1}{m^2+2}\right)^2 \tag{7.3}$$

is the scattering cross section of a Rayleigh sphere with radius $r$. $S$ is the time-averaged Poynting vector, $n$ is the index of refraction of the particle, $m = n/n_m$ is the relative index, and $k = 2\pi n_m/\lambda$ is the wave number of the light. Scattering force is proportional to the energy flux and points along the direction of propagation of the incident light. The gradient force is the Lorentz force acting on the dipole induced by the light field. It is given by

$$F_{grad} = \frac{\alpha}{2}\nabla\langle E^2 \rangle \tag{7.4}$$

where

$$\alpha = n_m^2 r^3 \left(\frac{m^2-1}{m^2+2}\right) \tag{7.5}$$

is the polarizability of the particle. The gradient force is proportional and parallel to the gradient in energy density (for $m > 1$). Stable trapping requires that the gradient force in the $-\hat{z}$ direction, against the direction of incident light, be greater than the scattering force. Increasing the NA decreases the focal spot size and increases the gradient strength [15].

To perform the proposed concept, a bright soliton pulse is introduced into the multi-stage nano ring resonators as shown in Figure 7.1, the input optical fields ($E_{in}$) of the bright and dark soliton pulses are given by an Eqs (7.6) and (7.7) as [13]

$$E_{in}(t) = A\,\text{sech}\left[\frac{T}{T_0}\right]\exp\left[\left(\frac{z}{2L_D}\right) - i\omega_0 t\right] \tag{7.6}$$

$$E_{in}(t) = A\tanh\left[\frac{T}{T_0}\right]\exp\left[\left(\frac{z}{2L_D}\right) - i\omega_0 t\right] \tag{7.7}$$

where $A$ and $z$ are the optical field amplitude and propagation distance, respectively. $T$ is a soliton pulse propagation time in a frame moving at the group velocity, $T = t - \beta_1 z$, where $\beta_1$ and $\beta_2$ are the coefficients of the linear and second order terms of Taylor expansion of the propagation constant. $L_D = T_0^2/|\beta_2|$ is the dispersion length of the soliton pulse. $T_0$ in equation is the initial soliton pulse width. Where $t$ is the soliton phase shift time, and the frequency shift of the soliton is $\omega_0$. This solution describes a pulse that keeps its temporal width invariance as it propagates and is thus called a temporal soliton. When a soliton peak intensity ($|\beta_2/\Gamma T_0^2|$) is given, then propagation time for input $T_0$ is known. For the soliton pulse in the micro ring device, a balance should be achieved between the dispersion length ($L_D$) and the nonlinear length ($L_{NL} = (1/\Gamma\phi_{NL})$, where $\Gamma = n_2 k_0$, is the length scale over which dispersive or nonlinear effects makes the beam becomes wider or narrower. For a soliton pulse, there is a balance between dispersion and nonlinear lengths, hence $L_D = L_{NL}$.

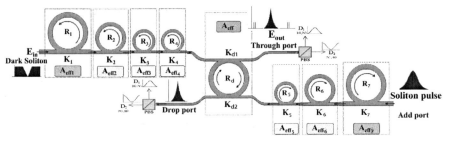

**FIGURE 7.1** A schematic diagram of the multi variable quantum tweezers generation system, where $R_i$: ring radii,. $\kappa_s$: coupling coefficients, $R_d$: add/drop filter radius, $K_{di}$: add/drop coupling coefficients, $A_{\text{eff},s}$: Effective core areas, PBS$_s$: polarizing beam-splitters [1].

We assume that the nonlinearity of the optical ring resonator is of the Kerr-type, i.e., the refractive index is given by

$$n = n_0 + n_2 I = n_0 + \left(\frac{n_2}{A_{\text{eff}}}\right) P \qquad (7.8)$$

where $n_0$ and $n_2$ are the linear and nonlinear refractive indexes, respectively. $I$ and $P$ are the optical intensity and optical power, respectively. The effective mode core area of the device is given by $A_{\text{eff}}$. For the microring and nanoring resonators, the effective mode core areas range from 0.10 to 0.50 µm² [16].

When a Gaussian pulse is input and propagated within a fiber ring resonator, the resonant output is formed, thus, the normalized output of the light field is the ratio between the output and input fields ($E_{\text{out}}(t)$ and $E_{\text{in}}(t)$) in each roundtrip, which can be expressed as [15]

$$\left|\frac{E_{\text{out}}(t)}{E_{\text{in}}(t)}\right|^2 = (1-\gamma)\left[1 - \frac{(1-(1-\gamma)x^2)\kappa}{(1-x\sqrt{1-\gamma}\sqrt{1-\kappa})^2 + 4x\sqrt{1-\gamma}\sqrt{1-\kappa}\sin^2\left(\frac{\varphi}{2}\right)}\right] \qquad (7.9)$$

Eq. (7.8) indicates that a ring resonator in the particular case is very similar to a Fabry-Pérot cavity, which has an input and output mirror with a field reflectivity, $(1 - \kappa)$, and a fully reflecting mirror. $k$ is the coupling coefficient, and $x = \exp(-\alpha L/2)$ represents a roundtrip loss coefficient, $\phi_0 = kLn_0$ and $\phi_{NL} = kL(n_2/A_{\text{eff}})P$ are the linear and nonlinear phase shifts, $k = 2\pi/\lambda$ is the wave propagation number in a vacuum, where $L$ and $\alpha$ are a waveguide length and linear absorption coefficient, respectively. In this chapter, the iterative method is introduced to obtain the results as shown in Eq. (7.9), similarly, when the output field is connected and input into the other ring resonators.

Figure 7.1 shows a schematic diagram of the multi variable quantum tweezers system, which two input fields at input and add ports. The dark soliton and bright soliton signals are input into input and add port respectively. The uplink carrier signal at 2 GHz is generated by using the dark soliton in a microring resonator system [13], where firstly the chaotic signal is generated with the rings $R_1 - R_4$, where the ring radii ($R_1 - R_7$) are between 5–12 µm, and the coupling coefficients are between 0.5–0.98.

The input optical field as shown in Eq. (7.6), i.e. a Gaussian pulse, is input into a nonlinear microring resonator. By using the appropriate parameters, the chaotic signal is obtained by using Eq. (7.8). To retrieve the signals from the chaotic noise, we propose to use the add/drop device with the appropriate parameters. This is given in details as followings. The optical outputs of a ring resonator add/drop filter can be given by the Eqs (7.10) and (7.11).

$$\left|\frac{E_t}{E_{in}}\right|^2 = \frac{(1-\kappa_1) - 2\sqrt{1-\kappa_1} \cdot \sqrt{1-\kappa_2}\, e^{-\frac{\alpha}{2}L} \cos(k_n L) + (1-\kappa_2) e^{-\alpha L}}{1+(1-\kappa_1)(1-\kappa_2)e^{-\alpha L} - 2\sqrt{1-\kappa_1} \cdot \sqrt{1-\kappa_2}\, e^{-\frac{\alpha}{2}L} \cos(k_n L)} \quad (7.10)$$

and

$$\left|\frac{E_d}{E_{in}}\right|^2 = \frac{\kappa_1 \kappa_2 e^{-\frac{\alpha}{2}L}}{1+(1-\kappa_1)(1-\kappa_2)e^{-\alpha L} - 2\sqrt{1-\kappa_1} \cdot \sqrt{1-\kappa_2}\, e^{-\frac{\alpha}{2}L} \cos(k_n L)} \quad (7.11)$$

where $E_t$ and $E_d$ represent the optical fields of the throughput and drop ports respectively. The transmitted output can be controlled and obtained by choosing the suitable coupling ratio of the ring resonator, which is well derived and described by reference [17]. Where $\beta = kn_{eff}$ represents the propagation constant, $n_{eff}$ is the effective refractive index of the waveguide, and the circumference of the ring is $L = 2\pi R$, here R is the radius of the ring. In the following, new parameters will be used for simplification, where $\phi = \beta L$ is the phase constant. The chaotic noise cancellation can be managed by using the specific parameters of the add/drop device, which the required signals at the specific wavelength band can be filtered and retrieved. $\kappa_1$ and $\kappa_2$ are coupling coefficient of add/drop filters, $k_n = 2\pi/\lambda$ is the wave propagation number for in a vacuum, and the waveguide (ring resonator) loss is $\alpha = 0.5$ dBmm$^{-1}$. The fractional coupler intensity loss is $\gamma = 0.1$. In the case of add/drop device [18], the nonlinear refractive index is neglected.

Figure 7.2 shows the simulation results of (a) dark soliton input with 2 W peak power, (b) the chaotic signal is generated within the first ring ($R_1$), whereas the ring radius used is 12 µm, with the coupling coefficient ($\kappa_1$) is 0.9713, (c) the radius of the second ring ($\kappa_2$) is 11.5 µm, with the coupling coefficient ($\kappa_2$) is 0.9723, (d) the radius of the third ring ($R_3$) is 10 µm, with the coupling coefficient ($\kappa_3$) is 0.9768 , the radius of the fourth ring ($R_4$) is 10 µm, and the coupling coefficient ($\kappa_4$) is 0.9768, (e) the input bright soliton with the center wavelength is at 1.500 µm, (f) the drop port output signals (multi optical tweezers), (g) the throughput port output signal amplitude and (h) the output signal at through port with the center wavelength is at 1.5 µm, and the radius of add/drop filter ($R_d$) is 50 µm, the coupling coefficient of add/drop filter is $\kappa_{d1} = \kappa_{d2} = 0.5$, which the through port signal free spectrum range is 0.002 µm.

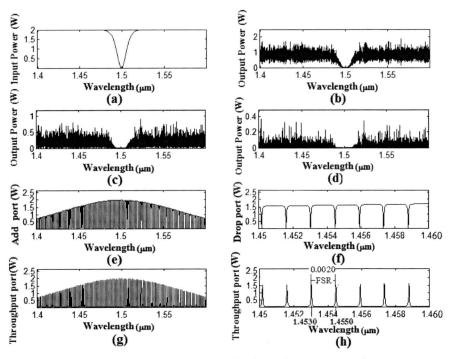

**FIGURE 7.2** Input and output signals of the dynamic optical tweezers.

In Figure 7.3, a polarization coupler that separates the basic vertical and horizontal polarization states corresponds to an optical switch between the first and second incoming pulses. We assume those horizontally polarized pulses with a temporal separation of $\Delta t$. The coherence time of the consecutive pulses, i.e. between two pulses is larger than $\Delta t$. Then the following state is created by Eq. (7.12) [2, 19, 20].

$$|\Phi\rangle_p = |1, H\rangle_s |1, H\rangle_i + |2, H\rangle_s |2, H\rangle_i \quad (7.12)$$

In the expression $|k, H\rangle$, $k$ is the number of time slots (1 or 2), which denotes the state of polarization, horizontal ($H$) or vertical ($V$), and the subscript identifies whether the state is the signal ($s$) or the idler ($i$) state. In Eq. (7.12), for simplicity we have omitted an amplitude term that is common to all product states. We employ the same simplification in subsequent equations. In this chapter, this two-photon state with $H$ polarization shown by Eq. (7.12) is input into the orthogonal polarization-delay circuit shown schematically in Figure 7.3. The delay circuit consists of a coupler and

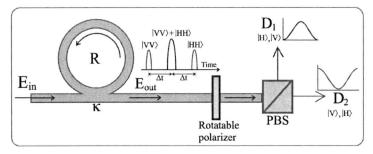

**FIGURE 7.3** A schematic diagram of the entangled photon generation system, where PBS: Polarizing Beam Splitter, $D_s$: Detectors.

the difference between the round-trip times of the micro ring resonator, which is equal to $\Delta t$. The microring is tilted by changing the round trip of the ring is converted into $V$ at the delay circuit output. That is the delay circuits convert;

$$|k,H\rangle + r|k,H\rangle + t_2 \exp(i\phi)|k+1,V\rangle + rt_2 \exp(i_2\phi)|k+2,H\rangle$$
$$+ r_2 t_2 \exp(i_3\phi)|k+3,V\rangle \qquad (7.13)$$

where $t$ and $r$ is the amplitude transmittances to cross and bar ports in a coupler. Then Eq. (7.8) is converted into the polarized state by the delay circuit as

$$|\Phi\rangle = [|1,H\rangle_s + \exp(i\varphi_s)|2,V\rangle_s] \times [|1,H\rangle_i + \exp(i\varphi_i)|2,V\rangle_i]$$
$$+ [|2,H\rangle_s + \exp(i\varphi_s)|3,V\rangle_s] \times [|2,H\rangle_i + \exp(i\varphi_i)|2,V\rangle_i]$$
$$= [|1,H\rangle_s|1,H\rangle_i + \exp(i\varphi_i)|1,H\rangle_s|2,V\rangle_i] + \exp(i\varphi_s)|2,V\rangle_s|1,H\rangle_i \qquad (7.14)$$
$$+ \exp[i(\varphi_s + \varphi_i)]|2,V\rangle_s|2,V\rangle_i + |2,H\rangle_s|2,H\rangle_i + \exp(i\varphi_i)|2,H\rangle_s|3,V\rangle_i$$
$$+ \exp(i\varphi_s)|3,V\rangle_s|2,H\rangle_i + \exp[i(\varphi_s + \varphi_i)]|3,V\rangle_s|3,V\rangle_i$$

By the coincidence counts in the second time slot, we can extract the fourth and fifth terms. As a result, we can obtain the following polarization entangled state as

$$|\Phi\rangle = |2,H\rangle_s|2,H\rangle_i + \exp[i(\varphi_s + \varphi_i)]|2,V\rangle_s|2,V\rangle_i \qquad (7.15)$$

We assume that the response time of the Kerr effect is much less than the cavity round-trip time. Because of the Kerr nonlinearity of the optical fiber, the strong pulses acquire an intensity dependent phase shift during propagation. The interference of light pulses at a coupler introduces the output beam, which is entangled. Due to the polarization states of light pulses are changed and converted while circulating in the delay circuit, where the polarization entangled photon pairs can be generated. The entangled photons of the nonlinear ring resonator are separated to be the signal and idler photon probability. The polarization angle adjustment device, as shown in Figure 7.3, is applied to investigate the orientation and optical output intensity.

In operation, a dark soliton input power of 2 W is input into the system as shown in Figure 7.1. The first ring radius ($R_1$) is 12 μm, and the coupling coefficient ($\kappa_1$) is 0.9713. The radius of the second ring ($R_2$) is 11.5 μm, where the coupling coefficient ($\kappa_2$) is 0.9723. The radius of the third ring ($R_3$) is 10 μm, the coupling coefficient ($\kappa_3$) is 0.9768, where the radius of the fourth ring ($R_4$) is 10 μm, and the coupling coefficient ($\kappa_4$) is 0.9768. The bright soliton with the center wavelength at 1.500 μm is input into the add port. The coupling coefficient of add/drop filter is $\kappa_{d1} = \kappa_{d2} = 0.5$. Figure 7.4 shows results of the multi quantum tweezers using the entangled photon states identification, where (a) is $|E|^2|V\rangle$, which represents the intensity of vertical polarization, where (b) $|E|^2|H\rangle$ is the intensity in the horizontal polarization. Where (c) and (d) are the multi quantum tweezers within the wavelength range between 1.5528 and 1.5548 μm, whereas the free spectrum range is 0.002 μm within the period of 100 ns.

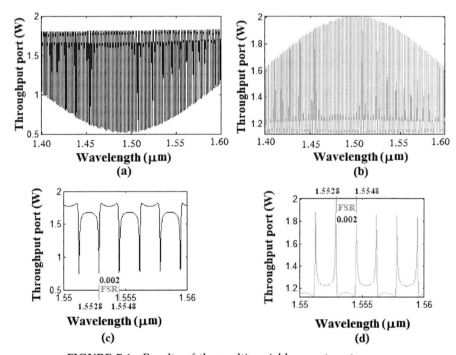

**FIGURE 7.4** Results of the multi variable quantum tweezers.

Figure 7.5 shows the results of the dynamic quantum tweezers (wells), where the tweezers intensity at the throughput port is (a) $|E|^2|V\rangle$, which is the vertical polarization intensity, (b) $|E|^2|H\rangle$ is the horizontal polarization intensity. For instance, these can be used to form the quantum tweezers for DNA size of 0.34 nm, and (c) is $|E|^2[|H\rangle+|V\rangle]$ is the vertical $|V\rangle$ and horizontal $|H\rangle$ polarization intensity respectively for dynamic quantum tweezers, which are performed by using the entangled photons as shown in Figures 7.1 and 7.5.

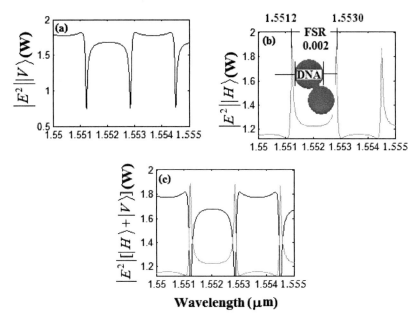

**FIGURE 7.5** The multi variable quantum tweezers with trapped DNAs.

## 7.3 Molecular Transporter Generation for Quantum-Molecular Transmission

Optical tweezers have been the good candidate when it can penetrate into the measurement target area. Since, optical tweezers have been recognized as a powerful tool for the use in many researches [22–28], for instance, the use for three-dimensional rotation of and translation (location manipulation) of nano-structures such as micro- and nano-particles as well as living micro-organisms. The benefit offered by optical tweezers is the ability to interact with nano-scaled objects in a non-invasive manner, i.e. there is no physical contact (non-demolition) with the sample, thus preserving many important characteristics of the sample, such as the manipulation of a cell with no harm to the cell. Optical tweezers are now widely used and they are particularly powerful in the field of microbiology to study cell–cell interactions, manipulate organelles without breaking the cell membrane and to measure adhesion forces between cells. In this chapter we describe a new concept of developing an optical tweezers source using a Gaussian pulse (laser pulse) to use in atomic spectroscopy. The developed tweezers have many potential applications in electron, ion, atom and molecule probing and manipulation as well as DNA probing, transportation and imaging.

We present the theoretical background in the physical model concept, where potential well can be formed by the barrier of optical filed, i.e. laser pulses. And also, the experimental of optical tweezers have been tested to verify the result of optical tweezers generated by using a Gaussian pulse in a fiber optic amplification system. The advantage is that the optical tweezers in this case is a moveable and tunable probe, which can be used in any location of the interesting target, which the resolution of the atom size is obtained.

Recently, several research works have shown that use of dark and bright soliton in various applications can be realized [29–34], where one of them has shown that the secured signals in the communication link can be retrieved by using a suitable an add/drop filter that is connected into the transmission line. The other promising application of a dark soliton signal [35] is for the large guard band of two different frequencies which can be achieved by using a dark soliton generation scheme and trapping a dark soliton pulse within a nanoring resonator [36]. Furthermore, the dark soliton pulse shows a more stable behavior than the bright solitons with respect to the perturbations such as amplifier noise, fiber losses, and intra-pulse stimulated Raman scattering [37]. It is found that the dark soliton pulses propagation in a lossy fiber, spreads in time at approximately half the rate of bright solitons. Recently, the localized dark solitons in the add/drop filter has been reported [38]. In this chapter, the use of dark and bright solitons propagating within the proposed ring resonator systems is investigated and described, where the use of suitable parameters based on the realistic device is discussed. The potential of using the generated dark soliton signals for single photon tweezers and molecular transporter, especially, for the hybrid quantum-molecular communication and transportation in the communication network, which is described in detail.

### 7.3.1 Transporter Generation

Bright and dark soliton pulses are introduced into the multi-stage nanoring resonators as shown in Figure 7.6, the input time dependent optical field ($E_{in}$) of the bright and dark soliton pulses input are given by an Eqs (7.16) and (7.17) [36], respectively.

$$E_{in}(t) = A\operatorname{sech}\left[\frac{T}{T_0}\right]\exp\left[\left(\frac{z}{2L_D}\right) - i\omega_0 t\right] \qquad (7.16)$$

and

$$E_{in}(t) = A\tanh\left[\frac{T}{T_0}\right]\exp\left[\left(\frac{z}{2L_D}\right) - i\omega_0 t\right] \qquad (7.17)$$

where $A$ and $z$ are the optical field amplitude and propagation distance, respectively. $T$ is a soliton pulse propagation time in a frame moving at the group velocity, $T = t - \beta_1 z$, where $\beta_1$ and $\beta_2$ are the coefficients of

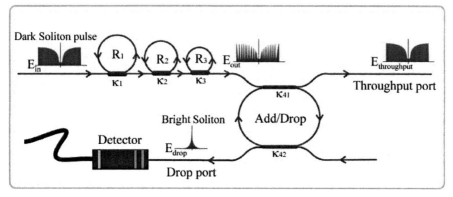

**FIGURE 7.6** A schematic diagram of a dark-bright soliton conversion system, where $R_s$ is the ring radii, $\kappa_s$ is the coupling coefficient, and $\kappa_{41}$ and $\kappa_{42}$ are the add/drop coupling coefficients.

the linear and second-order terms of Taylor expansion of the propagation constant. $L_D = T_0^2/|\beta_2|$ is the dispersion length of the soliton pulse. $T_0$ in equation is a soliton pulse propagation time at initial input (or soliton pulse width), where $t$ is the soliton phase shift time, and the frequency shift of the soliton is $\omega_0$. This solution describes a pulse that keeps its temporal width invariance as it propagates, and is thus called a temporal soliton. When a soliton peak intensity $(|\beta_2/\Gamma T_0^2|)$ is given, then $T_0$ is known. For the soliton pulse in the microring device, a balance should be achieved between the dispersion length ($L_D$) and the nonlinear length ($L_{NL} = 1/\Gamma \phi_{NL}$), where $\Gamma = n_2 k_0$ is the length scale over which dispersive or nonlinear effects makes the beam become wider or narrower. For a soliton pulse, there is a balance between dispersion and nonlinear lengths, hence $L_D = L_{NL}$. Similarly, the output soliton of the system in Figure 7.7 can be calculated by using Gaussian equations as given in the above case.

We assume that the nonlinearity of the optical ring resonator is of the Kerr-type, i.e., the refractive index is given by

$$n = n_0 + n_2 I = n_0 + \left(\frac{n_2}{A_{\text{eff}}}\right) P \qquad (7.18)$$

where $n_0$ and $n_2$ are the linear and nonlinear refractive indexes, respectively. $I$ and $P$ are the optical intensity and optical power, respectively. The effective mode core area of the device is given by $A_{\text{eff}}$. For the microring and nanoring resonators, the effective mode core areas range from 0.10 to 0.50 μm² [39, 40].

When a Gaussian pulse is input and propagated within a fiber ring resonator, the resonant output is formed, thus, the normalized output of the light field is the ratio between the output and input fields [$E_{\text{out}}(t)$ and $E_{\text{in}}(t)$] in each roundtrip, which can be expressed as [41]

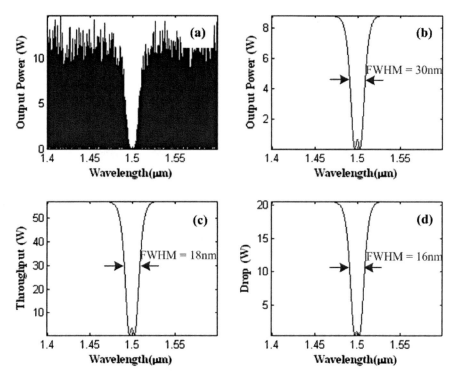

**FIGURE 7.7** The dynamic dark soliton (optical tweezers) occurs within add/drop tunable filter, where (a) add/drop signals, (b) dark–bright soliton collision, (c) optical tweezers at throughput port, and (d) optical tweezers at drop port.

$$\left|\frac{E_{\text{out}}(t)}{E_{\text{in}}(t)}\right|^2 = (1-\gamma)\left[1 - \frac{(1-(1-\gamma)\ x^2)\kappa}{(1-x\sqrt{1-\gamma}\sqrt{1-\kappa})^2 + 4x\sqrt{1-\gamma}\sqrt{1-\kappa}\sin^2\left(\frac{\phi}{2}\right)}\right] \quad (7.19)$$

Equation (7.19) indicates that a ring resonator in the particular case is very similar to a Fabry-Pérot cavity, which has an input and output mirror with a field reflectivity, $(1 - \kappa)$, and a fully reflecting mirror. $\kappa$ is the coupling coefficient, and $x = \exp(-\alpha L/2)$ represents a roundtrip loss coefficient, $\phi_0 = kLn_0$ and $\phi_{\text{NL}} = kL(n_2/A_{\text{eff}})P$ are the linear and nonlinear phase shifts, $k = 2\pi/\lambda$ is the wave propagation number in a vacuum, where $L$ and $\alpha$ are a waveguide length and linear absorption coefficient, respectively. In this chapter, the iterative method is introduced to obtain the results as shown in Eq. (7.4), and similarly, when the output field is fed into the other ring resonators.

The input optical field as shown in Eqs (7.16) and (7.17), i.e. a soliton pulse, is input into a nonlinear microring resonator. By using the appropriate parameters, the chaotic signal is obtained by using Eq. (7.19). To retrieve the signals from the chaotic noise, we propose to use the add/drop device with the appropriate parameters. This is given in details as followings. The optical outputs of a ring resonator add/drop filter can be given by the Eqs (7.20) and (7.21) [42].

$$\left|\frac{E_t}{E_{in}}\right|^2 = \frac{(1-\kappa_1) - 2\sqrt{1-\kappa_1}\cdot\sqrt{1-\kappa_2}e^{-\frac{\alpha}{2}L}\cos(k_n L) + (1-\kappa_2)e^{-\alpha L}}{1+(1-\kappa_1)(1-\kappa_2)e^{-\alpha L} - 2\sqrt{1-\kappa_1}\cdot\sqrt{1-\kappa_2}e^{-\frac{\alpha}{2}L}\cos(k_n L)} \quad (7.20)$$

and

$$\left|\frac{E_d}{E_{in}}\right|^2 = \frac{\kappa_1\kappa_2 e^{-\frac{\alpha}{2}L}}{1+(1-\kappa_1)(1-\kappa_2)e^{-\alpha L} - 2\sqrt{1-\kappa_1}\cdot\sqrt{1-\kappa_2}e^{-\frac{\alpha}{2}L}\cos(k_n L)} \quad (7.21)$$

where $E_t$ and $E_d$ represent the optical fields of the throughput and drop ports respectively. The transmitted output can be controlled and obtained by choosing the suitable coupling ratio of the ring resonator, which is well derived and described by reference [42]. Where $\beta = kn_{\text{eff}}$ represents the propagation constant, $n_{\text{eff}}$ is the effective refractive index of the waveguide, and the circumference of the ring is $L = 2\pi R$, here $R$ is the radius of the ring. In the following, new parameters will be used for simplification, where $\phi = \beta L$ is the phase constant. The chaotic noise cancellation can be managed by using the specific parameters of the add/drop device, in which the required signals at the specific wavelength band can be filtered and retrieved. $\kappa_1$ and $\kappa_2$ are coupling coefficient of add/drop filters, $k_n = 2\pi/\lambda$ is the wave propagation number for in a vacuum, and the waveguide (ring resonator) loss is $\alpha = 0.5$ dBmm$^{-1}$. The fractional coupler intensity loss is $\gamma = 0.1$. In the case of add/drop device, the nonlinear refractive index is neglected.

In operation, a dark soliton pulse with 50-ns pulse width with the maximum power of 0.65 W is input into the dark-bright soliton conversion system as shown in Figure 7.6. The suitable ring parameters are ring radii, where $R_1 = 10.0$ µm, $R_2 = 7.0$ µm, and $R_3 = 5.0$ µm. In order to make the system associate with the practical device [39, 41] the selected parameters of the system are fixed to $\lambda_0 = 1.50$ µm, $n_0 = 3.34$ (InGaAsP/InP). The effective core areas are $A_{\text{eff}} = 0.50, 0.25$, and $0.10$ µm$^2$ for a microring resonator (MRR) and nanoring resonator (NRR), respectively. The waveguide and coupling loses are $\alpha = 0.5$ dBmm$^{-1}$ and $\gamma = 0.1$, respectively, and the coupling coefficients $\kappa_s$ of the MRR are ranged from 0.05 to 0.90. The nonlinear refractive index is $n_2 = 2.2 \times 10^{-13}$ m$^2$/W. In this case, the waveguide loss used is 0.5 dBmm$^{-1}$. The input dark soliton pulse is chopped (sliced) into the smaller output signals of the filtering signals within the rings $R_2$ and $R_3$. We find that the output signals from $R_3$ are smaller than from $R_1$, which is more difficult to detect when it is used in the link. In fact, the multistage ring system is proposed due to the different core effective areas of the rings in the system, where the effective areas can be transferred from 0.50 to 0.10 µm$^2$ with some losses.

The soliton signals in $R_3$ are entered in the add/drop filter, where the dark-bright soliton conversion can be performed by using Eqs (7.20) and (7.21). The dynamic dark soliton control can be configured to be an optical dynamic tool known as an optical tweezers, where more details of optical tweezers can be found in references [43, 44]. After the bright soliton input is added into the system via add port as shown in Figure 7.7, the optical tweezers behavior is occurred. The parameters of system are used the same as the previous case. The bright soliton is generated with the central wavelength $\lambda_0 = 1.5$ µm, when the bright soliton propagating into the add/drop system, the dark-bright soliton collision in add/drop system is seen. The dark soliton valley dept, i.e. potential well is changed when it is modulated by the trapping energy (dark–bright solitons interaction) as shown in Figure 7.7. The dynamic dark soliton (optical tweezers) occurs within add/drop tunable filter, when the bright soliton is input into the add port with the central wavelength $\lambda_0 = 1.5$ µm. (a) add/drop signals, (b) dark–bright soliton collision, (c) optical tweezers at throughput port, and (d) optical tweezers at drop port. The recovery photon can be obtained by using the dark–bright soliton conversion, which is well analyzed by Sarapat et al. [34], where the trapped photon or molecule can be released or separated from the dark soliton pulse, in practice, in this case the bright soliton is become alive and seen.

### 7.3.2 Transporter Quantum State

Let us consider that the case when the optical tweezers output from the throughput port in Figures 7.6 and 7.7 is partially input into the quantum processor unit as shown in Figures 7.8 and 7.9. Generally, there are two pairs of possible polarization entangled photons forming within the ring device, which are represented by the four polarization orientation angles as [0°, 90°], [135° and 180°]. These can be formed by using the optical component called the polarization rotatable device and a polarizing beam splitter (PBS). In this concept, we assume that the polarized photon can be performed by using the proposed arrangement. Where each pair of the transmitted qubits can be randomly formed the entangled photon pairs. To begin this concept, we introduce the technique that can be used to create the entangled photon pair (qubits) as shown in Figure 7.9, a polarization coupler that separates the basic vertical and horizontal polarization states corresponds to an optical switch between the short and the long pulses. We assume those horizontally polarized pulses with a temporal separation of $\Delta t$. The coherence time of the consecutive pulses is larger than $\Delta t$. Then the following state is created by Eq. (7.22) [45].

$$|\Phi\rangle_p = |1, H\rangle_s |1, H\rangle_i + |2, H\rangle_s |2, H\rangle_i \qquad (7.22)$$

In the expression $|k, H\rangle$, $k$ is the number of time slots (1 or 2), which represents the state of polarization [horizontal $|H\rangle$ or vertical $|V\rangle$], and the subscript identifies whether the state is the signal ($s$) or the

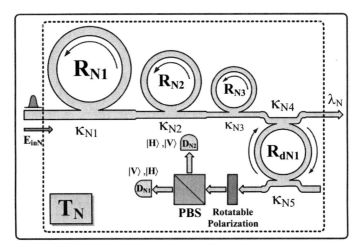

**FIGURE 7.8** A schematic diagram of a quantum tweezer generation system at the transmission unit ($T_N$), where $R_{NS}$: ring radii, $\kappa_{NS}$: coupling coefficients, $R_{dNS}$: an add/drop ring radius, can be used to be the received part, PBS: Polarizing Beam Splitter, $D_N$: Detectors.

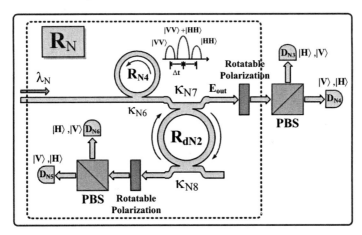

**FIGURE 7.9** A schematic diagram of an entangled photon pair manipulation within a ring resonator at the receiver unit ($R_N$). The quantum state is propagating to a rotatable polarizer and then is split by a Polarizing Beam Splitter (PBS) flying to detector $D_{N3}$ and $D_{N4}$.

idler ($i$) state. In Eq. (7.22), for simplicity, we have omitted an amplitude term that is common to all product states. We employ the same simplification in subsequent equations In this chapter. This two-photon state with $|H\rangle$ polarization shown by Eq. (7.22) is input into the orthogonal polarization-delay circuit shown schematically. The delay circuit consists of a coupler and the difference between the round-trip times of the microring resonator, which is equal to $\Delta t$. The microring is tilted by changing the round trip of the ring and it is converted into $|V\rangle$ at the delay circuit output. That is the delay circuits convert $|k, H\rangle$ to be

$$r|k,H\rangle + t_2 \exp(i\phi)|k+1,V\rangle + rt_2 \exp(i_2\phi)|k+2,H\rangle + r_2 t_2 \exp(i_3\phi)|k+3,V\rangle$$

where $t$ and $r$ are the amplitude transmittances to cross and bar ports in a coupler. Then Eq. (7.22) is converted into the polarized state by the delay circuit as

$$|\Phi\rangle = [|1,H\rangle_s + \exp(i\phi_s)|2,V\rangle_s] \times [|1,H\rangle_i + \exp(i\phi_i)|2,V\rangle_i]$$
$$+ [|2,H\rangle_s + \exp(i\phi_s)|3,V\rangle_s] \times [|2,H\rangle_i + \exp(i\phi_i)|2,V\rangle_i]$$
$$= [|1,H\rangle_s |1,H\rangle_i + \exp(i\phi_i)|1,H\rangle_s |2,V\rangle_i] + \exp(i\phi_s)|2,V\rangle_s |1,H\rangle_i \quad (7.23)$$
$$+ \exp[i(\varphi_s + \varphi_i)]|2,V\rangle_s |2,V\rangle_i + |2,H\rangle_s |2,H\rangle_i + \exp(i\varphi_i)|2,H\rangle_s |3,V\rangle_i$$
$$+ \exp(i\varphi_s)|3,V\rangle_s |2,H\rangle_i + \exp[i(\varphi_s + \varphi_i)]|3,V\rangle_s |3,V\rangle_i$$

By the coincidence counts in the second time slot, we can extract the fourth and fifth terms. As a result, we can obtain the following polarization entangled state as

$$|\Phi\rangle = |2,H\rangle_s |2,H\rangle_i + \exp[i(\phi_s + \phi_i)]|2,V\rangle_s |2,V\rangle_i \quad (7.24)$$

We assume that the response time of the Kerr effect is much less than the cavity round-trip time. Because of the Kerr nonlinearity of the optical device, the strong pulses acquire an intensity dependent phase shift during propagation. The interference of light pulses at a coupler introduces the output beam, which is entangled. Due to the polarization states of light pulses are changed and converted while circulating in the delay circuit, where the polarization entangled photon pairs can be generated. The entangled photons of the nonlinear ring resonator are separated to be the signal and idler photon probability. The polarization angle adjustment device is applied to investigate the orientation and optical output intensity, this concept is well described by the published work [45, 46]. The transporter states can be controlled and identified by using the quantum processing system as shown in Figures 7.8 and 7.9.

### 7.3.3 Multi Quantum-Molecular Transportation

By using the reasonable dark–bright soliton input power, the tunable optical tweezers can be controlled, which can provide the entangled photon as the dynamic optical tweezers probe. The smallest tweezer width of 16 nm is generated and achieved. In application, such a behavior can be used to confine the suitable size of light pulse or molecule, which can be employed in the same way of the optical tweezers. But in this case the terms dynamic probing is come to be a realistic function, therefore, the transportation of the trapped atom/molecule/photon by a single photon is plausible. For simplicity, the entangled photons power is attenuated to be a single photon before the detection, therefore, the separation between photon and molecule is employed the same way of a single photon detection scheme. This means that the detection of the transported single atom/molecule can be configured by using the single photon detection method. Thus, the transported molecule/atom with long distance link via quantum-molecular transporter is realized. Furthermore, the secured hybrid quantum-molecular communication can be implemented within the existed transmission link.

The transmitter unit (as shown in Figure 7.8) can be used to generate the quantum codes within the series of micro ring resonators and the cloning unit, which is operated by the add/drop filter ($R_{dN1}$). The receiver unit can be used to detect the quantum bits via the optical link, which can be obtained via the end quantum processor and the reference states can be recognized by using the cloning unit, which is operated by the add/drop filter ($R_{dN2}$) as shown in the schematic diagram in Figure 7.9. The remaining part of a system is the parallel processing system using the multi transporters multiplexing via an optical multiplexer as shown in the schematic diagram in Figure 7.10. The multi transporters are allowed to form and transmit via the optical link, where the transporters with different wavelengths ($\lambda_N$) can be generated and obtained which is available for multi molecules transportation. Moreover, the molecule identification can be recognized and confirmed by using the transporter quantum state.

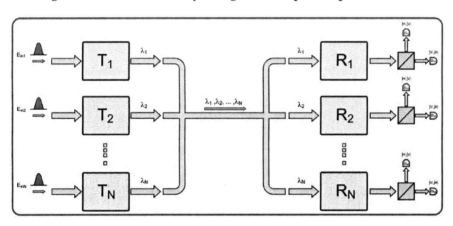

**FIGURE 7.10** A schematic diagram of a system of the transporters transmission system via an optical link, where $R_N$: the receiver part, $T_N$: the transmitter part.

## 7.4 Conclusion

We have demonstrated that some interesting results can be obtained when the laser pulse is propagated within the nonlinear optical ring resonator, especially, in microring and nanoring resonators, which can be used to perform many applications. For instance, the broad spectrum of a monochromatic source with the reasonable power can be generated and achieved. By using a dark soliton, it can be converted to be a bright soliton by using the ring resonator system incorporating the add/drop multiplexer, which can be configured as a dynamic optical tweezers. The use of quantum tweezers for quantum-molecular communication/cryptography via a single photon based technology in the communication link is plausible.

The multi variable quantum tweezers array can be generation by using a microring resonator system for use of quantum-molecular and atomic tweezers transmission, whereas the free spectrum range obtained is between 2.0–5.0 nm, which is shown the quantum tweezers array function. In applications, the smallest and simple quantum tweezers system can be employed, where the quantum states, entangle photon and multi tweezers can be generated, controlled and trapping, then transmitted via the transmission lines, which is allowed to use for multi molecules or atoms transportation via the optical wireless link. Finally, we have claimed this system is the novel design for multi variable quantum tweezers array by using dark–bright soliton control within an add/drop multiplexer, which is a possible technique that can form the DNA with quantum states identification.

We propose a novel system of a quantum-molecular transportation using the multi optical tweezers, whereas the transportation of molecules in the communication system can be performed. Initially, the generated transporter can be formed by the dark soliton, it can be tuned and attenuated to be a single photon by the bright soliton control and transmitted into the link. The molecule transportation states can be identified by using a single photon state of the transporter, which can be formed in the transmission line. The proposed fabricated material used is InGaAsP/InP, which can be formed the device that can provide the required output signals. The design system consists of a nonlinear microring/nanoring resonator system incorporating an add/drop filter and a quantum signal processor. In applications, the use of the proposed system incorporating a quantum processor can be performed as the secured molecular communication in the network.

# REFERENCES

[1] N. Pornsuwancharoen, M. Tasakorn, S. Julajaturasirarath, K. Chaiyawong, and P.P. Yupapin, "Multi-variable quantum tweezers generation using photon entanglement", *Nano Comm. Networks*, 1, 131–137 (2010).

[2] B. Piyatamrong, K. Kulsirirat, Techithdeera, S. Mitatha, and P.P. Yupapin, "Dynamic potential well generation and control using double resonators incorporating in an add/drop filter", *Mod. Phys. Lett. B*, vol. 24, pp. 3071–3080 (2010).

[3] T. Threepak, X. Luangvilay, S. Mitatha, and P.P. Yupapin, "Novel quantum-molecular transporter and networking via a wavelength router", *Microw. Opt. Technol. Lett.*, 52(6), 1353–1357 (2010).

[4] A. Ashkin, J.M. Dziedzic, J.E. Bjorkholm, and S. Chu, "Observation of a single-beam gradient force optical trap for dielectric particles", *Opt. Lett.*, 11, 288–290 (1986).

[5] R.L. Eriksen, V.R. Daria, and J. Gluckstad, "Fully dynamic multiple-beam optical tweezers", *Opt. Express*, 10, 597–602 (2002).

[6] P.J. Rodrigo, V.R. Daria, and J. Gluckstad, "Real-time interactive optical micromanipulation of a mixture of high- and low-index particles", *Opt. Express*, 12, 1417–1425 (2004).

[7] J. Liesener, M. Reicherter, T. Haist, and H.J. Tiziani, "Multi-functional optical tweezers using computer generated holograms", *Opt. Commun.*, 185, 77–82 (2000).

[8] J.E. Curtis, B.A. Koss, and D.G. Grier, "Dynamic holographic optical tweezers", *Opt. Commun.*, 207, 169–175 (2002).

[9] W.J. Hossack, E. Theofanidou, J. Crain, K. Heggarty, and M. Birch, "High-speed holographic optical tweezers using a ferroelectric liquid crystal microdisplay", *Opt. Express*, 11, 2053–2059 (2003).

[10] V. Boyer, R.M. Godun, G. Smirne, D. Cassettari, C.M.Chandrashekar, A.B. Deb, Z.J. Laczik, and C.J. Foot, "Dynamic manipulation of Bose-Einstein condensates with a spatial light modulator", *Phys. Rev., A* 73, 031402(R) (2006).

[11] A.V. Carpentier, J. Belmonte-Beitia, H. Michinel, and V.M. Perez-Garcia, "Laser tweezers for atomic solitons", *J. Mod. Opt.*, 55(17), 2819–2829 (2008).

[12] V. Milner, J.L. Hanssen, W.C. Campbell, and M.G. Raizen, "Optical billiards for atoms", *Phys. Rev. Lett.*, 86, 1514–1516 (2001).

[13] P.T. Korda, M.B. Taylor, and D.G. Grier, "Kinetically locked-in colloidal transport in an array of optical tweezers", *Phys. Rev. Lett.*, 89, 128301 (2002).

[14] S. Mithata, N. Pornsuwancharoen, and P.P. Yupapin, "A simultaneous short wave and millimeter wave generation using a soliton pulse within a nano-waveguide", *IEEE Photon. Technol. Lett.*, 21, 932–934 (2009).

[15] A. Ashkin, "Forces of a single-beam gradient laser trap on a dielectric sphere in the ray optics regime", *J. Biophys.*, 61, 569–82 (1992).

[16] K. Svoboda and S.M. Block, "Biological Applications of Optical Forces", *Annu. Rev. Biophs. Biomol. Struct.*, 247–282 (1994).

[17] Y. Kokubun, Y. Hatakeyama, M. Ogata, S. Suzuki, and N. Zaizen, "Fabrication technologies for vertically coupled microring resonator with multilevel crossing busline and ultracompact-ring radius", *IEEE J. Sel. Top. Quantum Electron.*, 11, 4–10 (2005).

[18] P.P. Yupapin and W. Suwancharoen, "Chaotic signal generation and cancellation using a microring resonator incorporating an optical add/drop multiplexer", *Opt. Commun.*, 280, 343–350 (2007).

[19] P.P. Yupapin, P. Saeung, and C. Li, "Characteristics of complementary ring-resonator add/drop filters modeling by using graphical approach", *Opt. Commun.*, 272, 81–86 (2007).

[20] P.P. Yupapin, "Generalized quantum key distribution via microring resonator for mobile telephone networks", *Optik-International Journal for Light and Electron Optics*, 121 (5), 422–425 (2010).

[21] C.H. Bennett, G. Brassard, C. Crepeau, R. Jozsa, A. Peres, and W.K. Wootters, "Teleporting an unknown quantum state via dual classical an Eistein-Podolsky-Rosen", *Phys. Rev. Lett.*, 70, 1895–1899 (1993).

[22] K. Svoboda and S.M. Block, "Biological applications of optical forces", *Annu. Rev. Biophys. Biomol. Struct.*, 23, 247–285 (1994).

[23] S.M. Block, "Making light work with optical tweezers". *Nature*, 360, 493–495 (1992).

[24] R.M. Simmons, J.T. Finer, S. Chu, and J.A. Spudich, "Quantitative measurements of force and displacement using an optical trap". *J. Biophys*, 70(4), 1813–1822 (1996).

[25] M.J. Lang, C.L. Asbury, J.W. Shaevitz, and S.M. Block, "An automated two-dimensional optical force clamp for single molecule studies", *J. Biophys*, 83(1), 491–501 (2002).

[26] J. Pine and G. Chow, "Moving live dissociated neurons with an optical tweezer", *IEEE Transaction on Biomedical Engineering*, 56(4), 1184–1188 (2009).

[27] A. Ashkin, J.M. Dziedzic, and T. Yamane, "Optical trapping and manipulation of single cells using infrared laser beams", *Nature*, 330, 769–771 (1987).

[28] J. Conia, B.S. Edwards, and S. Voelkel, "The micro-robotic laboratory: optical trapping and scissing for the biologist", *J. Clin. Lab. Anal.*, 11, 28–38 (1997).

[29] N. Pornsuwancharoen, U. Dunmeekaew, and P.P. Yupapin, "Multi-soliton generation using a microring resonator system for DWDM based soliton communication", *Microw. Opt. Technol. Lett.*, 51(5), 1374–1377 (2009).

[30] P.P. Yupapin, N. Pornsuwanchroen, and S. Chaiyasoonthorn, "Attosecond pulse generation using nonlinear microring resonators", *Microw. Opt. Technol. Lett.*, 50(12), 3108–3011 (2008).

[31] N. Pornsuwancharoen and P.P. Yupapin, "Generalized fast, slow, stop, and store light optically within a nanoring resonator", *Microw. Opt. Technol. Lett.*, 51(4), 899–902 (2009).

[32] N. Pornsuwancharoen, S. Chaiyasoonthorn, and P.P. Yupapin, "Fast and slow lights generation using chaotic signals in the nonlinear microring resonators for communication security", *Opt. Eng.*, 48(1), 50005-1–5 (2009).

[33] P.P. Yupapin and N. Pornsuwancharoen, "Proposed nonlinear microring resonator arrangement for stopping and storing light", *IEEE Photon. Technol. Lett.*, 21, 404–406 (2009).

[34] K. Sarapat, N. Sangwara, K. Srinuanjan, P.P. Yupapin, and N. Pornsuwancharoen, "Novel dark-bright optical solitons conversion system and power amplification", *Opt. Eng.*, 48, 045004–1–7 (2009).

[35] S. Mitatha, N. Pornsuwancharoen, and P.P. Yupapin, "A simultaneous short-wave and millimeter-wave generation using a soliton pulse within a nano-waveguide", *IEEE Photon. Technol. Lett.*, 21, 932–934 (2009).

[36] G.P. Agrawal, *Nonlinear Fiber Optics*, New York: Academic Press, 4th edition (2007).

[37] M.E. Heidari, M.K. Moravvej-Farshi, and A. Zariffkar, "Multichannel wavelength conversion using fourth-order soliton decay", *J. Lightwave Technol.*, 25, 2571–2578 (2007).

[38] A. Charoenmee, N. Pornsuwancharoen, and P.P. Yupapin, "Trapping a dark soliton pulse within a nanoring resonator", *International Journal of Light and Electron Optics*, vol. 121(18), 1670–1673 (2010).

[39] Y. Kokubun, Y. Hatakeyama, M. Ogata, S. Suzuki, and N. Zaizen, "Fabrication technologies for vertically coupled microring resonator with multilevel crossing busline and ultracompact-ring radius", *IEEE J. Sel. Top. Quantum Electron.* 11, 4–10 (2005).

[40] Y. Su, F. Liu, and Q. Li, "System performance of slow-light buffering, and storage in silicon nano-waveguide", *Proc. SPIE* 6783, 67832P (2007).

[41] P.P. Yupapin, P. Saeung, and C. Li, "Characteristics of complementary ring-resonator add/drop filters modeling by using graphical approach", *Opt. Comm.*, 272, 81–86 (2007).

[42] P.P. Yupapin and W. Suwancharoen, "Chaotic signal generation and cancellation using a microring resonator incorporating an optical add/drop multiplexer", *Opt. Comm.*, 280/2, 343–350 (2007).

[43] L. Yuan, Z. Liu, J. Yang, and C. Guan, "Twin-core fiber optical tweezers", *Opt. Express*, 16, 4559–4566 (2008).

[44] N. Malagninoa, G. Pescea, A. Sassoa, and E. Arimondo, "Measurements of trapping efficiency and stiffness in optical tweezers", *Opt. Comm.*, 214, 15–24 (2002).

[45] S. Suchat, W. Khannam, and P.P. Yupapin, "Quantum key distribution via an optical wireless communication link for telephone network", *Opt. Eng. Lett.*, 46(10), 100502–1–5(2007).

[46] P.P. Yupapin and S. Suchat, "Entangle photon generation using fiber optic Mach-Zehnder interferometer incorporating nonlinear effect in a fiber ring resonator", *Nanophotonics* (JNP), 1, 13504–1–7(2007).

# 8

# Nanosensors

### CHAPTER OUTLINE

- Introduction
- Network Sensors using a PANDA Ring Resonator Type
- Self-calibration in a Fiber Optic Sensing System
- Conclusion
- References

## 8.1 Introduction

Nano-science and technology has become the common subject that introduces many related research areas today. One of them is the nano-scale measurement resolution, i.e. nano metrology which can be used to support and confirm the small scale regime of the observed physical quantities. To date, there are many techniques that can be performed the nano-scale measurement resolutions and standards [1–4], however, the searching for new measurement standards and techniques still remains. Recently, Yupapin and Pornsuwancharoen [5] have shown the interesting result that light pulse can be stretched or compressed and stored within a tiny device known as "nano-waveguide". Several research works have also shown the interesting results that light can be stored within the micro- cavities [6], micro-sphere [7] and nano-waveguide [8]. The promising concept is that white light, i.e. continuous light spectra can be generated, amplified and stored within a nano-waveguide, which is allowed to raise the concept of storing a narrow light pulse, i.e. a narrow spectral width. This is suitable for high resolution measurement in terms of optical resonant wavelength, whereas the measurement resolution of picometers(pm) or femtometer(fm) can be achieved. Many earlier works of soliton applications in either theory or experimental works are found in a soliton application book by Hasegawa et al. [9]. Many of the soliton related concepts in fiber optic are discussed by Agarwal [10]. The problems of soliton–soliton interactions [11], collision [12], rectification [13] and dispersion management [14] are required to solve and addressed. In practice, the soliton–soliton interaction would affect the dense wavelength division multiplexing (DWDM), however, this problem can be solved by using the suitable free spectrum range arrangement, which can be designed [15]. To present this concept, the use of optical soliton in nano-ring resonator is formed with the large bandwidth signal, which can be compressed, filtered and stored within a nano-waveguide. Finally, the different resonant signals with different wavelengths are stored via the specific nano-waveguides in the array waveguide, which is available for sensing applications. The changes in physical quantities relating to the device parameters are observed and seen on the spectrum analyzer. The measurements of the related parameters with respect to the change in each wavelength can be performed by the distributed or multiplexed sensors. Quantum metrology using the present system is also described.

### 8.1.1 Operating Principle

To perform the proposed concept, a bright soliton pulse is introduced into the multi-stage nano ring resonators as shown in Figure 8.1, the input optical field ($E_{in}$) of the bright soliton pulse input is given by an Eq. (8.1) [5] as

$$E_{in}(t) = A \operatorname{sech}\left[\frac{T}{T_0}\right] \exp\left[\left(\frac{z}{2L_D}\right) - i\omega_0 t\right] \quad (8.1)$$

where $A$ and $z$ are the optical field amplitude and propagation distance, respectively. $T$ is a soliton pulse propagation time in a frame moving at the group velocity, $T = t - \beta_1^* z$, where $\beta_1$ and $\beta_2$ are the coefficients of the linear and second order terms of Taylor expansion of the propagation constant. $L_D = T_0^2/|\beta_2|$ is the dispersion length of the soliton pulse. $T_0$ in equation is the initial soliton pulse width, where $t$ is the soliton phase shift time, and the frequency shift of the soliton is $\omega_0$. This solution describes a pulse that keeps its temporal width invariance as it propagates and is thus called a temporal soliton. When a soliton peak intensity $(|\beta_2/\Gamma T_0^2|)$ is given, then $T_0$ is known. For the soliton pulse in the microring device, a balance should be achieved between the dispersion length ($L_D$) and the nonlinear length ($L_{NL} = 1/\Gamma\phi_{NL}$), where $\Gamma = n_2 * k_0$, is the length scale over which dispersive or nonlinear effects makes the beam becomes wider or narrower. For a soliton pulse, there is a balance between dispersion and nonlinear lengths, hence $L_D = L_{NL}$.

When light propagates within the nonlinear material (medium), the refractive index ($n$) of light within the medium is given by

$$n = n_0 + n_2 I = n_0 + \left(\frac{n_2}{A_{eff}}\right) P \qquad (8.2)$$

where $n_0$ and $n_2$ are the linear and nonlinear refractive indexes, respectively. $I$ and $P$ are the optical intensity and optical power, respectively. The effective mode core area of the device is given by $A_{eff}$. For the microring and nano ring resonators, the effective mode core areas range from 0.50 to 0.1 $\mu m^2$ [16], where they found that fast light pulse can be slowed down experimentally after input into the nano ring.

When a soliton pulse is input and propagated within a microring resonator as shown in Figure 8.1, which consists of a series microring resonators. The resonant output is formed, thus, the normalized output of the light field is the ratio between the output and input fields ($E_{out}(t)$ and $E_{in}(t)$) in each roundtrip, which can be expressed as [5, 15]

$$\left|\frac{E_{out}(t)}{E_{in}(t)}\right|^2 = (1-\gamma)\left[1 - \frac{(1-(1-\gamma)x^2)\kappa}{(1-x\sqrt{1-\gamma}\sqrt{1-\kappa})^2 + 4x\sqrt{1-\gamma}\sqrt{1-\kappa}\sin^2\left(\frac{\phi}{2}\right)}\right] \qquad (8.3)$$

The close form of Eq. (8.3) indicates that a ring resonator in the particular case is very similar to a Fabry-Pérot cavity, which has an input and output mirror with a field reflectivity, $(1 - \kappa)$, and a fully reflecting mirror. $\kappa$ is the coupling coefficient, and $x = \exp(-\alpha L/2)$ represents a roundtrip loss coefficient, $\phi_0 = kLn_0$ and $\phi_{NL} = kLn_2|E_{in}|^2$ are the linear and nonlinear phase shifts, $k = 2\pi/\lambda$ is the wave propagation number in a vacuum, where $L$ and $\alpha$ are a waveguide length and linear absorption coefficient, respectively. In this work, the iterative method is introduced to obtain the results as shown in Eq. (8.3), similarly, when the output field is connected and input into the other ring resonators.

## 8.1.2 Distributed Spatial Sensors

In operation, the localized soliton pulses can be stored within the microring device as shown in Figure 8.1. In Figure 8.2, results obtained have shown that the multi-soliton pulses can be generated and localized within the device (ring $R_3$), where the ring radii are $R_1 = 10$ μm, $R_2 = 7$ μm, $R_3 = 4$ μm, where $R_3 = R_4$ ($R_d = R_4$). The obtained results of the multi-soliton pulse generation are; (a) input soliton, (b) large bandwidth signals (c) temporal solitons, and (d) localized solitons. The key advantage of the proposed system is the multi-soliton pulses, which is available for multiplex/distributed sensing applications since, a soliton communication has been recognized as a good candidate for long distance communication. To generate a single soliton with slightly difference in wavelength, the input soliton pulse is chopped (sliced) into

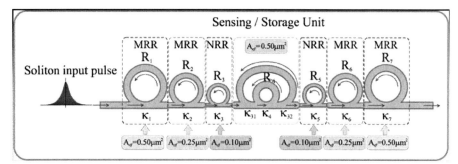

**FIGURE 8.1** A schematic diagram of an upstream and downstream soliton generation system with a storage unit, where $R_s$: ring radii, $k_s$: coupling coefficients, MRR: microring resonator, NRR: nano ring resonator.

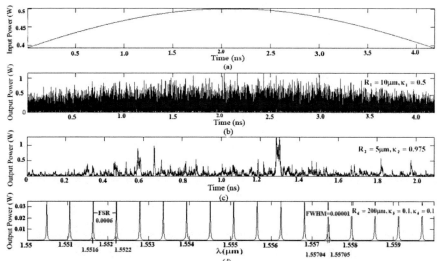

**FIGURE 8.2** Results of the multi-soliton pulse generation, where (a) input soliton, (b) large bandwidth signals, (c) temporal soliton, (d) spatial soliton. The free spectrum range is 600 pm, and the FWHM of the pulse is 10 pm [18].

a smaller signal spreading over the spectrum as shown in Figure 8.3(b), which is shown that the large bandwidth signal is generated within the first ring device. Figures 8.3(c) and (d) shows the compressing in spectral width of the output signals, with the parameters used are $R_2 = 7$ μm, $R_3 = R_4 = R_5 = 5$ μm. In operation, the upstream and downstream conversion of soliton generation can be performed using the system as shown in Figure 8.1. Furthermore, the generated soliton/signal can be stored within a nano-waveguide (ring $R_4$), which is confirmed by Yupapin and Poensuwancharoen [5]. Whereas the trapping pulse is circulated within the nano-waveguide (stopping/storing pulse), which is available to detect, i.e. it can be slowed down and detected by any available detector they have also found that the light pulse energy recovery can be obtained by connecting into the nano ring device. By using an Eq. (8.3), the output light pulse within a ring $R_4$ is obtained, where the main parameters that can provide the constant coupling energy are $K_{31}$, $K_{32}$ and the input power.

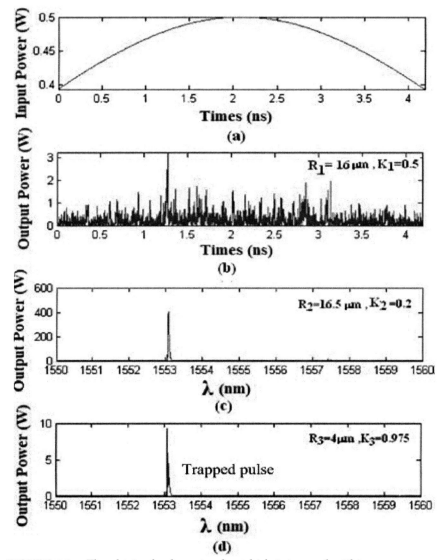

**FIGURE 8.3** The obtained coherent pulse which is trapped within a nano-waveguide, where (a) an input soliton pulse, large bandwidth signals, and (c) and (d) the compressing spectral width signals [5].

The tunable spatial solitons can be obtained by using the array waveguide as shown in Figure 8.4, which means that the multi-solitons

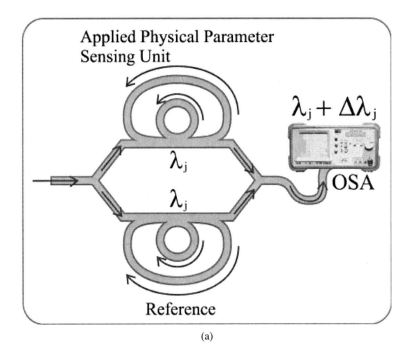

(a)

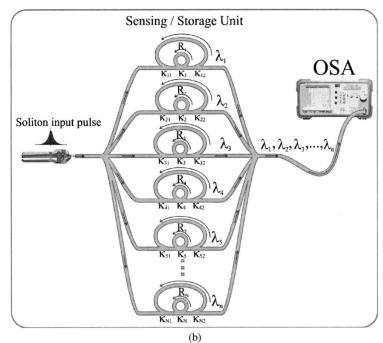

(b)

**FIGURE 8.4** Nano ring storage array waveguide system: (a) a sensing unit, (b) a distributed sensing system, where $R_i$: ring radii, $\kappa_i$: coupling coefficient, $\kappa_{ij}$: coupling loss, $\lambda_n$: signal output with wavelength $\lambda_n$, OSA: Optical Spectrum Analyzer.

can be stored within the ring $R_3$ (or $R_4$). The regeneration of the stored signals can be performed by using the ring radii $R_5$, $R_6$ and $R_7$, respectively. However, the coupling losses are included within the power distribution. The induced change within the ring $R_4$ occurs and is detected relating to the applied physical quantities. The sensing system is as shown in Figure 8.4, whereas the signal detection device can be optical spectrum analyzer, wavelength tuner, wavelength filter, or the recovery ring resonator.

### 8.1.3 Distributed Quantum Sensors

In general, the input pulse can be a single pulse or pulse trains, where the output pulses after some roundtrips with random polarization are in the form as shown in Figure 8.5. $|H\rangle$ and $|V\rangle$ represent the horizontal and vertical polarization components, respectively. To begin this concept, firstly, light pulse is input and chopped to form many pulses by the chaotic behavior within the microring resonator. Secondly, we introduce the technique that can be used to create the entangled photon pair (qubits) as shown in Figure 8.5 [19], polarized light can be formed the basic vertical and horizontal polarization states corresponds to the input short and long pulses (different time). We assume those horizontally polarized pulses with a temporal separation of $\Delta t$. The coherence time of the consecutive pulses is larger than $\Delta t$. When the coupled mode is formed by the external environment then the following state is created by Eq. (8.4).

$$|\Phi\rangle_p = |1, H\rangle_s |1, H\rangle_i + |2, H\rangle_s |2, H\rangle_i \tag{8.4}$$

In the expression $|k, H\rangle$, $k$ is the number of time slots (1 or 2), where $|H\rangle$ and $|V\rangle$ denote the state of polarization in horizontal and vertical components, respectively, and the subscript identifies whether the state is the signal (s) or the idler (i) state. In Eq. (8.4), for simplicity we have omitted an amplitude term that is common to all product states. We employ the same simplification in subsequent equations in this chapter. This two-photon state with $|H\rangle$ polarization shown by Eq. (8.4) is input into the orthogonal polarization-delay circuit shown schematically in Figure 8.1. The delay circuit consists of a coupler and the difference between the round-trip times of the microring resonator, which is equal to $\Delta t$. The microring is tilted by changing the round trip of the ring is converted into $|V\rangle$ at the delay circuit output. That is the delay circuits convert.

$$|k, H\rangle \text{ to } r|k, H\rangle + t_2 \exp(i\phi)|k+1, V\rangle + rt_2 \exp(i_2\phi)|k+2, H\rangle + r_2 t_2 \exp(i_3\phi)|k+3, V\rangle$$

where $t$ and $r$ are the amplitude transmittances to cross and bar ports in a coupler. Then Eq. (8.4) is converted into the polarized state by the delay circuit as

$$|\Phi\rangle = [|1,H\rangle_s + \exp(i\phi_s)|2,V\rangle_s] \times [|1,H\rangle_i + \exp(i\phi_i)|2,V\rangle_i]$$
$$+ [|2,H\rangle_s + \exp(i\phi_s)|3,V\rangle_s] \times [|2,H\rangle_i + \exp(i\phi_i)|2,V\rangle_i]$$
$$= [|1,H\rangle_s|1,H\rangle_i + \exp(i\phi_i)|1,H\rangle_s|2,V\rangle_i] + \exp(i\phi_s)|2,V\rangle_s|1,H\rangle_i \quad (8.5)$$
$$+ \exp[i(\phi_s + \phi_i)]|2,V\rangle_s|2,V\rangle_i + |2,H\rangle_s|2,H\rangle_i + \exp(i\phi_i)|2,H\rangle_s|3,V\rangle_i$$
$$+ \exp(i\phi_s)|3,V\rangle_s|2,H\rangle_i + \exp[i(\phi_s + \phi_i)]|3,V\rangle_s|3,V\rangle_i$$

By the coincidence counts in the second time slot, we can extract the fourth and fifth terms. As a result, we can obtain the following polarization entangled state as

$$|\Phi\rangle = |2,H\rangle_s|2,H\rangle_i + \exp[i(\phi_s + \phi_i)]|2,V\rangle_s|2,V\rangle_i \quad (8.6)$$

We assume that the response time of the Kerr effect is much less than the cavity round-trip time. Because of the Kerr nonlinearity of the optical device [1–4], the strong pulses acquire an intensity dependent phase shift during propagation. The interference of light pulses at a coupler introduces the output beam, which is entangled. Due to the polarization states of light pulses are changed and converted while circulating in the delay circuit, where the polarization entangled photon pairs can be generated. The entangled photons of the nonlinear ring resonator are separated to be the signal and idler photon probability. The polarization angle adjustment device is applied to investigate the orientation and optical output intensity, where the compensation, i.e. the measurement is performed.

When polarized light propagates in optical device, the change in birefringence is introduced. This means the change in the phase of the entangled photon pair occurs. The transversal walk-off produces a shift between the ordinary and extraordinary mode, while the longitudinal walk-off introduces a time delay between horizontally and vertically polarized photons. The amount of the walk-off depends on the location where the photon-pairs are created within the device. This position is completely random due to the coherent nature of light in the optical device. To compensate the longitudinal timing-walk off effect, a polarization controller is recommended to ensure that the polarization rotation is the same on both photons from the entangled pair. Additionally the compensator device is used to change the relative phase, $\phi$, of the states of the polarized light. Because of the change in birefringence, the tilting of the compensator allows to apply a phase shift to the entangled states of the two entangled photons, which is given by Eq. (8.7) as [1]

$$|\psi\rangle_{12} = \frac{1}{\sqrt{2}}\left(|H\rangle_1 \otimes |V\rangle_2 + e^{i\phi}|V\rangle_1 \otimes |H\rangle_2\right) \quad (8.7)$$

In applications, the walk-off entangled state parameters involving in the measurement are related to the changes in the applied physical parameters such as force, stress, strain, heat, and pressure etc. and the optical device properties. However, the interested parameters in this proposed system are concerned the optical birefringence parameters, which can be given by

$$\Delta\phi = \frac{2\pi(n_x - n_y)L_w}{\lambda} \quad (8.8)$$

where $\Delta n = (n_x - n_y)$ is the optical birefringence, $L_w$ is the entangled state walk-off length, and $\lambda$ is the light source wavelength.

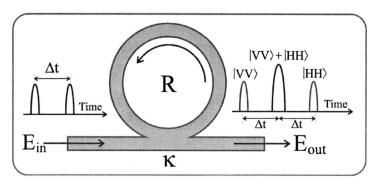

**FIGURE 8.5** A schematic diagram of polarized photon generated within a microring resonator and the polarized entangled photon components.

In principle, the movement of the signal and idler from the optimum location introduces the walk-off length change by the external or physical quantities, which can be compensated, i.e. measured. This effect is presented by the measurement parameter known as birefringence. The entangled photon visibility is randomly produce a pair of signal and idler, which was well described by reference [1]. When the chaotic signal is generated then passing through the rotatable polarizer and polarizing beam splitter as shown in Figure 8.6. The required azimuth angle is adjusted to obtain the specific orientation angle via the rotatable polarizer. The random entangled photon pair is split via a PBS and detected by the two detectors. In general, the entangled photon pairs may be formed by two different forms of the orientation angles (0°, 90°) or (135°, 180°), which is represented by light

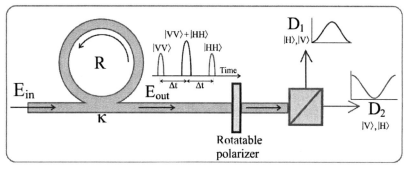

**FIGURE 8.6** A schematic diagram of the entangled photon generation system; PBS: Polarizing Beam Splitter, $D_s$: Detectors.

traveling into the optical components as described earlier. The entangled photon visibility is seen when the azimuth angle is rotated between 0 and 180°, where each peak power of the entangled photon pair is formed by each value of the maximum peak power at the specific orientation angle. Further applications, the induced change due to the external disturbance on the stored light pulse within the sensing waveguide affects the change in the entangled photon phase shift, which is recovered by using the walk-off compensation. This is related to the change in the applied physical quantities such as force, temperature or strain on the sensing device. The proposed system is also available for the large sensing area, whereas the large number of sensing heads is required, i.e. distributed sensors.

## 8.2 Network Sensors using a PANDA Ring Resonator Type

We propose a new system of microring sensing transducer using a PANDA ring resonator type, in which the sensing unit is consisted of an optical add/drop filter and two nanoring resonators, where one ring is placed as a transducer (sensing unit), the other ring is set as a reference ring. In operation, the external force is assumed to exert on the sensing ring resonator. The method of finite difference time domain (FDTD) via the computer programming called Opti-wave is used to simulate the sensing behaviors. The obtained results have shown that the change in wavelength due to the change in sensing ring radii is seen, in which the wavelength shift of 1 nm resolution is achieved. The behavior of light within a PANDA ring resonator is also analyzed and reviewed.

### 8.2.1 General Review

In the recent years, the integrated nonlinear optical device using a microring resonator has been widely investigated in both theory and experiment [19–21]. One of the interesting results is the use of a specific model of a ring resonator known as a PANDA ring resonator [22], which can be a good candidate for nanoscale sensing applications. Optical sensors have been implemented and widely used in various applications, for instance, in medicine, microbiology, communication, particle physics, automotive, environmental safety and defence [23–25]. The use of nanoscale measurement with more efficiency systems has been reported by several research groups [26, 27]. However, the searching of new device and technology remains, which can be fulfilled by the specific requirement of sensing applications, especially, in the nanoscale resolution regime. The use of nanoscale measurement and self-calibrations and more efficiency system has been presented by Yupapin et al. [28, 29], in which the interesting results are the self-calibration sensor base on ring resonator. Moreover, the distributed or multiplexed system is also available for a large area of sensing application by using the nano-scale sensing transducer devices vie multi wavelength router [29, 30]. Recently, the use of a new form of a ring resonator called a PANDA ring resonator has shown the interesting aspect of applications [31, 32]. The authors have shown that such a proposed form of a ring resonator can establish the new concept of dark-bright soliton collision, whereas the use of random encoding, optical vortices (tweezers) and optical/quantum gate can be generated. In this chapter, we propose the other aspect of a PANDA ring resonator, in which the system of a nano-scale sensing transducer based on a PANDA ring resonator type is proposed. The sensing system is functioned by mean of the change of a ring radius due to a load cell or other physical parameters, in which the change in optical path length of light is caused by the same way of an interferometer [33, 34], while the other ring radius remains constant (reference). The sensing and the reference signals are analyzed, simulated and compared. Simulation results obtained have shown that the system can be employed to be the nano-scale sensing transducer. However, the measurement limitation occurs due to wavelength meter resolution and material elongation limit, in which the measurement resolution of 1 nm is noted in this work. Lastly, the distributed or multiplexed sensing applications are also available using the nano-scale sensing transducer devices vie multi- wavelength router, which are discussed in detail.

### 8.2.2 Principle and Method

To form the simulation sensing performance, the microring material used is InGaAsP/InP, with the refractive index is $n_0 = 3.34$ [35–37]. The schematic diagram of a sensing transducer using a PANDA ring resonator is as shown in Figure 8.7. The system consists of three microring resonators, where the first ring is positioned as a reference ring, with radius $R_1 = 1.550$ μm.

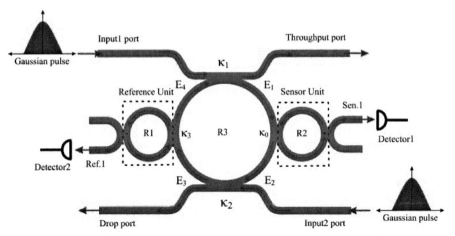

**FIGURE 8.7** A schematic diagram of a nano-scale sensing transducer using PANDA ring resonator.

The second ring $R_2$ is the sensing ring, the radius is varied from 1.550–1.558, and the third ring is used to form the interference signal between reference and sensing rings, with the radius $R_3 = 3.10$ μm. In operation, the change in sensing ring radius is caused by the change in the shift in signals circulated in the interferometer ring ($R_3$), in which the interference signals are seen. The change in optical path length which is related to the change of the external parameters is measured.

Two identical beams of monochromatic optical field ($E_{in}$) of Gaussian pulse with the center wavelength 1.550 μm are launched into in the system at the input port and the add port, which is given by

$$E_{in}(t) = E_0 \exp[-\alpha L + j\phi_0(t)] \quad (8.9)$$

where $L = 2\pi R$ is a propagation distance (waveguide length), $\alpha$ is an attenuation and $\phi_0$ is the phase constant. When light propagates within the nonlinear material (medium), by the Kerr nonlinear effect within the ring devices the refractive index ($n$) of light within the medium is given by

$$n = n_0 + n_2 I = n_0 + n_2 \left(\frac{P}{A_{eff}}\right) \quad (8.10)$$

where $n_0$ and $n_2$ are the linear and nonlinear refractive indexes, respectively. $I$ and $P$ are the optical intensity and optical power, respectively. The effective mode core area of the device is given by $A_{eff}$.

The resonance output is formed, thus, the normalized output of the light field is the ratio between the output ant input fields [$E_{out}(t)$ and $E_{in}(t)$] in each roundtrip, which is given by [19, 22]

$$\left|\frac{E_{out}(t)}{E_{in}(t)}\right|^2 = (1-\gamma) \cdot \left[1 - \frac{(1-(1-\gamma)x^2)\kappa}{\left(1-x\sqrt{1-\gamma}\sqrt{1-\kappa}\right)^2 + 4x\sqrt{1-\gamma}\sqrt{1-\kappa}\sin^2\left(\frac{\phi}{2}\right)}\right] \quad (8.11)$$

The optical output of ring resonator add/drop filter for the through and drop port can be given by Eqs (8.12) and (8.13), respectively [22]

$$\left|\frac{E_{t1}}{E_{in}}\right|^2 = \frac{(1-\kappa_1) - 2\sqrt{1-\kappa_1}\cdot\sqrt{1-\kappa_2}\cdot e^{-\frac{\alpha}{2}L}\cos(k_n L) + (1-\kappa_2)e^{-\alpha L}}{1+(1-\kappa_1)(1-\kappa_2)e^{-\alpha L} - 2\sqrt{1-\kappa_1}\cdot\sqrt{1-\kappa_2}e^{-\frac{\alpha}{2}L}\cos(k_n L)} \quad (8.12)$$

$$\left|\frac{E_{t2}}{E_{in}}\right|^2 = \frac{\kappa_1\kappa_2 \cdot e^{-\frac{\alpha}{2}L}}{1+(1-\kappa_1)(1-\kappa_2)e^{-\alpha L} - 2\sqrt{1-\kappa_1}\cdot\sqrt{1-\kappa_2}e^{-\frac{\alpha}{2}L}\cos(k_n L)} \quad (8.13)$$

where $E_{t1}$ and $E_{t2}$ represent the optical fields of the through and drop ports, respectively. $x = \exp(-\alpha L/2)$ is a roundtrip loss coefficient, $k_n = 2\pi/\lambda$ is the wave propagation number in vacuum, $n_{eff}$ is an effective refractive index, $\phi = kn_{eff}L$ is the phase constant, $\gamma$ is the fractional coupler intensity loss,

$\kappa$ is the coupling coefficient, and $\beta$ is a complex coefficient. The signals of both rings $R_1$ and $R_2$ are observed at the point Ref.1 ($E_{R1}$) and Sen.1 ($E_{S1}$) respectively as shown in Figure 8.7 and the mathematical form of those signals are also analyzed, which can be expressed as

$$\left|\frac{E_{S1}}{E_{in}}\right|^2 = \left[\frac{-(1-\gamma_S)\kappa_S}{1-Z_2(1-\gamma_S)(1-\kappa_S)}\right] \times$$
$$\left[\frac{j \cdot Z_3\sqrt{(1-\gamma_C)\kappa_C}\left(1+Z_3^2\beta_1\sqrt{(1-\gamma_C)(1-\kappa_C)}\right)}{1-Z_3^4\beta_1\beta_2(1-\gamma_C)(1-\kappa_C)}\right]^2 \quad (8.14)$$

$$\left|\frac{E_{R1}}{E_{in}}\right|^2 = \left[\frac{-(1-\gamma_R)\kappa_R}{1-Z_1(1-\gamma_R)(1-\kappa_R)}\right] \times$$
$$\left[\frac{j \cdot Z_3\sqrt{(1-\gamma_C)\kappa_C}\left(1+Z_3^2\beta_2\sqrt{(1-\gamma_C)(1-\kappa_C)}\right)}{1-Z_3^4\beta_1\beta_2(1-\gamma_C)(1-\kappa_C)}\right]^2 \quad (8.15)$$

where $E_{S1}$ and $E_{R1}$ represent the sensing and reference signal respectively, $\gamma_S$ and $\gamma_R$ are the fractional coupler intensity loss in sensing and reference unit, $\kappa_S$ and $\kappa_R$ are the coupling coefficient in sensing and reference unit,

$$Z_1 = \exp\left(\frac{-\alpha}{8}\frac{L_1}{2} - jk_n\frac{L_1}{2}\right)$$

$$Z_2 = \exp\left(\frac{-\alpha}{8}\frac{L_2}{2} - jk_n\frac{L_2}{2}\right)$$

and

$$Z_3 = \exp\left(\frac{-\alpha}{8}\frac{L_3}{4} - jk_n\frac{L_3}{4}\right)$$

are loss coefficient and $\beta$ is a complex coefficient which are described by

$$\beta_1 = \left[\frac{\sqrt{(1-\gamma_3)(1-\kappa_3)}+(1-\gamma_3)\cdot e^{\frac{-\alpha L_1}{4}-jk_nL_1}}{1-\sqrt{(1-\gamma_3)(1-\kappa_3)}\cdot e^{\frac{-\alpha L_1}{4}-jk_nL_1}}\right] \quad (8.16)$$

$$\beta_2 = \left[\frac{\sqrt{(1-\gamma_0)(1-\kappa_0)}+(1-\gamma_0)\cdot e^{\frac{-\alpha L_2}{4}-jk_nL_2}}{1-\sqrt{(1-\gamma_0)(1-\kappa_0)}\cdot e^{\frac{-\alpha L_2}{4}-jk_nL_2}}\right] \quad (8.17)$$

The power output $P$ at all ports is expressed by

$$P = |E|^2 \quad (8.18)$$

To compare the reference and sensing signals, we set $\gamma_0 = \gamma_3$, $\kappa_0 = \kappa_3$ so $\beta_1 = \beta_2$ and then set $\gamma_S = \gamma_R$, where finally $E_{S1}$ and $E_{R1}$ are both identical if $L_1 = L_2$. Then $E_{S1}$ is varied while $L_2$ is changed by mean of varying $R_2$ with respect to $E_{R1}$, in which $R_2$ remains constant. According to the finite difference time domain method (FDTD), the hold system is analyzed, whereas all the parameters are simulated via the computer programming called Opti-wave. The simulation steps are 40,000 iterations and the peak spectrum at point Ref.1 and Sen.1 are set as reference and sensing signals respectively, as shown in Figures 8.8 and 8.9. The change in optical path length between sensing and reference signals is compared, in which the induced change by the external parameters is measured.

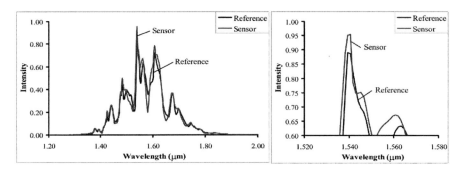

**FIGURE 8.8** The relationship between intensity and wavelength of sensing ($E_{S1}$) and reference signals ($E_{R1}$), with the radius $R_1$ (Reference) and $R_2$ (Sensor) = 1.550 μm.

This measurement is formed by the comparison of the shift in wavelength (optical path length), which is called self-calibration, and the difference between center peak wavelengths $\Delta\lambda$ is obtained by

$$\Delta\lambda = \lambda_2 - \lambda_1 \tag{8.19}$$

where $\lambda_1$ and $\lambda_2$ are the peak wavelengths of Ref.1 and Sen.1, respectively. The relationships between intensity and wavelength-shift are plots as shown in Figure 8.9, which is set as a self-calibration sensing transducer.

### 8.2.3 Distributed Network Sensors Using a Microring Sensing Transducer

In simulation, the sensing ring radius $R_2$ is varied from 1.550 to 1.558 μm, in which the optical path length is also changed, and an interferometer system is formed [30, 31]. Both signals, sensing and reference signals are observed and compared, and the self-calibration transducer is performed [26, 27]. We assume that the load cell or others sensing parameters is applied on the second ring $R_2$, whereas stress and strain are introduced on the sensing unit by mean of the elastic modulus of the materials. This is caused by

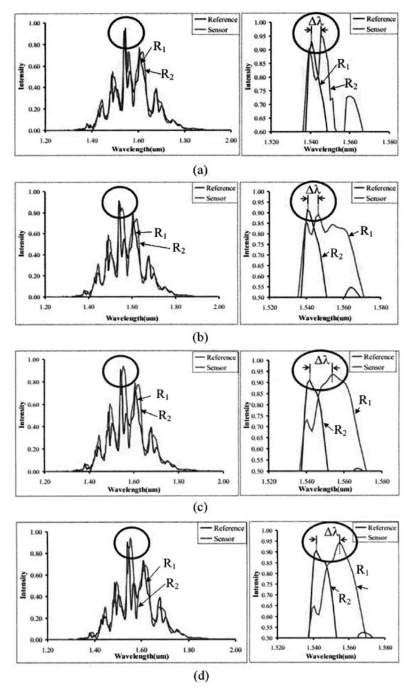

**FIGURE 8.9** The relationship between intensity and wavelength of sensing ($E_{S1}$) and reference signals ($E_{R1}$), with the radius $R_1$ = 1.550 μm and $R_2$ is varying from 1.552–1.558 μm, (a) $R_2$ = 1.552 μm, (b) $R_2$ = 1.554 μm, (c) $R_2$ = 1.556 μm, and (d) $R_2$ = 1.558 μm.

the difference in peak spectrum of both signals, which is described by the Eq. (8.20).

$$Y_0 = \frac{F/A}{\Delta L/L} = \frac{\text{Stress}}{\text{Strain}} \qquad (8.20)$$

and the relationship between force and the difference length is described by

$$F = \left(\frac{Y_0 A_0}{L_0}\right) \cdot \Delta L \qquad (8.21)$$

where $F$ is the applied force, $Y_0$ is the Young's Modulus, $A_0$ is the initial cross-section area, $L_0$ is the initial length and $\Delta \lambda$ is the difference in length. According to the properties of InGaAsP/InP material [32–34], the relationship between force and wavelength shift is plotted as shown in Figure 8.10.

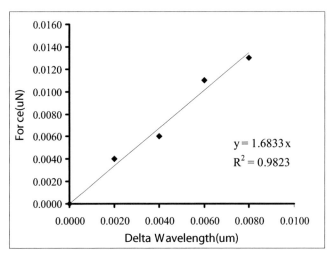

**FIGURE 8.10** Graph of the linear relationship between Force and the Wavelength-shift.

By using Eq. (8.21) and the simulation results in Figures 8.8 and 8.9, the relationship between force and wavelength-shift of the reference and sensing signals is as shown in Figure 8.10. In this case study, the applied micro/nano force ranges are between 0.000 and 0.016 µN. We find that a sensing range in terms of wavelength-shift ($\Delta \lambda$) within the resolution of 1 nm is achieved. Figure 8.3 shows the relationship between the varying ring radius $R_2$ and the wavelength-shift ($\Delta \lambda$) by comparing the sensing and reference signals, where the self-calibration sensing transducer is formed. By using the least square curve fitting, the linearity relationship between the applied force and wavelength shift with $R_2 = 0.9823$ is formed, which is shown in good linearity for sensing application.

From the above reasons, the network sensors using a PANDA ring resonator can be performed as shown in Figure 8.11. Two optical fields $E_{in}$

of Gaussian pulse can propagate into the system at the input1 port and the input2 port as shown in Figure 8.7. By using a PANDA ring resonator, where can be measure of the delta wavelength between of sensing and reference unit. This means that each node device acts as self-calibration sensing transducer suitable to perform the measurements in the nano-scale regime such as force, stress and temperature. Except the network sensors sensing information still can be transmitted via a wavelength router.

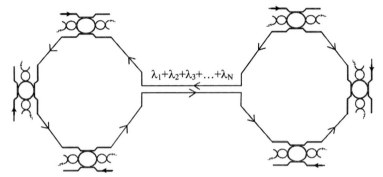

**FIGURE 8.11** Distributed network sensor system.

## 8.3 Self-calibration in a Fiber Optic Sensing System

In this section, we present a concept of a self calibration of the classical and quantum parameters measurement using a fiber optic system. The measurement of the change in phase of the optimum entangled states visibility, is performed in term of a walk-off length, i.e. birefringence. The two applied physical parameters (i.e. force/stress and temperature) on the sensing fiber can be simultaneously measured, whereas the self calibration with respect to the entangled photon walk-off length can be performed. The scheme of the entangled photons generation in fiber optic is reviewed and the walk-off on the polarization entangled states presented. The potential of self-calibration and simultaneous measurement using an interferometric sensing technique and fiber grating sensor are proposed and discussed. The entangled photon walk-off compensation when the sensing fiber was in the thermal controlled environment, is presented and discussed.

### 8.3.1 General Review

Recently, Yupapin and Suchat [38, 39] have demonstrated that the use of a fiber optic ring resonator could be used to generate a pair of the entangled photons. The advantage of such a system is that there is no optical pumping part and the component included in the system (i.e. an all fiber optic scheme), which is a remarkably simple arrangement and is easy to implement in the practical applications. However, the problem of the fiber optic property known as a fiber birefringence could affect the optimum entangled state visibility after traveling within a length of the fiber. Recently, Trojek et al. [40] have analyzed the timing-walk-off on the entangled photons in fiber optic, which could be compensated by using the phase retardation device (phase shifter). Such a device could be in the form of a bulky or fiber optic. To shift the polarization orientation angle, therefore, the polarization controller device is recommended to use for adjusting and preserving the entangled states along the fiber optic length. The entangled states walk-off due to the effects of the fiber optic properties, especially, the polarization mode dispersion i.e. fiber birefringence and its related parameters that affect the timing-walk-off on the entangled photons are analyzed and discussed. The feasibility of using such concept with the systems in applications is described in detail.

### 8.3.2 Operation Principle

When the entangled photons generate and enter the fiber optic system, the amount of the walk-off on the entangled photons depends on the location where the photon pairs are within the fiber. This position is completely random due to the coherent nature of light in fiber optic. To compensate the longitudinal timing-walk off effect, a polarization controller is recommended to ensure that the polarization rotation is the same on both photons from the entangled pair. Additionally the compensator fiber is used to change the relative phase $\phi$ of the states of the polarized light. Because of the change in birefringence, the tilting of the compensator allows to apply a phase shift to the entangled states of the two photons, which are given by Eq. (8.22) [40],

$$|\psi\rangle_{12} = \frac{1}{\sqrt{2}}\left(|H\rangle_1 \otimes |V\rangle_2 + e^{i\phi}|V\rangle_1 \otimes |H\rangle_2\right) \quad (8.22)$$

To begin this concept, firstly, a light pulse is input and chopped to form many pulses by the chaotic behavior within the microring resonator. Secondly, we introduce a technique that can be used to create the entangled photon pair (qubits) as shown in Figure 8.12. Polarized light can be formed

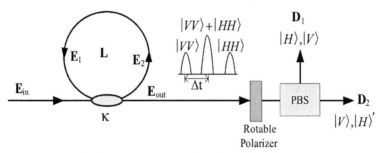

**FIGURE 8.12** A schematic diagram of the entangled photon generation system: PBS: Polarizing Beam Splitter, $D_s$: Detectors [43].

from the basic vertical and horizontal polarization states that correspond to the input of short and long pulses (different time). We assume that those horizontally polarized pulses have a temporal separation of $\Delta t$. The coherence time of the consecutive pulses is larger than $\Delta t$. When the coupled polarization mode is introduced by the external environment, then the following state is created by Eq. (8.23).

$$|\Phi\rangle_p = |1,H\rangle_s |1,H\rangle_i + |2,H\rangle_s |2,H\rangle_i \qquad (8.23)$$

In the expression $|k,H\rangle$, $k$ is the number of time slots (1 or 2), where $|H\rangle$ and $|V\rangle$ denote the state of polarization in horizontal and vertical components, respectively, and the subscript identifies whether the state is in the signal (s) or the idler (i) state. In Eq. (8.23), for simplicity we have omitted an amplitude term that is common to all product states. We employ the same simplification in subsequent equations in this article. This two-photon state with $|H\rangle$ polarization, shown by Eq. (8.23), is input into the orthogonal polarization-delay circuit shown schematically in Figure 8.12. The delay circuit consists of a coupler and the difference between the round-trip times of the microring resonator, which is equal to $\Delta t$. The light signal within a microring resonator is coupled and converted into $|V\rangle$ at the delay circuit output. That is the delay circuits convert.

$$|k,H\rangle \text{ to } r|k,H\rangle + t_2 \exp(i\phi)|k+1,V\rangle + rt_2 \exp(i_2\phi)|k+2,H\rangle \\ + r_2 t_2 \exp(i_3\phi)|k+3,V\rangle$$

where $t$ and $r$ are the amplitude transmittances of the cross and bar ports in the coupler, respectively. Then Eq. (8.23) is converted into the polarized state by the delay circuit as

$$\begin{aligned}|\Phi\rangle &= [|1,H\rangle_s + \exp(i\phi_s)|2,V\rangle_s] \times [|1,H\rangle_i + \exp(i\phi_i)|2,V\rangle_i] \\ &\quad + [|2,H\rangle_s + \exp(i\phi_s)|3,V\rangle_s] \times [|2,H\rangle_i + \exp(i\phi_i)|2,V\rangle_i] \\ &= [|1,H\rangle_s |1,H\rangle_i + \exp(i\phi_i)|1,H\rangle_s |2,V\rangle_i] + \exp(i\phi_s)|2,V\rangle_s |1,H\rangle_i \\ &\quad + \exp[i(\phi_s + \phi_i)]|2,V\rangle_s |2,V\rangle_i + |2,H\rangle_s |2,H\rangle_i + \exp(i\phi_i)|2,H\rangle_s |3,V\rangle_i \\ &\quad + \exp(i\phi_s)|3,V\rangle_s |2,H\rangle_i + \exp[i(\phi_s + \phi_i)]|3,V\rangle_s |3,V\rangle_i \end{aligned} \qquad (8.24)$$

by the coincidence counts in the second time slot, we can extract the fourth and fifth terms. As a result, we can obtain the following polarization entangled state as

$$|\Phi\rangle = |2,H\rangle_s |2,H\rangle_i + \exp[i(\phi_s + \phi_i)]|2,V\rangle_s |2,V\rangle_i \qquad (8.25)$$

we assume that the response time of the Kerr effect is much less than the cavity round-trip time. Because of the Kerr nonlinearity of the optical device, strong pulses acquire an intensity dependent phase shift during propagation. The interference of light pulses at a coupler introduces the superposition of light, which is entangled. Due to the polarization, states of light pulses are changed and converted while circulating in the delay circuit where the polarization entangled photon pairs can be generated. The entangled photons of the nonlinear ring resonator are separated to be the signal and idler photons. A polarization angle adjustment device is used to investigate the orientation and optical output intensity where the compensation, i.e. the measurement, is performed.

Generally, there are two pairs of possible polarization entangled photons, which are represented by the four polarization orientation angles as [0°, 90°], [135° and 180°]. They can be formed by using a quarter wave plate after the polarization rotation device and polarizing beam splitter. However, in practice, a continuously variable entangled photon can be generated, which introduces the possibility that a random entangled photon pair can be formed from the photon visibility. To generate the polarized photons, a rotatable polarizer is included in the system after the ring device as shown in Figure 8.13. The output polarized state is controlled by a rotatable polarizer, where the input azimuth angle is changed to obtain the required state. The randomly polarized light is formed from the specific angle by controlling the rotatable polarizer and PBS. The two polarization modes ($|H\rangle$ and $|V\rangle$) are randomly formed) and detected by the two detectors. By rotating the azimuth angles from 0–180°, the polarization entangled photon visibility is plotted and seen.

In applications, the walk-off entangled state parameters involving in the measurement are related to the changes in the applied physical parameters such as force, stress, strain, heat, pressure, etc. However, the interested parameters in these proposed systems are concerned the fiber optic birefringence related parameters, which can be given by

$$\Delta\phi = \frac{2\pi(n_x - n_y)L_w}{\lambda} \tag{8.26}$$

where $\Delta n = (n_x - n_y)$ is the fiber optic birefringence, $L_w$ is named as the entangled states walk-off length, and $\lambda$ is the light source wavelength.

We begin with the first proposed sensing system which is as shown in Figure 8.13. A pair of the entangled photons is formed after light pulses circulating in a fiber ring resonator (EPR source), and the polarization controller applied. Such a system of an optical fiber interferometer incorporating the entangled states generation setup is as shown schematically. The entangled photons are generated by the first part of the setup, which was well confirmed by Yupapin and Suchat [38]. This generates two pairs of the entangled photons which one is entered into the sensing and the other into the reference arm where both arms are coated to obtain the maximum reflected powers. There are only the signal and idler photons entered into a 2 × 2 coupler with the 50/50 coupling ratio and detected by two detectors ($D_1$ and $D_2$). The difference between the round-trip times of two arms is set at $\Delta t$. As a result, we can obtain the following polarization entangled state, which is given by [41]

$$|\Phi\rangle = |2, H\rangle_s |2, H\rangle_i + \exp[i(\phi_s + \phi_i)]|2, V\rangle_s |2, V\rangle_i \tag{8.27}$$

The subscripts $(s, i)$ identify whether the state is the signal $(s)$ or the idler $(i)$ state. The center wavelength of the signal is reflected and interfered being detected by detectors $D_1$ and $D_2$ and seen by using the interferometric technique. The entangled photons are interfered and recombined by a polarization combiner after reflected back from the fiber ends of the same coupler. The applied physical parameter that performs on the sensing arm will change the fiber birefringence, which will be recovered by rotating the Polarization Controller (PC). The change in rotation angle then can be related to the change in fiber birefringence, i.e. physical parameter, which is given in Eq. (8.27).

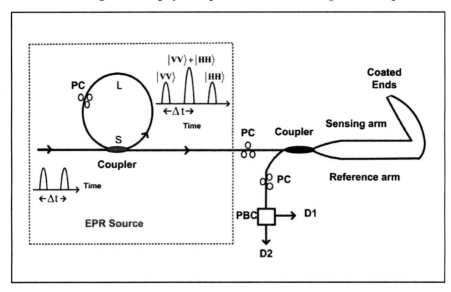

**FIGURE 8.13** A schematic diagram of the quantum interferometric system. LD: Laser Diode, PCs: polarization controllers, $D_s$: Detectors, PBC: Polarization Beam Combiner. S and L are stands for short and long fiber lengths [43].

### 8.3.3 Entangled Photon States Walk-off Compensation

For example, the entangled photons probability is as shown in term of the optical output intensity, i.e. entangled photon visibility. It was generated by using the first part of the system, when the phase difference $\phi = 0°$ is as shown in Figure 8.14(a) [38]. This is the optimum entangled photon visibility. Figure 8.14(b) presents the optical intensity at the output of the polarization output, when the phase difference of the signal peak and the delay peak of the nonlinear fiber ring resonator is $\phi = 40°$. The interference signal of the center wavelength signal can be presented in the form of the output intensity, i.e. visibility, when the PBC is replaced by a beamsplitter and visibility is plotted. This is named as a general interferometric technique. The interference of the entangled photons is measured by the change in phase of the optimum entangled photon visibility, where the self-calibration and simultaneous measurement between the classical (optical visibility) and quantum (entangled photon visibility) with respect to the same physical parameter can be realized.

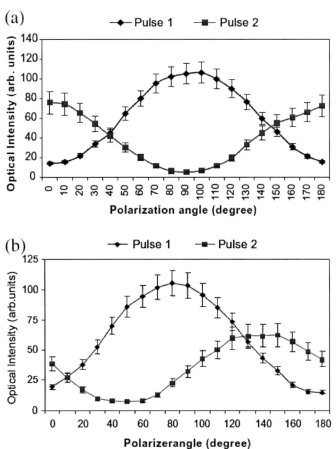

**FIGURE 8.14** Graphs of the measured optical signals: (a) $\phi = 0°$ and (b) $\phi = 45°$ at room temperature, where pulse 1 and pulse 2 were detected by $D_1$ and $D_2$ respectively [43].

For instance, the measurement relationship between the physical parameter and fiber birefringence, i.e. rotation angle was experimentally measured. The measurement data was obtained by using detectors $D_1$ and $D_2$. The result obtained is as shown in Figure 8.15, which presents the optical intensity of the polarization output, where the phase difference was obtained by changing temperature on the sensing arm, which was in the heat chamber. At each temperature values, the polarization angle was adjusted to obtain the optimum visibility, i.e. the change in phase or birefringence was measured. To recover the change in phase of the entangled states by the applied physical parameter, i.e. change in temperature. The adjusting on the Polarization Controller (PC) before the polarizing beam splitter was operated. Then the measured orientation angle was noted and plotted relatively to the change in fiber birefringence ($\Delta n$ i.e. physical parameter) as shown.

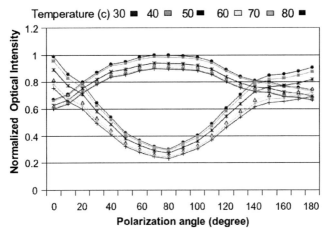

**FIGURE 8.15** A graph of the entangled photon visibility, when the sensing arm is in the temperature control unit [43].

In principle, to obtain the optimal visibility of the entangled photons after walk-off compensation, the polarization orientation was adjusted to obtain the optimum entangled photon visibility, the amount of phase shift was recorded with respect to each applied temperature which was obtained by adjusting the output PC, where the change in phase shift ($\Delta\phi$) or birefringence was measured. The plot of visibility of the change on the entangled states with temperatures from 30°C to 80°C is as shown in Figure 8.15. The shift in phase with temperature is seen, which means the walk-off compensation to obtain the optimum visibility was operated.

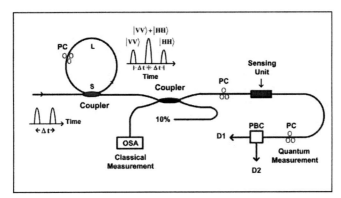

**FIGURE 8.16** A schematic diagram of the simultaneous classical-quantum measurement system. LD: laser diode, PCs: polarization controllers, $D_s$: detectors, OSA: Optical Spectrum Analyzer, PBC: Polarization Beam Combiner. S and L are stands for short and long fiber lengths [43].

The other system of this proposal is as shown in Figure 8.16. The entangled photons enter into a 90:10 fiber coupler, where more input power (90%) is entered into a sensing part (fiber Bragg grating). The Bragg grating resonance, which is the center wavelength of light back reflected from a Bragg grating depends on the effective index of refraction of the core and the periodicity of the grating. The effective index of refraction, as well as the periodic spacing between the grating planes, will be affected with changes in strain and temperature. The shift in the Bragg grating center wavelength due to strain and temperature changes is given by

$$\Delta\lambda_{Bragg} = 2\left(\Lambda\frac{\partial n}{\partial l} + n\frac{\partial \Lambda}{\partial l}\right)\Delta l + 2\left(\Lambda\frac{\partial n}{\partial T} + n\frac{\partial \Lambda}{\partial T}\right)\Delta T \qquad (8.28)$$

where $T$ is temperature and $l$ is length of strain effect and $\Lambda$ is the periodicity of the reflective index variation and $n$ is the effective refractive index of the core. The first term in Eq. (8.28) represents the strain effect on an optical fiber. This corresponds to a change in the grating spacing and the strain optic induced change in the refractive index. The second term in Eq. (8.28) represents the temperature effect on an optical fiber. A shift in the Bragg wavelength due to thermal expansion changes the grating spacing and changes the index of refraction. This fractional wavelength shift for temperature change $\Delta T$ may be written as

$$\Delta\lambda_{Bragg} = \lambda_{Bragg}(\alpha - \zeta)\Delta T \qquad (8.29)$$

where $\alpha = (1/\Lambda)(\partial\Lambda/\partial T)$ is the thermal expansion coefficient of the fiber ($\approx 0.55 \times 10^{-6}$ for silica). The quantity $\zeta = (1/n)(\partial n/\partial T)$ represents the thermo-optic coefficient and it is approximately equal to $8.6 \times 10^{-6}$ for germanium-doped silica core fiber.

In operation, the measurement of the change in the entangled states can be formed by the change in the optimum entangled sate visibility, and the change in walk-off length or phase, i.e. birefringence is measured. Then the fiber optic sensing self-calibration system using the entangled photon in fiber optic ring resonator is realized, where the applied physical parameters can be simultaneously measured and the self-calibration applied. In the system, a pair of the entangled photons (signal and idler) is randomly generated via a fiber ring resonator, which is now easy to generate. Then the optical signals with the entangled photons enter the system and sensing unit, i.e. fiber grating. The center wavelength signal is blocked by the fiber grating property and reflected back and detected by the OSA. The change in Bragg wavelength and the entangled photon states walk-off are measured relatively to the change in physical parameter on the sensing unit, which means the simultaneous measurement and the self-calibration between the change in the entangled photon states walk-off, and Bragg wavelength respectively. The change in phase, i.e. birefringence of the entangled photon visibility is obtained. For instance, the entangled states of a pair of photon with and without the thermal effects are shown in Figures 8.14 and 8.16 respectively. When the sensing unit is put into a thermal control environment, the physical parameter such as pressure, strain or force, the thermal sensitivity can be measured both by Eq. (8.26), the sensitivity of temperature and the change in the optimum entangled photon visibility with respect to temperature is as shown in Figures 8.15, where the applied temperatures were ranged from 30°C to 80°C.

To obtain the optimum measurement performance with the weak coupled into a reference arm. The resonant signal in fiber grating (sensing unit) is reflected back to the same coupler with the coupling ratio 90:10 to the optical spectrum analyzer. The optical isolator is also recommended to protect the optical feedback into the operating source. The signal and idler (quantum signal) are entered into a fiber grating and then enter a polarization combiner though the detectors. The simultaneous measurement is performed with respect to the same applied physical parameter.

The measurement resolution is one of the advantages of these schemes. For example, if we set the parameters of Eq. (8.30) as followings. Given light source wavelength is 800 nm, then we have the relation of Eq. (8.29) which is expressed by

$$\Delta\phi = 3.92 \times 10^6 (n_x - n_y) L_w \tag{8.30}$$

we set $L_w$ is a constant value, the linear relationship between the fiber birefringence and the entangled photons phase shift is measured. To obtain the maximum entangled photon visibility as the initial states before the measurement, the polarization device is rotated to obtain the optimum photon visibility. The resolution of measurement depends on the angle rotation device resolution that can be retried by the change in sensing birefringence, i.e. $\Delta n$. When one of the physical parameters such as strain or heat is applied on the fiber grating, the measurement between the entangled states walk-off length and Bragg wavelength can be made simultaneously, the self-calibration of the system between the interested physical parameter, Bragg wavelength and walk-off phase shift is also validity.

## 8.4 Conclusion

Firstly, we have proposed the interesting concept where the small scale measurement in the nano scale resolution regime is plausible, using the nano-waveguide transducer, whereas the distributed system is also available for large area of sensing applications. The measurement of physical quantities such as nano force, nano temperature and nano strain can be performed using the proposed system. The major advantage is the sensing device can be from the storing signal which is capable to perform the long period of measurement, while the very narrow optical pulse has shown the potential of using such system for nano scale measurement and beyond. Finally, the quantum measurement concept is also plausible, which is allowed to obtain the increasing in measurement resolution. By using the polarization concept, the measurement of force under zero gravity perturbation is also available.

Secondly, we have proposed the system of a nano-scale sensing transducer using PANDA ring resonator to make the benefit of the accuracy of measurements in nano-scale range. The system has the measuring regions with in the range of 1 nm, in which the self-calibration of the measurement of two sensing and reference signal can be compared without any additional optical part or other addition unit. The calibration is allowed by using the change in wavelength between sensing and reference signals, which is existed within the system. The other advantage is that the remote measurement is also available due to the use of the integrated optic device, which has shown the potential of broad applications in the near future.

Finally, we have proposed the new concept of using fiber optic systems to make the benefit of the simultaneous measurement of quantum and classical parameters. Such proposed schemes have also shown the potential of broad applications in the near future. In conclusion, the self-calibration of the measurement of two physical parameters can be performed without any additional optical part or sensing unit. The calibration is allowed by using the entangled photon, which is existed within the sensing system. The other advantage is that the remote measurement is also available due to the use of the fiber optic link. The potential of the applications in quantum interferometer and simultaneous measurement sensing systems are proposed and discussed in detail. The walk-off compensation of the entangled photon pair in the thermal effects is discussed.

# REFERENCES

[1] P.P. Yupapin, "Fiber optic sensing applications using the entangled state walk-off compensation", *Int. Light and Electron Opt.*, 120, pp. 265–267 (2009).

[2] P. Yabosdee, P. Phophithirankarn, and P.P. Yupapin, "A new concept of nano strain monitoring using strain perturbation", *Int. Light and Electron Opt.*, vol. 121(5), pp. 442–445 (2010).

[3] N. Pornsuwancharoen, P. Phiphithirankarn, P.P. Yupapin, and J. Ali, "Pulse polarization entangled photon generated by chaotic signal in a nonlinear microring resonator for birefringence based sensing applications", *Opt. and Laser Technol.*, vol. 46(6), pp. 788–793 (2009).

[4] P.P. Yupapin and P. Yabosdee, Optimum entangled photon generated by microring resonator for new generation interferometry use", *Int. Light and Electron Opt.*, vol. 121(4), pp. 389–393 (2010).

[5] P.P. Yupapin and N. Pornsuwancharoen, "Proposed nonlinear microring resonator arrangement for stopping and storing light", *IEEE Photon. Technol. Lett.*, vol. 21, pp. 404–406 (2009).

[6] M.F. Yanik and S. Fan, "Stopping light all optically", *Phys. Rev. Lett.*, 92, p. 083901 (2004).

[7] Y. Zhao, H.W. Zhao, X.Y. Zhang, B. Yuan, and S. Zhang, "New mechanisms of slow light and their applications", *Opt. and Laser Technol.*, vol. 21(4), pp. 515–525 (2009).

[8] N. Pornsuwancharoen and P.P. Yupapin, "Generalized fast, slow, stop, and store light optically within a nanoring resonator", *Microw. Opt. Technol. Lett.*, vol. 51, pp. 899–902 (2009).

[9] A. Hasegawa (ed.), *Massive WDM and TDM Soliton Transmission Systems*, Netherlands: Kluwer Academic Publishers (2000).

[10] G.P. Agarwal, *Nonlinear Fiber Optics*, 4th edition, New York: Academic Press (2007).

[11] Yu. A. Simonov and J.A. Tjon, "Soliton–soliton interaction in confining models", *Phys. Lett., B*, vol. 85, pp. 380–384 (1979).

[12] J.K. Drohm, L.P. Kok, Yu. A. Simonov, J.A. Tjon, and A.I. Veselov, "Collision and rotation of solitons in three space-time dimensions", *Phys. Lett., B*, vol. 101, pp. 204–208 (1981).

[13] T. Iizuka and Y. S. Kivshar, "Optical gap solitons in nonresonant quadratic media", *Phys. Rev. E*, vol. 59, pp. 7148–7151 (1999).

[14] R. Ganapathy, K. Porsezian, A. Hasegawa, and V.N. Serkin, "Soliton interaction under soliton dispersion management", *IEEE Quantum Electron.*, vol. 44, pp. 383–390 (2008).

[15] P.P. Yupapin, P. Saeung, and C. Li, "Characteristics of complementary ring-resonator add/drop filters modeling by using graphical approach", *Opt. Commun.*, vol. 272, pp. 81–86 (2007).

[16] Y. Su, F. Liu, and Q. Li, "System performance of slow-light buffering and storage in silicon nano-waveguide", *Proc. SPIE*, 6783, p. 67832P (2007).

[17] Y. Kokubun, Y. Hatakeyama, M. Ogata, S. Suzuki, and N. Zaizen, "Fabrication technologies for vertically coupled microring resonator with multilevel crossing busline and ultracompact-ring radius", *IEEE J. of Selected Topics in Quantum Electron.*, vol. 11, pp. 4–10 (2005).

[18] N. Sangwara, N. Pornsuwancharoen, and P.P. Yupapin, "Soliton pulses generation and filtering using microring resonators for DWDM based soliton communication", *Int. J. of Light and Electron Opt.*, vol. 121(14), pp. 1263–1267 (2010).

[19] Y.G. Boucher and P. Feron, "Generalized transfer function: A simple model applied to active single-mode, microring resonators", *Opt. Commun.*, vol. 282, pp. 3940–3947 (2009).

[20] Y. Dumeige, C. Arnaud, and P. Fe'ron, "Combining FDTD with coupled mode theories for bistability in microring resonators", *Opt. Commun.*, vol. 250, pp. 376–383 (2005).

[21] T.T. Maia, F. Hsiaoa, C. Leea, W. Xianga, C. Chenc, and W.K. Choia, "Optimization and comparison of photonic crystal resonators for silicon microcantilever sensors", *Sensors and Actuators A: Physical*, vol. 165(1), pp. 16–25 (2011).

[22] K. Uomwech, K. Sarapat, and P.P. Yupapin, "Dynamic modulated Gaussian pulse propagation whthin the double PANDA ring resonator", *Microw. and Opt. Technol. Lett.*, vol. 52, no. 8, pp. 1818–1821 (2010).

[23] Y. J. Rao, D.J. Webb, D.A. Jackson, L. Zhang, and I. Bennion, "Optical in-fiber Bragg grafting sensor systems for medical applications", *Biomedical Optics*, vol. 3, pp. 38–44 (1998).

[24] A Grillet, D. Kinet, J. Witt, M. Schukar, K. Krebber, F. Pirotte, and A. Depre, "Optical fiber sensors embedded into medical textiles for healthcare monitoring", *IEEE J. Sensors*, vol. 8, pp. 1215–1222 (2008).

[25] Yi Cui, Qingqiao Wei, Hongkun Park, and Charles M. Lieber, "Nanowire nanosensors for highly sensitive and selective detection of biological and chemical species", *Science*, vol. 293, pp. 1289–1292 (2001).

[26] M.A. Haque and M.T.A. Saif, "Application of MEMS force sensors for in situ mechanical characterization of nano-scale thin films in SEM and TEM", *Sensors and Actuators, A*, vol. 97–98, pp. 239–245 (2002).

[27] Z. Djinovic, M. Tomic, and A. Vujanic, "Nanometer scale measurement of wear rate and vibrations by fiber-optic white light interferometry", *Sensors and Actuators, A*, vol. 123–124, pp. 92–98 (2005).

[28] P. Yabosdee, K. Srinuanjan, and P.P. Yupapin, "Proposal of the nanosensing device and system using a nano-waveguide transducer for distributed sensors", *Opt. Int. J. Light and Electron. Opt.*, vol. 121, no. 23, pp. 2117–2121 (2010).

[29] S. Suchat, W. Khunnam, and P.P. Yupapin, "Self-calibration in a fiber optic sensing system using the walk-off compensation", *Measurement*, vol. 42, pp. 1263–1267 (2009).

[30] N. Pornsuwancharoen, "Multi-soliton generation with extremely narrow free spectrum range for multiplexed sensors via optical network", *Optik*, vol. 121, 1285–1289 (2010).

[31] K. Kulsirirat, W. Techithdeera, and P.P. Yupapin, "Dynamic potential well generation and control using double resonators incorporating in an add/drop filter", *Mod. Phys. Lett. B*, vol. 24, no. 32, pp. 3071–3080 (2010).

[32] T. Phatharaworamet, Chat Teeka, R. Jomtarak, S. Mitatha, and P.P. Yupapin, "Random binary code generation using dark-bright soliton conversion control within a PANDA ring resonator", *IEEE J. Lightwave Technology*, vol. 28, no. 19, pp. 2804–2809 (2010).

[33] N.K. Berger, "Measurement of subpicosecond optical waveforms using a resonator-based phase modulator", *Opt. Commun.*, vol. 283, pp. 1397–1405 (2010).

[34] P. Hua, B.J. Luff, G.R. Quigley, J.S. Wilkinson, and K. Kawaguchi, "Integrated optical dual Mach–Zehnder interferometer sensor", *Sensors and Actuators B*, vol. 87, pp. 250–257 (2002).

[35] M. Levinshtein, S. Rumyantsev, and M. Shur, *Handbook Series on Semiconductor Parameters*, vol. 1, London: World Scientific, pp. 147–168 (1996).

[36] M. Levinshtein, S. Rumyantsev, and M. Shur, *Handbook Series on Semiconductor Parameters*, vol. 2, London: World Scientific, pp. 153–179 (1999).

[37] T.P. Pearsall, *GaInAsP Alloy Semiconductors*, John Wiley and Sons, New York (1982).

[38] P.P. Yupapin and S. Suchat, "Entangled photon generation using a fiber optic Mach-Zehnder interferometer incorporating the nonlinear effect in a fiber ring resonator", *Nanophotonics*, vol. 1, p. 013504 (2007).

[39] S. Suchat, W. Khunnam, and P.P. Yupapin, "The entangled photon states recovery using a fiber ring resonator incorporating an erbium-doped fiber amplifier," *Opt. Eng.*, vol. 47, no. 6, p. 100502 (2008).

[40] P. Trojek, Ch Schmid, M. Bourennane, and H. Weinfurter, "Compact source for polarization entangled photon pairs", *Opt. Exp.*, vol. 12, pp. 276–281 (2004).

[41] H. Takesue K. Inoue, O. Tadanaga, Y. Nishida, and M. Asobe, "Generation of pulsed polarization-entangled photon pairs in a 1.55 micrometer band with a periodically poled lithium niobate waveguide and an orthogonal polarization delay circuit", *Opt. Lett.* vol. 30, 293–295 (2005).

[42] W. Khunnam and P.P. Yupapin, "An investigation of the entangled photon walk-off compensation generated by a fiber ring resonator", *Int. J. of Light and Electron Opt.*, vol. 121(4), pp. 389–393 (2010).

[43] S. Suchat, W. Khunnam, and P.P. Yupapin, "Self-calibration in a fiber optic sensing system using the entangled photon states walk-off compensation", *Measurement*, vol. 42, pp. 1263–1267 (2009).

# 9
# Optical and Quantum Computing

**CHAPTER OUTLINE**
- Introduction
- Quantum Controlled-NOT (CNOT) Gate
- Quantum SWAP Gate
- All-optical Logic Gate
- Dark-Bright Soliton Conversion Mechanism
- Optical XOR/XNOR Logic Gate Operation
- Operation Principle of Simultaneous All-optical Logic Gates.
- OOK Generation
- Conclusion
- References

## 9.1 Introduction

Quantum computing has attracted widely attention for its factoring power and efficient simulation of quantum dynamics. Many efforts have been made in building quantum computers with various physical systems and optical qubits which are regarded as a prominent candidate for their robustness against decoherence. An important theoretical breakthrough in the field was the Knill-Laflamme-Milburn (KLM) protocol [1], a circuit-based approach using single-photon sources, single-photon detectors, and linear optical elements. A two-qubit gate could be realized in an asymptotically deterministic way, as the number of photons forming an entangled state for teleportation in the protocol grows to infinity [1, 2]. It opens up the possibility of building any quantum logic gate which can be decomposed into two-qubit and single-qubit gates theoretically [3]. The prohibitively large overhead cost of a two-qubit gate in the KLM protocol, however, necessitates various improvements. Most progress works follow in the direction of one way computation [4], an approach imprinting circuits on a particular class of entangled states (cluster states) through measurements. Though it is possible to create cluster states with realistic optical methods [5], the generation of such multiple-entangled states is still not efficient with available techniques, imposing a bottleneck on the practical implementation. Beyond linear optics, a near deterministic controlled-NOT (CNOT) gate based on weak nonlinearities [6–11] has been proposed and it suggests a way for deterministic quantum computation [12], quantum SWAP gate in an optical cavity [13] and quantum Hadamard gate [14]. However in the realistic quantum computation it will still require considerable resource to perform a gate involving more than two qubits if one decomposes a complicated quantum circuit into the basic CNOT and single-qubit gates. In this chapter, we propose an architecture for quantum CNOT, SWAP and Hadamard gates operating on photonic circuits encoded as the nonlinear combinations of dark-bright soliton conversion modes, e.g. $|0\rangle \equiv |D\rangle$ and $|1\rangle \equiv |B\rangle$, where $D$ and $B$ are dark and bright soliton. Finally, the optical gate using the dark-bright soliton conversion states are also described and discussed.

## 9.2 Quantum Controlled-NOT (CNOT) Gate

The proposed model of quantum controlled-NOT gate can be realized by using dark-bright optical solitons, in which the zero logical state $|0\rangle$ (or dark soliton input pulse: $|D\rangle$) and the logical one $|1\rangle$ (bright soliton input pulse: $|B\rangle$) are encoded into an optical Mach–Zehnder Interferometer (MZI). The mathematical forms that are used to describe the electric fields based on MZI are as shown in Eqs (9.1)–(9.3).

$$E_{11} = \frac{1}{2} In + j\frac{1}{2} A$$

$$E_{21} = \frac{1}{2} A + j\frac{1}{2} In$$

(9.1)

$$E_{12} = E_{11} e^{-j\phi}$$

$$E_{22} = E_{21} e^{-\frac{\alpha}{2} L_1}$$

(9.2)

$$Th = \frac{1}{2} E_{12} + j\frac{1}{2} E_{22}$$

$$Drop = \frac{1}{2} E_{22} + j\frac{1}{2} E_2$$

(9.3)

where *In* and *A* are the input an added (controlled) fields respectively. $E_{11}$ and $E_{21}$ are the fields that split into two arms of MZI passing through the first coupler as shown in Figure 9.1. $E_{12}$ and $E_{22}$ are the fields before merging into the second coupler. Th is the output field of quantum CNOT gate, and Drop is the target field. $L_1$ is the lower arm with a length of 8 μm, $\phi$ is the nonlinear phase shift depending on the *In* and *A*, and *j* is the complex conjugate.

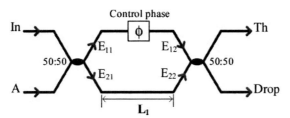

**FIGURE 9.1** A schematic diagram of quantum CNOT gate based on MZI.

When *In* field of the dark soliton $|D\rangle$ or bright soliton $|B\rangle$ is input into *In* and *A* ports, the fields are split into two parts when passing through the first 3 dB (50:50) coupler as shown in Figure 9.2. The field $E_{12}$ at the upper arm of MZI is the controlled phase shifted signal after passing through $E_{11}$. Finally, the fields are combined again at the second 3 dB coupler. The

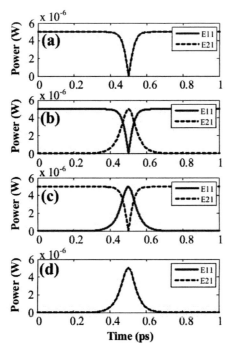

**FIGURE 9.2** Result of quantum Hadamard gate, where (a) logical input $|DD\rangle$, (b) logical input $|DB\rangle$, (c) logical input $|BD\rangle$, and (d) logical input $|BB\rangle$.

output results at the Th port are the required quantum CNOT gate output signals as shown in Figure 9.3(a)–(d). From Figure 9.2(a)–(d), we found that the outputs of quantum Hadamard gate are $|DD\rangle \rightarrow |DD\rangle$, $|DB\rangle \rightarrow |DB\rangle$, $|BD\rangle \rightarrow |BD\rangle$ and $|BB\rangle \rightarrow |BB\rangle$, respectively.

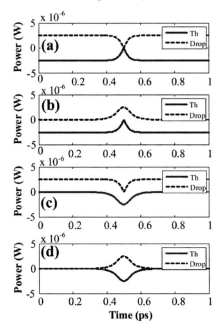

**FIGURE 9.3** Results of quantum CNOT gate operation, where (a) logical input $|DD\rangle$, (b) logical input $|DB\rangle$, (c) logical input $|BD\rangle$, and (d) logical input $|BB\rangle$.

The simulation result of quantum CNOT gate (quantum converter gate) is the through port (Th) output. From Figure 9.3(a)–(d), we found that the outputs of quantum CNOT gate based on MZI are $|DD\rangle \rightarrow |BD\rangle$, $|DB\rangle \rightarrow |BB\rangle$, $|BD\rangle \rightarrow |DD\rangle$ and $|BB\rangle \rightarrow |DB\rangle$, respectively. This means that such device allows the dark soliton $|D\rangle$ or bright soliton $|B\rangle$ being phase shift controlled by the input and added fields. If no phase shift part is introduced, the device is caused the required transmission, for instance,

$|DD\rangle \rightarrow |DD\rangle \rightarrow |DD\rangle$,

$|DB\rangle \rightarrow (|DB\rangle - |BD\rangle)/\sqrt{2} \rightarrow |DB\rangle$,

$|BD\rangle \rightarrow (|DB\rangle + |BD\rangle)/\sqrt{2} \rightarrow |BD\rangle$, and

$|BB\rangle \rightarrow |BB\rangle \rightarrow |BB\rangle$

where the first and second arrow correspond to the action of the first and second quantum Hadamard gate, respectively.

Figure 9.4 shows the switching output of quantum CNOT gate based on MZI. We found that the switching time is faster when input and added signals are same logical input such as $|DD\rangle$ or $|BB\rangle$ as shown in Figure 9.4(a) and (d). When input and added signals are $|DB\rangle$ the switching time of output at Drop port is faster than the Th port, whereas the input and added signals are $|BD\rangle$ the switching time of output at Th port is faster than the Drop port.

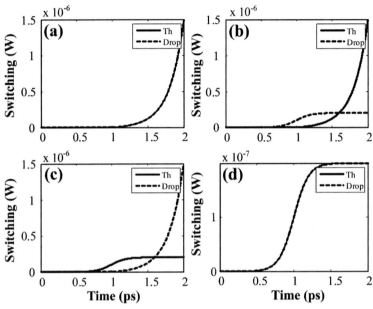

**FIGURE 9.4** Switching time of quantum CNOT gate when (a) logical input $|DD\rangle$, (b) logical input $|DB\rangle$, (c) logical input $|BD\rangle$, and (d) logical input $|BB\rangle$.

## 9.3 Quantum SWAP Gate

The proposed model of quantum SWAP gate based on add/drop photonic circuit is as shown in Figure 9.5. Qubits can be performed by using dark-bright optical solitons, in which the zero logical state: $|D\rangle$ and the logical one: $|B\rangle$ are formed in the same way of the previous case.

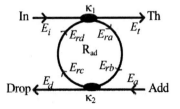

**FIGURE 9.5** A schematic diagram of SWAP gate based on add/drop photonic circuit.

The relative phase of $\pi/2$ is formed by the signals coupled into the ring and input bus. The signal coupled to the drop and through ports acquire a phase of $\pi$ with respect to the input port signal. This means that if we engineer the coupling coefficients appropriately, the field coupled to the through port on resonance would completely extinguish the resonant wavelength, and all power would be coupled into the drop port, which will be shown later in this section.

$$E_{ra} = -j\kappa_1 E_i + \tau_1 E_{rd} \tag{9.4}$$

$$E_{rb} = \exp(j\omega T/2)\exp(-\alpha L/4)E_{ra} \tag{9.5}$$

$$E_{rc} = \tau_2 E_{rb} - j\kappa_2 E_a \tag{9.6}$$

$$E_{rd} = \exp(j\omega T/2)\exp(-\alpha L/4)E_{rc} \tag{9.7}$$

$$E_t = \tau_1 E_i - j\kappa_1 E_{rd} \tag{9.8}$$

$$E_d = \tau_2 E_a - j\kappa_2 E_{rb} \tag{9.9}$$

where $E_i$ is the input field, $E_a$ is the added(control) field, $E_t$ is the throughput field, $E_d$ is the dropped field, $E_{ra} \ldots E_{rd}$ are the fields in the ring at the points $a \ldots d$. $\kappa_1$ is the field coupling coefficient between the input bus and the ring, $\kappa_2$ is the field coupling coefficient between the ring and the output bus, $L$ is the circumference of the ring, $T$ is the time taken for one round trip, and $\alpha$ is the power loss in the ring per unit length. We assume that lossless coupling, i.e., $\tau_{1,2} = \sqrt{1-\kappa_{1,2}^1}$. $T$ is the round-trip time; $T = Ln_{\text{eff}}/c$. Solving for output power/intensity at drop port and through port are given by

$$|E_d|^2 = \left| \frac{-\kappa_1\kappa_2 A_{1/2}\Phi_{1/2}}{1-\tau_1\tau_2 A\Phi} E_i + \frac{\tau_2 - \tau_1 A\Phi}{1-\tau_1\tau_2 A\Phi} E_a \right|^2 \tag{9.10}$$

$$|E_t|^2 = \left| \frac{\tau_2 - \tau_1 A\Phi}{1-\tau_1\tau_2 A\Phi} E_i + \frac{-\kappa_1\kappa_2 A_{1/2}\Phi_{1/2}}{1-\tau_1\tau_2 A\Phi} E_a \right|^2 \tag{9.11}$$

where $A_{1/2} = \exp(-\alpha L/4)$ (the half-round-trip amplitude), $A = A_{1/2}^2$, $\Phi_{1/2} = \exp(-j\omega T/2)$ (the half-round-trip phase contribution), and $\Phi = \Phi_{1/2}^2$.

The parameters of add/drop device are fixed to be 300 nm radius, non-linear refractive index, $n_2$, $1.3 \times 10^{-13}$ cm$^2$/W [15] InGaAsP/InP waveguide, coupling ratio $\kappa_1 = \kappa_2 = 0.5$.

From Figure 9.6(a)–(d), we found that the outputs of quantum SWAP gate are $|DD\rangle \rightarrow |DD\rangle$, $|DB\rangle \rightarrow |BD\rangle$, $|BD\rangle \rightarrow |DB\rangle$, and $|BB\rangle \rightarrow |BB\rangle$, respectively.

## 9.4 All-optical Logic Gate

To date, many researchers have demonstrated the interesting techniques that can be used to realize the various optical logic functions (i.e. AND, NAND, OR, XOR, XNOR, NOR) by using different schemes, including thermo-optic effect in two cascaded microring resonator [16], quantum

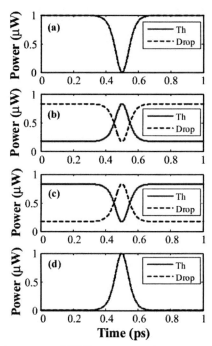

**FIGURE 9.6** Result of quantum SWAP gate operation, where (a) logical input $|DD\rangle$, (b) logical input $|DB\rangle$, (c) logical input $|BD\rangle$, and (d) logical input $|BB\rangle$.

dot [17, 18], semiconductor optical amplifier (SOA) [19–26], TOAD-based interferometer device [27], nonlinear effects in SOI waveguide [28, 29], nonlinear loop mirror [30, 31], DPSK format [32, 33], local nonlinear in MZI [34], photonic crystal [35, 36], error correction in multipath differential demodulation [37], fiber optical parametric amplifier [38], multimode interference in SiGe/Si [39], polarization and optical processor [40], and injection-locking effect in semiconductor laser [41]. However, the searching of new techniques remain, there is some room for new techniques that can be used to be the good candidate. Finally, we propose the use of the simultaneous arbitrary two-input logic XOR/XNOR and all-optical logic gates based on dark-bright soliton conversion within the add/drop optical filter system. The advantage of the scheme is that the random codes can be generated simultaneously by using the dark-bright soliton conversion behavior, in which the coincidence dark and bright soliton can be separated after propagating into a $\omega/2$ phase retarder, which can be used to form the security codes. Moreover, this is a simple and flexible scheme for an arbitrary logic switching system, which can be used to form the advanced complex logic circuits. The proposed scheme is based on a 1 bit binary comparison XOR/XNOR scheme that can be compared to any 2 bits, i.e., between 0 and 0 (dark-dark solitons), 0 and 1 (dark-bright solitons), 1 and 0 (bright-dark solitons) or 1 and 1 (bright-bright solitons), which will be detailed in the next section.

## 9.5 Dark-Bright Soliton Conversion Mechanism

In operation, dark-bright soliton conversion using a ring resonator optical channel dropping filter (OCDF) is composed of two sets of coupled waveguides, as shown in Figure 9.7(a) and (b), where for convenience, Figure 9.7(b) is replaced by Figure 9.7(a). The relative phase of the two output light signals after coupling into the optical coupler is $\pi/2$ before coupling into the ring and the input bus, respectively. This means that the signals coupled into the drop and through ports are acquired a phase of $\pi$ with respect to the input port signal. In application, if we design the coupling coefficients appropriately, the field coupled into the through port on resonance would completely extinguish the resonant wavelength, which they are given by

$$E_{ra} = -j\kappa_1 E_i + \tau_1 E_{rd} \quad (9.12)$$

$$E_{rb} = \exp(j\omega T/2)\exp(-\alpha L/4) E_{ra} \quad (9.13)$$

$$E_{rc} = \tau_2 E_{rb} - j\kappa_2 E_a \quad (9.14)$$

$$E_{rd} = \exp(j\omega T/2)\exp(-\alpha L/4) E_{rc} \quad (9.15)$$

$$E_t = \tau_1 E_i - j\kappa_1 E_{rd} \quad (9.16)$$

$$E_d = \tau_2 E_a - j\kappa_2 E_{rb} \quad (9.17)$$

where $E_i$ is the input field, $E_a$ is the add (control) field, $E_t$ is the through field, $E_d$ is the drop field, $E_{ra}\ldots E_{rd}$ are the fields in the ring at points $a\ldots d$, $\kappa_1$ is the field coupling coefficient between the input bus and ring, $\kappa_2$ is the field coupling coefficient between the ring and output bus, $L$ is the

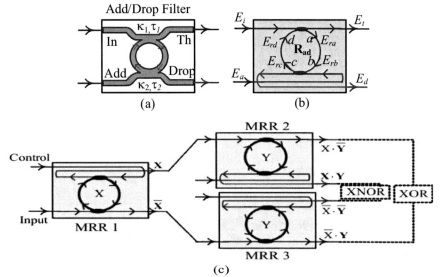

**FIGURE 9.7** A schematic diagram of a simultaneous optical logic XOR and XNOR gate.

circumference of the ring, $T_1$ is the time taken for one round trip (round-trip time), and $\alpha$ is the power loss in the ring per unit length. We assume that this is the lossless coupling, i.e., $\tau_{1,2} = \sqrt{1 - \kappa_{1,2}^1}$, $T = Ln_{\text{eff}}/c$.

The output power/intensities at the drop and through ports are given by

$$|E_d|^2 = \left| \frac{-\kappa_1 \kappa_2 A_{1/2} \Phi_{1/2}}{1 - \tau_1 \tau_2 A \Phi} E_i + \frac{\tau_2 - \tau_1 A \Phi}{1 - \tau_1 \tau_2 A \Phi} E_a \right|^2 \tag{9.18}$$

$$|E_t|^2 = \left| \frac{\tau_2 - \tau_1 A \Phi}{1 - \tau_1 \tau_2 A \Phi} E_i + \frac{-\kappa_1 \kappa_2 A_{1/2} \Phi_{1/2}}{1 - \tau_1 \tau_2 A \Phi} E_a \right|^2 \tag{9.19}$$

where $A_{1/2} = \exp(-\alpha L/4)$ (the half-round-trip amplitude), $A = A_{1/2}^2$, $\Phi_{1/2} = \exp(-j\omega L/2)$ (the half-round-trip phase contribution), and $\Phi = \Phi_{1/2}^2$.

The input and control fields at the input and add ports are formed by the dark-bright optical soliton as shown in Eqs (9.20)–(9.21),

$$E_{\text{in}}(t) = E_0 \tanh\left[\frac{T}{T_0}\right] \exp\left[\left(\frac{z}{2L_D}\right) - i\omega_0 t\right] \tag{9.20}$$

$$E_{\text{in}}(t) = E_0 \operatorname{sech}\left[\frac{T}{T_0}\right] \exp\left[\left(\frac{z}{2L_D}\right) - i\omega_0 t\right] \tag{9.21}$$

where $E_0$ and $z$ are the optical field amplitude and propagation distance, respectively. $T = t - \beta_1 z$, where $\beta_1$ and $\beta_2$ are the coefficients of the linear and second-order terms of Taylor expansion of the propagation constant. $L_D = T_0^2/|\beta_2|$ is the dispersion length of the soliton pulse. $T_0$ in equation is a soliton pulse propagation time at initial input (or soliton pulse width), where $t$ is the soliton phase shift time, and the frequency shift of the soliton is $\kappa_0$. When the optical field is entered into the nanoring resonator as shown in Figure 9.8, where the coupling coefficient ratio $\kappa_1:\kappa_2$ are 50:50, 90:10, 10:90. By using (a) dark soliton is input into input and control ports, (b) dark and bright soliton are used for input and control signals, (c) bright and dark soliton are used for input and control signals, and (d) bright soliton is used for input and control signals. The ring radii $R_{\text{ad}} = 5$ μm, $A_{\text{eff}} = 0.25$ μm², $n_{\text{eff}} = 3.14$ (for InGaAsP/InP), $\alpha = 5$ dB/mm, $\gamma = 0.1$, $\lambda_0 = 1.51$ μm.

## 9.6 Optical XOR/XNOR Logic Gate Operation

The proposed architecture is schematically shown in Figure 9.7(c). A continuous optical wave with a wavelength of $\lambda$ is formed by an optical dark-bright soliton pulse train X using MRR 1, in which the optical pulse trains that appear at the through and drop ports of MRR 1 are $\overline{X}$ and X, respectively ($\overline{X}$ is the inverse of X or dark-bright conversion). It is assumed that the input optical dark-bright soliton wave is directed to the drop port

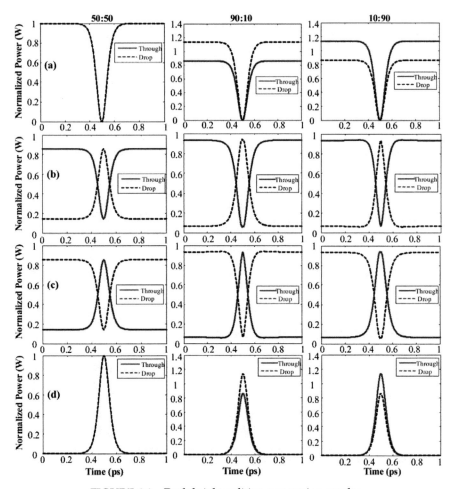

**FIGURE 9.8** Dark-bright solition conversion results.

when the optical signal is 1 (dark soliton pulse). In other words, the MRR 1 resonates at $\lambda$ when the input dark soliton pulse is applied.

If the optical pulse train $X$ is fed into MRR 2 solely from its input port and is formed by an optical pulse train $Y$ bit by bit using MRR 2, where we assume that no signal is fed into MRR 2 from its add port, in which the optical pulse trains that appear at the through and drop ports of MRR 2 will be $X \cdot \overline{Y}$ and $X \cdot Y$, respectively, whereas the aforementioned assumption is provided. The symbol represents the logical operation AND here.

If the optical pulse train $\overline{X}$ is fed into MRR 3 from its input port solely and is formed by an optical pulse train $Y$ bit by bit using MRR 3, where we assume that no signal is fed into MRR 3 from its input port, in which the optical pulse trains that appear at the through and drop ports of MRR 3 will be $\overline{X} \cdot Y$ and $\overline{X} \cdot \overline{Y}$, respectively. If the optical pulse trains $X$ and $\overline{X}$ are fed into MRR 2 and MRR 3 from its input ports simultaneously [see Figure 9.7(c)], in which the optical pulse trains $X \cdot \overline{Y} + \overline{X} \cdot Y$ and $X \cdot Y + \overline{X} \cdot \overline{Y}$ are achieved at the through and drop ports of MRR 2 and MRR 3, respectively [see Figure 9.1(c)]. The symbol + represents the logical operation OR, which is implemented through the multiplexing function of MRR 2 and MRR 3. It is well known that the XOR and XNOR operations can be calculated by using the formulas $X \oplus Y = X \cdot \overline{Y} + \overline{X} \cdot Y$ and $X \otimes Y = X \cdot Y + \overline{X} \cdot \overline{Y}$, where the capital letters represent logical variables and the symbols $\oplus$ and $\otimes$ represent the XOR and XNOR operators, respectively. Therefore, the proposed architecture can be used as an XOR and XNOR calculator.

By using the dynamic performance of the device, shown in Figure 9.9, two pseudo-random binary sequence $2^4-1$ signals at 100 Gbit/s are converted to two optical signals bit by bit according to the rule presented above and then applied to the corresponding MRRs. Clearly, a logic 1 is obtained when the applied optical bright soliton pulse signals, and a logic 0 is obtained when the applied optical dark soliton pulse signals are generated. Therefore, the device performs the XOR and XNOR operation correctly.

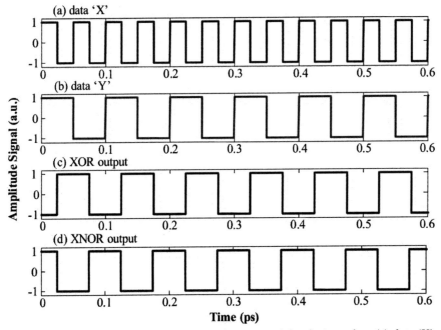

**FIGURE 9.9** Output results dynamics performance of the device, when (a) data 'X', (b) data [Y], (c) all-optical XOR and (d) XNOR logic gates.

The proposed simultaneous all-optical logic XOR and XNOR gates device is as shown in Figure 9.7(c) The input and control light pulse trains are input into the first add/drop optical filter (MRR 1) using the dark solitons (logic '0') or the bright solitons (logic '1'). Firstly, the dark soliton is converted to be dark and bright soliton via the add/drop optical filter, which they can be seen at the through and drop ports with $\pi$ phase shift [42], respectively. By using the add/drop optical filters (MRR 2 and MRR 3), both input signals are generated by the first stage add/drop optical filter. Next, the input data "Y" with logic "0" (dark soliton) and logic "1"(bright soliton) are added into both add ports, the dark-bright soliton conversion with $\pi$ phase shift is operated again. For large scale [Figure 9.7(c)], results obtained are simultaneously seen by $D_1$, $D_2$, $T_1$, and $T_2$ at the drop and through ports for optical logic XNOR and XOR gates, respectively.

In simulation, the add/drop optical filter parameters are fixed for all coupling coefficients to be $\kappa_s = 0.05$, $R_{ad} = 5$ µm, $A_{eff} = 0.25$ µm² [30], $\alpha = 5$ dBmm$^{-1}$ for all add/drop optical filters in the system. Results of the simultaneous optical logic XOR and XNOR gates are generated by using dark-bright soliton conversion with wavelength center at $\lambda_0 = 1.51$ µm, pulse width 45 fs. In Figure 9.10, simulation result of the simultaneous output optical logic gate is seen when the input data logic "00" is added, whereas the obtained output optical logic is "0001" [see Figure 9.10(a)]. Similarly, when the simultaneous output optical logic gate input data logic "01" is added, the output optical logic "0010" is formed [see Figure 9.10(b)]. Next, when the output optical logic gate input data logic "10" is added, the output optical logic "1000" is formed [see Figure 9.10(c)]. Finally, when the output optical logic input data logic "11" is added, we found that the output optical logic "0100" is obtained [see Figure 9.10(d)]. The simultaneous optical logic gate output is concluded in Table 9.1. We found that the output data logic in the drop ports, $D_1$, $D_2$ are optical logic

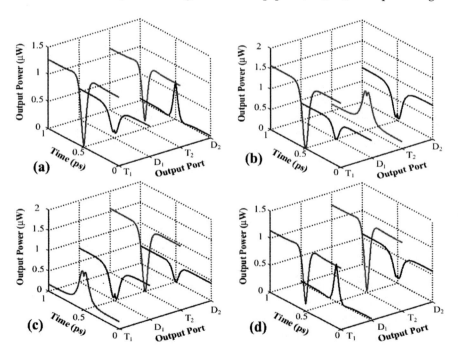

**FIGURE 9.10** The output logic XOR/XNOR gates when the input logic states are (a) 'DD', (b) 'DB', (c) 'BD', and (d) 'BB', respectively.

XNOR gates, whereas the output data logic in the through ports, $T_1$ and $T_2$ are optical logic XOR gates, the switching time of 55.1 fs is noted.

Table 9.1  Conclusion output optical logic XOR and XNOR gates

| Input data | | Output logic | | | | | |
|---|---|---|---|---|---|---|---|
| X | Y | $(T_1)$ $X \cdot Y$ | $(D_1)$ $X \cdot Y$ | $(T_2)$ $\overline{X} \cdot Y$ | $(D_2)$ $\overline{X} \cdot Y$ | XOR $X \cdot \overline{Y} + \overline{X} \cdot Y$ | XNOR $X \cdot Y + \overline{X} \cdot \overline{Y}$ |
| D | D | D | D | D | B | D | B |
| D | B | D | D | B | D | B | D |
| B | D | B | D | D | D | B | D |
| B | B | D | B | D | D | D | B |

'D (Dark soliton)' = logic '0', 'B (Bright soliton)' = logic '1'.

## 9.7 Operation Principle of Simultaneous All-optical Logic Gates

The configuration of the proposed simultaneous all-optical logic gates is as shown in Figure 9.11. The input and control light ("A") pulse trains in the first add/drop optical filter (No. "01") are the dark soliton (logic '0'). In the first stage of the add/drop filter, the dark-bright soliton conversion is seen at the through and drop ports with $\pi$ phase shift, respectively. In the second stage (Nos "11" and "12"), both inputs are generated by the first stage of the add/drop optical filter, in which the input data "B" with

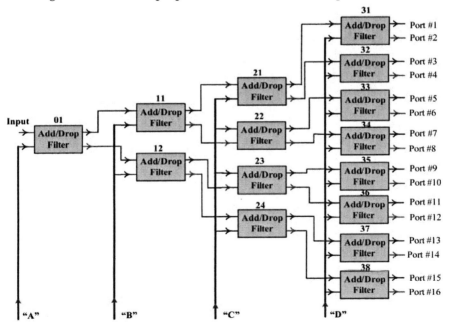

FIGURE 9.11. A schematic diagram of the proposed all-optical logic gate.

logic "0" (dark soliton) and logic "1" (bright soliton) are added into both add ports. The outputs of second stage are dark-bright soliton conversion with $\pi$ phase shift again. In the third stage of the add/drop optical filter (Nos "21" to "24"), the input data "C" with logic "0" (dark soliton) and logic "1" (bright soliton) are inserted into all final stage add ports. In the final stage of the add/drop optical filter (No. "31" to "38"), the input data "D" with logic "0" (dark soliton) and logic "1" (bright soliton) are inserted into all final stage add ports and outputs numbers 1 to 16 and shown simultaneously all-optical logic gate.

Table 9.2 Conclusion output results of all-optical logic gates

| "A" | Output Port No. | | | | | | | | | | | | | | | |
|---|---|---|---|---|---|---|---|---|---|---|---|---|---|---|---|---|
| | 1 | 2 | 3 | 4 | 5 | 6 | 7 | 8 | 9 | 10 | 11 | 12 | 13 | 14 | 15 | 16 |
| DARK ('0') | 0 | 0 | 0 | 0 | 0 | 0 | 0 | 0 | 0 | 0 | 0 | 0 | 0 | 0 | 0 | 0 |
| | 0 | 1 | 0 | 1 | 0 | 1 | 0 | 1 | 0 | 1 | 0 | 1 | 0 | 1 | 0 | 1 |
| | 0 | 0 | 1 | 0 | 0 | 0 | 1 | 0 | 0 | 0 | 1 | 0 | 0 | 0 | 1 | 0 |
| | 0 | 1 | 1 | 1 | 0 | 1 | 1 | 1 | 0 | 1 | 1 | 1 | 0 | 1 | 1 | 1 |
| | 0 | 0 | 0 | 0 | 1 | 0 | 0 | 0 | 0 | 0 | 0 | 0 | 1 | 0 | 0 | 0 |
| | 0 | 1 | 0 | 1 | 1 | 1 | 0 | 1 | 0 | 1 | 0 | 1 | 1 | 1 | 0 | 1 |
| | 0 | 0 | 1 | 0 | 1 | 0 | 1 | 0 | 0 | 0 | 1 | 0 | 1 | 0 | 1 | 0 |
| | 0 | 1 | 1 | 1 | 1 | 1 | 1 | 1 | 0 | 1 | 1 | 1 | 1 | 1 | 1 | 1 |
| BRIGHT ('1') | 0 | 0 | 0 | 0 | 0 | 0 | 0 | 0 | 1 | 0 | 0 | 0 | 0 | 0 | 0 | 0 |
| | 0 | 1 | 0 | 1 | 0 | 1 | 0 | 1 | 1 | 1 | 0 | 1 | 0 | 1 | 0 | 1 |
| | 0 | 0 | 1 | 0 | 0 | 0 | 1 | 0 | 1 | 0 | 1 | 0 | 0 | 0 | 1 | 0 |
| | 0 | 1 | 1 | 1 | 0 | 1 | 1 | 1 | 1 | 1 | 1 | 1 | 0 | 1 | 1 | 1 |
| | 0 | 0 | 0 | 0 | 1 | 0 | 0 | 0 | 1 | 0 | 0 | 0 | 1 | 0 | 0 | 0 |
| | 0 | 1 | 0 | 1 | 1 | 1 | 0 | 1 | 1 | 1 | 0 | 1 | 1 | 1 | 0 | 1 |
| | 0 | 0 | 1 | 1 | 1 | 0 | 1 | 0 | 1 | 0 | 1 | 0 | 1 | 0 | 1 | 0 |
| | 0 | 1 | 1 | 1 | 1 | 1 | 1 | 1 | 1 | 1 | 1 | 1 | 1 | 1 | 1 | 1 |

☐ AND, ▨ XNOR, ■ XOR, ▥ NAND

The simulation parameters of the add/drop optical filters are fixed for all coupling coefficients to be $\kappa_s = 0.05$, $R_{ad} = 5\mu m$, $A_{eff} = 0.25\ \mu m^2$ [15, 42], and $\alpha = 5$ dBmm$^{-1}$ for all add/drop optical filter system. Simulation results of the simultaneous all-optical logic gates are generated by using the dark-bright soliton conversion at wavelength center $\lambda_0 = 1.51\ \mu m$, pulse width 45 fs. In Figure 9.12 compared Figure 9.13, the simulation result of simultaneous output optical logic gate when the input and control signals ('A') are dark solitons. Input data 'BCD' are (a) 'DDD', (b) 'DDB', (c) 'DBD', (d) 'DBB', (e) 'BDD', (f) 'BDB', (g) 'BBD' and (h) 'BBB', respectively. Results of all outputs are concluded as shown in Table 9.2 for all-optical logic gates.

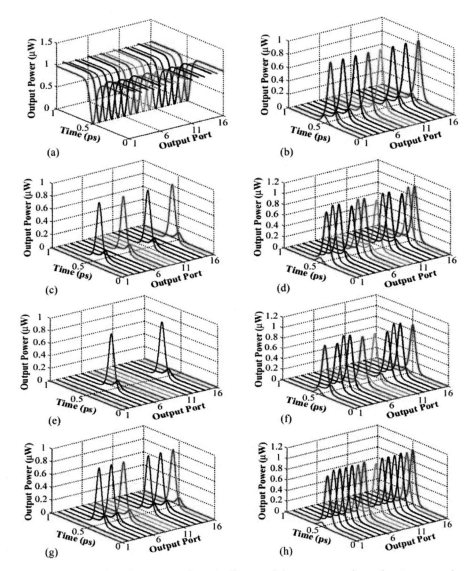

**FIGURE 9.12** Simulation results of all-optical logic gate when the input and control signals ('$A$') are dark soliton pulses. The inputs '$BCD$' are (a) '$DDD$', (b) '$DDB$', (c) '$DBD$', (d) '$DBB$', (e) '$BDD$', (f) '$BDB$', (g) '$BBD$' and (h) '$BBB$', respectively.

## 9.8 OOK Generation

In this section, we present an interesting result of nonlinear light pulse propagation within a Mach-Zehnder Infeterometer (MZI) which can be used to extend the existed On Off Keying (OOK) techniques. The goal of this chapter is OOK generation based on MZI incorporating a pumped

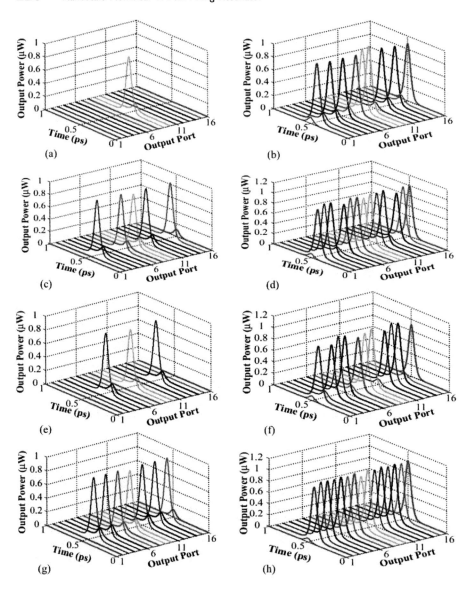

**FIGURE 9.13** Simulation results of all-optical logic gate, when the dark soliton input and control signals ('A') are bright soliton pulses. The inputs 'BCD' are (a) 'DDD', (b) 'DDB', (c) 'DBD', (d) 'DBB', (e) 'BDD', (f) 'BDB', (g) 'BBD' and (h) 'BBB', respectively.

nonlinear ring resonators system. We first analyze the principles of a phase modulation scheme using MZI incorporating the triple nonlinear ring resonators, which can be fabricated and used in practical communications. After that, we focus on the recent modulation schemes, where the all-optical on off keying and the phase shift control for phase shaped binary transmission (PSBT) are discussed in detail. The novelty of this work is that the nonlinear ring resonators are used incorporating a MZI, where the extended switching generation can be achieved and seen.

Today, most of the optical communication links use the well known modulation format called OOK (On Off Keying) [44–52], which is due to the increasing in bit rates, power and number of DWDM (Dense Wavelength Division Multiplexing) channels, where the OOK format can reach the communication requirement. However, the various modulation schemes have already been used in the electrical domain during last decade, but they have not been applied in optical schemes. Recently the use of new modulation formats in optical communications has been considered and compared for increasing the tolerance of the optical link to impairments such as chromatic dispersion, PMD (polarization mode dispersion) or nonlinearity (Kerr effect) [53]. Among the various modulation formats, we present here the differential phase shift keying (DPSK) and the phase shaped binary transmission (PSBT) schemes, whereas DPSK presents a better robustness to optical nonlinearities than the classical OOK, particularly for the cross phase modulation (XPM) in DWDM systems [46, 52, 54]. Moreover, it has also been shown that DPSK format has better performances due to PMD degradations than the classical OOK [52, 54]. The disadvantage of the DPSK is that a direct detection (DD) at the end of the optical link is not possible, since DPSK is a phase modulation. Thus, an interferometric demodulation stage must be inserted in front of the photo-detector. This stage is an "Add and Delay" structure, which is composed by a MZI [52, 54–58].

In this section, we first describe the principles of light propagation in the proposed system, where the nonlinear behavior of light within the nonlinear ring resonators (NRRs) can be used to analyze for PSBT modulation. After that, the MZI structure is detailed. However, the nonlinear ring resonator has been used [59–62] for phase shifted generate by couple to one arm of a MZI [59–60]. In this structure, one arm presents an optical delay line equal to the bit duration. This MZI converts an optical DPSK to an intensity-modulated (IM) signal, which is followed by a DD. The PSBT format is encoded from a DPSK: first a DPSK signal is generated and after that a MZI structure converts the DPSK to be an intensity-modulated signal: the PSBT. MZI characteristics for the two applications are slightly different, as we will see in the next sections. The novelty of this work is that the nonlinear light pulses generated by using the triple nonlinear ring resonators in one arm of a Mach-Zehnder interferometer can be used to enhance (amplify) the light pulse output signals, where the nonlinear outputs are generated by using the nonlinear coefficient refractive index $n_2 - 2.2 \times 10^{-13}$ m$^2$/W.

Figure 9.14 shows the proposal for on-off keying model using nonlinear index of refraction in three nonlinear ringresonator coupled to one arm of Mach-Zehnder interferometer. In this figure there are similar NRRs with the field-dependent absorption and index of refraction coefficients.

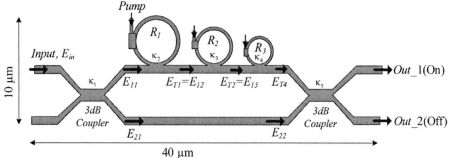

**FIGURE 9.14** A schematic diagram of OOK system, system size is $10 \times 40\ \mu m^2$.

When the input field, $E_{in}$, passes through 3 dB coupler with coupling coefficient ratio, $\kappa_1$, 50:50, then light is split into two ways, which can be expressed as

$$E_{11} = \sqrt{1-\gamma_1}\sqrt{1-\kappa_1}\,E_{in} \qquad (9.22)$$

$$E_{21} = j\sqrt{1-\gamma_1}\sqrt{\kappa_1}\,E_{in} \qquad (9.23)$$

According to the linear coupling theory, the following relations can connect input-output fields for each nonlinear ring resonators (NRRs) as shown in Figure 9.15, which can be expressed by [60]

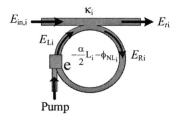

**FIGURE 9.15** A schematic diagram of single NRR ($i = 2, 3, 4$).

$$E_{Ri} = \sqrt{1-\gamma_i}\sqrt{1-\kappa_i}\,E_{Li} + j\sqrt{1-\gamma_i}\sqrt{\kappa_i}\,E_{in_i} \qquad (9.24)$$

$$E_{t_i} = \sqrt{1-\gamma_i}\sqrt{1-\kappa_i}\,E_{in_i} + j\sqrt{1-\gamma_i}\sqrt{\kappa_i}\,E_{Li} \qquad (9.25)$$

where $\gamma_i$ and $\kappa_i$ are the coupler loss and coupling coefficient in each NRR, respectively. Since the NRR length and the nonlinear index of refraction are small, the nonlinear Schrödinger equation (NLS) can be used to solve light propagation through the NRRs, the solution is given by [60]

$$E_{Li} = E_{Ri}\exp\left(-\frac{\alpha}{2}L_i - \gamma_{NL_i}|E_{Ri}|^2 L_i\right) \qquad (9.26)$$

where $\alpha$, $L_i = 2\pi R_i$ and $\gamma_{NL_i}$ are the NRR loss, NRR length ($R_i$ is ring radius) and the nonlinear coefficient including the nonlinear index of refraction and two-photon absorption phenomenon, respectively. By using the basic concepts in nonlinear optics, the following relation can be used for the above mentioned, the nonlinear coefficient is given by

$$\gamma_{NL_i} = \frac{\beta}{2} - j\frac{\omega_0}{c} n_2 \tag{9.27}$$

where $\beta$, $\omega_0$, C and $n_2$ are the two-photon absorption coefficient, incident light frequency, speed of light in free space and the nonlinear index of refraction coefficient, respectively. The following relation describes the nonlinear phenomenon in the NRRs as

$$\begin{aligned}\tilde{\alpha} &= \alpha + \beta |E|^2 \\ \tilde{n} &= n + n_2 |E|^2\end{aligned} \tag{9.28}$$

where both of the absorption coefficient and index of refraction includes linear and nonlinear parts and the following relations can be used for obtaining these variables in terms of the optical third order susceptibility as

$$\begin{aligned}n_2 &= \frac{3}{8n} \text{Re}\left[\chi^{(3)}\right] \\ \beta &= \frac{3\omega_0}{4nC} \text{Im}\left[\chi^{(3)}\right]\end{aligned} \tag{9.29}$$

Using Eqs (9.24)–(9.26) and some mathematical manipulations, the following transmission functions can be obtained for each NRRs as

$$T = \frac{E_{t_i}}{E_{in_i}} = \sqrt{1-\gamma_i}\sqrt{1-\kappa_i} - \frac{\kappa_i(1-\gamma_i)\exp\left(-\frac{\alpha}{2}L_i - \phi_{NL_i}\right)}{1 - \sqrt{1-\gamma_i}\sqrt{1-\kappa_i}\exp\left(-\frac{\alpha}{2}L_i - \phi_{NL_i}\right)} \tag{9.30}$$

where $\phi_{NL_i} = \gamma_{NL_i} L_i |E_{Ri}|^2$ is defined as a nonlinear phase shift.

For the obtained result in Eq. (9.30), the phase difference (effective phase from single NRR) can be found as follows:

$$\phi_{eff} = \phi_i = \tan^{-1}\left[\frac{-\kappa_i(1-\gamma_i)e^{-1/2(\alpha+\beta|E_{Ri}|^2)L_i} \times \sin\left((\omega_0/C)n_2 L_i |E_{Ri}|^2\right)}{A+B+D}\right] \tag{9.31}$$

where

$$A = (2-\kappa_i)(1-\gamma_i)e^{-1/2(\alpha+\beta|E_{Ri}|^2)L_i}\cos\left((\omega_0/C)n_2 L_i |E_{Ri}|^2\right)$$

$$B = \sqrt{(1-\kappa_i)(1-\gamma_i)} \quad \text{and} \quad D = (1-\gamma_i)\sqrt{(1-\kappa_i)(1-\gamma_i)}e^{-(\alpha+\beta|E_{Ri}|^2)L_i}$$

The obtained result can be simplified to the following formula, which is assumed by $\gamma_i = \alpha = \beta = 0$.

$$\phi_{eff} = \phi_i = \tan^{-1}\left[\frac{-\kappa_i \sin\left(\frac{\omega_0}{C} n_2 L_i |E_{Ri}|^2\right)}{2\sqrt{1-\kappa_i} - (2-\kappa_i)\cos\left(\frac{\omega_0}{C} n_2 L_i |E_{Ri}|^2\right)}\right] \quad (9.32)$$

In real system $\gamma_i$, $\alpha$, and $\beta$ are not equal to zero as shown, $\alpha = 0.5$ dBmm$^{-1}$, $\gamma = 0.1$ and $\beta = 2 \times 10^{-11}$, respectively.

And, the output power at light propagation through the first NRR is given by

$$\frac{P_{t1}}{P_{in}} = \left|\frac{E_{t1}}{E_{in}}\right|^2 \quad (9.33)$$

The electric field of light propagation through the second NRR is given by

$$\frac{E_{t2}}{E_{in}} = \frac{\sqrt{1-\gamma_1}\sqrt{1-\gamma_2}\sqrt{1-\gamma_3}\sqrt{1-\kappa_1}\left(\begin{array}{c}\sqrt{1-\kappa_2}\sqrt{1-\kappa_3} - \sqrt{1-\gamma_2}\sqrt{1-\kappa_3}\,e^{-\frac{\alpha}{2}L_1-\phi_{NL,1}} \\ -\sqrt{1-\gamma_3}\sqrt{1-\kappa_2}\,e^{-\frac{\alpha}{2}L_2-\phi_{NL,2}} \\ +\sqrt{1-\gamma_2}\sqrt{1-\gamma_3}\,e^{-\frac{\alpha}{2}(L_1+L_2)-(\phi_{NL,1}+\phi_{NL,2})}\end{array}\right)}{\left(\begin{array}{c}1+\sqrt{1-\gamma_2}\sqrt{1-\kappa_2}\,e^{-\frac{\alpha}{2}L_1-\phi_{NL,1}} - \sqrt{1-\gamma_3}\sqrt{1-\kappa_3}\,e^{-\frac{\alpha}{2}L_2-\phi_{NL,2}} \\ +\sqrt{1-\gamma_2}\sqrt{1-\gamma_3}\sqrt{1-\kappa_2}\sqrt{1-\kappa_3}\,e^{-\frac{\alpha}{2}(L_1+L_2)-(\phi_{NL,1}+\phi_{NL,2})}\end{array}\right)} \quad 9.34)$$

where $\sqrt{1-\gamma_1}$, $\sqrt{1-\gamma_2}$, $\sqrt{1-\gamma_3}$, $\sqrt{1-\gamma_4}$ and $\sqrt{1-\gamma_5}$ are coupler losses in each coupler and $\sqrt{1-\kappa_1}$, $\sqrt{1-\kappa_2}$, $\sqrt{1-\kappa_3}$, $\sqrt{1-\kappa_4}$ and $\sqrt{1-\kappa_5}$ are coupler separates in each coupler, respectively.

The output power of light propagation through the second NRR is given by

$$\frac{P_{t2}}{P_{in}} = \left|\frac{E_{t2}}{E_{in}}\right|^2 \quad (9.35)$$

The electric field of light propagation through third NRR the relation input-output field is

$$\frac{E_{t3}}{E_{in}} = \frac{A\begin{pmatrix} \sqrt{1-\kappa_2}\sqrt{1-\kappa_3}\sqrt{1-\kappa_4} \\ -\sqrt{1-\kappa_3}\sqrt{1-\kappa_4}\,e^{-\frac{\alpha}{2}L_1-\phi_{NL,1}} \\ -\sqrt{1-\kappa_2}\sqrt{1-\kappa_4}\,e^{-\frac{\alpha}{2}L_2-\phi_{NL,2}} \\ +\sqrt{1-\kappa_4}\,e^{-\frac{\alpha}{2}(L_1+L_2)-(\phi_{NL,1}+\phi_{NL,2})} \\ -\sqrt{1-\kappa_2}\sqrt{1-\kappa_3}\,e^{-\frac{\alpha}{2}L_3-\phi_{NL,3}} \\ +\sqrt{1-\kappa_3}\,e^{-\frac{\alpha}{2}(L_1+L_3)-(\phi_{NL,1}+\phi_{NL,3})} \\ +\sqrt{1-\kappa_2}\,e^{-\frac{\alpha}{2}(L_2+L_3)-(\phi_{NL,2}+\phi_{NL,3})} \\ +\sqrt{1-\kappa_3}\sqrt{1-\kappa_4}\,e^{-\frac{\alpha}{2}(L_1+L_2+L_3)-(\phi_{NL,1}+\phi_{NL,2}+\phi_{NL,3})} \end{pmatrix}}{\begin{pmatrix} 1+\sqrt{1-\gamma_2}\sqrt{1-\kappa_2}\,e^{-\frac{\alpha}{2}L_1-\phi_{NL,1}} \\ -\sqrt{1-\gamma_3}\sqrt{1-\kappa_3}\,e^{-\frac{\alpha}{2}L_2-\phi_{NL,2}} - \sqrt{1-\gamma_4}\sqrt{1-\kappa_4}\,e^{-\frac{\alpha}{2}L_3-\phi_{NL,3}} \\ +\sqrt{1-\gamma_2}\sqrt{1-\gamma_3}\sqrt{1-\kappa_2}\sqrt{1-\kappa_3}\,e^{-\frac{\alpha}{2}(L_1+L_2)-(\phi_{NL,1}+\phi_{NL,2})} \\ +\sqrt{1-\gamma_3}\sqrt{1-\gamma_4}\sqrt{1-\kappa_3}\sqrt{1-\kappa_4}\,e^{-\frac{\alpha}{2}(L_2+L_3)-(\phi_{NL,2}+\phi_{NL,3})} \\ -\sqrt{1-\gamma_2}\sqrt{1-\gamma_3}\sqrt{1-\gamma_4}\sqrt{1-\kappa_2}\sqrt{1-\kappa_3}\sqrt{1-\kappa_4} \\ \times e^{-\frac{\alpha}{2}(L_1+L_2+L_3)-(\phi_{NL,1}+\phi_{NL,2}+\phi_{NL,3})} \end{pmatrix}} \quad (9.36)$$

where $A = \sqrt{1-\gamma_1}\sqrt{1-\gamma_2}\sqrt{1-\gamma_3}\sqrt{1-\gamma_4}\sqrt{1-\kappa_1}$.

The output power of light propagation through the third NRR is

$$\frac{P_{t3}}{P_{in}} = \left|\frac{E_{t3}}{E_{in}}\right|^2 \quad (9.37)$$

When light propagation through the second 3 dB, upper and lower MZI arms that On Off Keying (OOK) or DPSK are controlled the optical pump in each NRR that yield

$$\begin{pmatrix} E_{out\_1}(On) \\ E_{out\_2}(Off) \end{pmatrix} = \sqrt{1-\gamma_5}\begin{pmatrix} \sqrt{1-\kappa_5} & j\sqrt{\kappa_5} \\ j\sqrt{\kappa_5} & \sqrt{1-\kappa_5} \end{pmatrix}\begin{pmatrix} E_{T3} \\ E_{22} \end{pmatrix} \quad (9.38)$$

For phase shaped binary transmission (PSBT), we generated input light pulse through NRRs that the difference phase shift is equal to $\pi$, as shown in Figure 9.16.

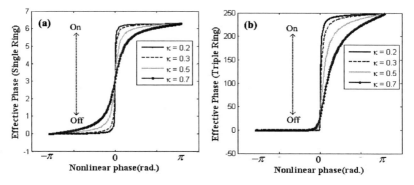

**FIGURE 9.16** Simulation results of effective phase, where (a) single NRR vs. nonlinear phase and (b) triple NRR vs. nonlinear phase. ($\alpha$ = 0.5 dBmm$^{-1}$, $\gamma$ = 0.1, $n_2$ = 2.2 × 10$^{-13}$ m$^2$/W, $n_0$ = 3.34, $\lambda$ = 1.55 µm, $\beta$ = 0).

In operation, all-optical OOK generated maximum power of 3 mW Gaussian modulated CW is input into the OOK system, as shown in Figure 9.14. The suitable parameters of NRRs are used, such as NRR radii where $R_1$ = 1.5 µm, $R_2$ = 1.0 µm and $R_3$ = 0.775 µm. In order to make the system associate with the practical device [63, 64], the selected parameters of the system are fixed at $\lambda_0$ = 1.55 µm and 1.31 µm, $n_0$ = 3.34 (GaInAsP/InP waveguide). The effective core areas are $A_{eff}$ = *0.10* µm$^2$ for NRRs. The waveguide and coupling loses are $\alpha$ =0.5 dBmm$^{-1}$ and $\gamma$ = 0.1, respectively, and the coupling coefficients $\kappa_i$ of the NRRs fixed 0.5 and $\beta$ = 2 × 10$^{-11}$ [60]. As for the numerical simulation of all-optical OOK, PSBT and DPSK, all our numerical work has been carried out by using commercially available simulation software-the OptiFDTD simulation package [65] which is based explicitly on the model described above [66]. However, more parameters are used as shown in Figure 9.14. The nonlinear refractive index is $n_2$ = 2.2 × 10$^{-13}$ m$^2$/W. In this case, the waveguide loss used is 0.5 dBmm$^{-1}$ and the size of the waveguide desired system is 10 × 40 µm$^2$.

Figure 9.16 shows the variation of the effective phase due to nonlinear phase ($\phi_{NL} = (\omega_0/C)n_2(2\pi R_i)|E_{Ri}|^2$), we assumed that $\beta$ = 0, the realized $\beta$ is not equal to zero as shown in next section. The coupling coefficient is decreased which is described as the slope of variation. We found that the effective phase of all-optical PSBT generation is changed rapidly when the coupling coefficient of NRRs is fixed at 0.2, in which the small coupling coefficient can obtain the fast switching on-off power, and in this case only PSBT is necessary for on-off keying (OOK) in each NRR. Figure 9.17 shows the output signal when the input power is 3 mW with center wavelength at 1.55 µm. Result of OOK modulation using the Gaussian CW modulated signal is as shown in Figure 9.18. The used parameters are wavelength center $\lambda_0$ = 1.31 µm and 3 mW input power. The on-off state occurs within

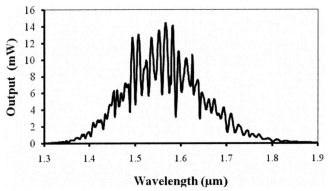

**FIGURE 9.17** Result of the output signal when the input power is 3 mW with center wavelength at 1.55 μm.

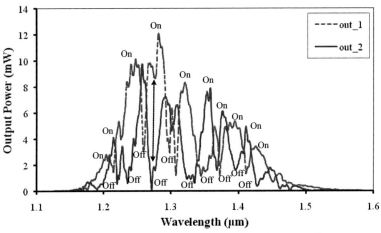

**FIGURE 9.18** Result of OOK generated at wavelength center $\lambda_0$ = 1.31 μm, the input power is 3 mW.

the upper (out_1) and lower (out_2) MZI arms, with the different phase shift is equal to π. Results of OOK modulation at the wavelength center $\lambda_0$ = 1.55 μm and 3 mW input power is as shown in Figure 9.19. The on-off state occurred within the upper (out_1) and lower (out_2) MZI arms, with the different phase shift is equal to π, however in this case the output power of out_1 and out_2 is generated rapidly compared with the result of Figure 9.18, in this result the obtained delay time is 1.2 fs. The operation of the proposed circuit is transient in time as shown in Figure 9.20. The comparison of OOK generated at wavelength center $\lambda_0$ = 1.55 μm and $\lambda_0$ = 1.31 μm is as shown in Figure 9.21, where we found that the OOK result at wavelength 1.55 μm is switching faster than the result of 1.31 μm, therefore, the switching OOK generated at wavelength center 1.55 μm can form the higher capacity packet on-off state appearance. The upper limit of the circuit in frequency domain (response) is 3.5 THz, which is obtained and shown in Figure 9.22.

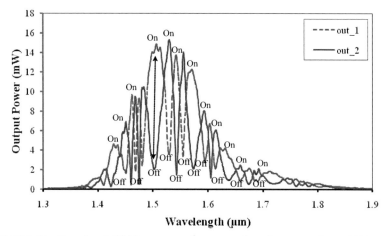

**FIGURE 9.19** Result of OOK generated at wavelength center $\lambda_0 = 1.55$ μm, the input power is 3 mW.

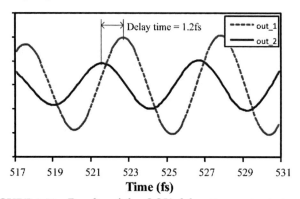

**FIGURE 9.20** Results of the OOK delay time output signals.

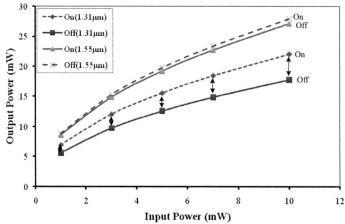

**FIGURE 9.21** Results of OOK generated at wavelength center, $\lambda_0 = 1.55$ μm and $\lambda_0 = 1.31$ μm, respectively.

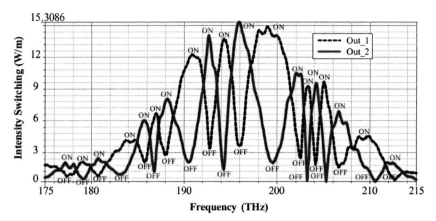

**FIGURE 9.22** OOK result is generated in the frequency domain, the input power is 3 mW.

## 9.9 Conclusion

Basically, in the first chapter the simple deterministic quantum CNOT and SWAP gates operation using the nonlinear photonic circuits are presented. Two optical qubits are generated by using dark-bright soliton conversion states: $|D\rangle$ and $|B\rangle$, in which the quantum controlled-NOT (CNOT) gate was controlled by the nonlinear phase shifter in MZI photonic circuit. The photonic add/drop device was also operated to perform the quantum SWAP gate.

Then we proposed the novel technique that can be used to simultaneously generate the optical logic XOR/XNOR gates using dark-bright soliton conversion within the add/drop optical filter system. By using the dark-bright soliton conversion concept, results obtained have shown that the input logic '0' and control logic '0' can be formed by using the dark soliton (D) pulse trains. We also found that the simultaneous optical logic XOR/XNOR can be seen randomly at the drop and through ports, respectively, which is shown the potential of application for large scale use, especially, for security code application requirement.

Finally, we have demonstrated that the OOK modulation format by using NRRs coupled into one arm of a MZI system could be performed. The solution of the nonlinear Schrödinger equation is

$$E_{Li} = E_{Ri} \exp\left(-\frac{\alpha}{2}L_i - \gamma_{NL_i}\left|E_{Ri}\right|^2 L_i\right)$$

which describes the nonlinear properties in each nonlinear ring resonator by using the term $|E_{Ri}|^2$ that circulates in each NRR. It is used to enhance and amplify the output signals. When the input light pulse is input through a 3 dB coupler of a MZI, the coupling power is partially circulated through $R_1$, where it is circulated and combined with the pump light within $R_1$. The output of rings $R_2$ and $R_3$ are obtained in the similar manner.

The feasibility of the device by comparison to already fabricated devices with the same radius NRR radii, where $R_1 = 1.5$ μm, $R_2 = 1.0$ μm and $R_3 = 0.775$ μm. The parameter details are given by reference [63]. Our proposed system is the extended system of a Mach–Zehnder interferometer combined with ring resonators. It was fabricated by Rabus [64], where the system size was 700 × 2500 μm$^2$. It is larger than the system in this chapter where the system size is 10 × 40 μm$^2$. Moreover, our system is combined with the triple nonlinear ring resonators. Two different results at the center wavelength 1.31 μm and 1.55 μm are compared, where they are dominated by the nonlinear refractive indices and two-photon absorption coefficients within the NRRs. We found that the OOK generated at 1.55 μm is shown the fastest switching, where the delay time of OOK is 1.2 fs. We have also presented the principles of MZI operation in PSBT-based systems, where the characterization of useful parameters required for the DPSK demodulation or PSBT encoding. DPSK and PSBT have been highlighted as the suitable modulation formats for optical transmissions. The DPSK modulation format presents the better performance for transmission than the conventional OOK, justifying its utilization. However, the DPSK requires the passive MZIs for interferometric demodulation. In application, the use of DWDM (Dense Wavelength Division Multiplexing) can be employed to obtain the multi-wavelength OOK, which may be available for high capacity packet switching.

# REFERENCES

[1] E. Knill, R. Laflamme, and G.J. Milburn, "A scheme for efficient quantum computation with linear optics", *Nature (London)* 409, 46–52 (2001).

[2] T.B. Pittman, B.C. Jacobs, and J.D. Franson, "Demonstration of non-deterministic quantum logic operations using linear optical elements", *Phys. Rev. Lett.* 88, 257902 (2002).

[3] M.A. Nielsen and I.L. Chuang, *Quantum Computation and Quantum Information*. Cambridge: Cambridge University Press (2000).

[4] R. Raussendorf and H.J. Briegel, "A one-way quantum computer", *Phys. Rev. Lett.* 86, 51885191 (2001).

[5] P. Kok *et al.*, "Linear optical quantum computing with photonic qubits", *Rev. Mod. Phys.* 79, 135–174 (2007).

[6] K. Nemoto and W.J. Munro, "Nearly deterministic linear optical controlled-NOT gate", *Phys. Rev. Lett.* 93, 250502 (2004).

[7] C. Bonato, F. Haupt, S.S.R. Oemrawsingh, J. Gudat, D. Ding, M.P. Exter, and D. Bouwmeester, "CNOT and Bell-state analysis in the weak-coupling cavity QED regime", *Phys. Rev. Lett.* 104, 160503 (2010).

[8] S. Saito, T. Tilma, S.J. Devitt, K. Nemoto, and K. Semba, "Experimentally realizable controlled NOT gate in a flux qubit/resonator system", *Phys. Rev. B* 80, 224509 (2009).

[9] L.P. Deng, H. Wang, and K. Wang, "Quantum CNOT gates with orbital angular momentum and polarization of single-photon quantum logic", *J. Opt. Soc. Am. B*, 24, 2517–2520 (2007).

[10] J. Ghosh and M.R. Geller, "Controlled-NOT gate with weakly coupled qubits: Dependence of fidelity on the form of interaction", *Phys. Rev. A* 81, 052340 (2010).

[11] L. Isenhower, E. Urban, X.L. Zhang, A.T. Gill, T. Henage, T.A. Johnson, T.G. Walker, and M. Saffman, "Demonstration of a neutral atom controlled-NOT quantum gate", *Phys. Rev. Lett.*, 104, 010503 (2010).

[12] D.W. Leung, I.L. Chuang, F. Yamaguchi, and Y. Yamamoto, "Efficient implementation of coupled logic gates for quantum computation", *Phys. Rev. A*, 61, 042310 (2000).

[13] G.W. Lin, X.B. Zou, M.Y. Ye, X.M. Lin, and G.C. Guo, "Quantum SWAP gate in an optical cavity with an atomic cloud", *Phys. Rev. A* 77, 064301 (2008).

[14] D.G. Angelakis, M.F. Santos, V. Yannaopapas, and A. Ekert, "A proposal for the implementation of quantum gates with photonic-crystal waveguides", *Phys. Lett. A* 362, 377–380 (2007).

[15] Y. Kokubun, Y. Hatakeyama, M. Ogata, S. Suzuki, and N. Zaizen, "Fabrication technologies for vertically coupled microring resonator with multilevel crossing busline and ultracompact-ring radius", *IEEE J. Sel. Top. Quantum Electron.*, 11, 4–10 (2005).

[16] L. Zhang, R. Ji, L. Jia, L. Yang, P. Zhou, Y. Tian, P. Chen, Y. Lu, Z. Jiang, Y. Liu, Q. Fang, and M. Yu, "Demonstration of directed XOR/XNOR logic gates using two cascaded microring resonators", *Opt. Lett.*, 35(10), 1620–1622 (2010).

[17] S. Ma, Z. Chen, H. Sun, and K. Dutta, "High speed all optical logic gates based on quantum dot semiconductor optical amplifiers", *Opt. Express*, 18(7), 6417–6422 (2010).

[18] T. Kawazoe, K. Kobayashi, K. Akahane, M. Naruse, N. Yamamoto, and M. Ohtsu, "Demonstration of nanophotonic NOT gate using near-field optically coupled quantum dots", *Appl. Phys. B*, 84, 243–246 (2006).

[19] J. Dong, X. Zhang, and D. Huang, "A proposal for two-input arbitrary Boolean logic gates using single semiconductor optical amplifier by picoseconds pulse injection", *Opt. Express*, 17(10), 7725–7730 (2009).

[20] J. Dong, X. Zhang, J. Xu, and D. Huang, "40 Gb/s all-optical logic NOR and OR gates using a semiconductor optical amplifier: Experimental demonstration and theoretical analysis", *Opt. Commun.* 281, 1710–1715 (2008).

[21] B.C. Han, J. L. Yu, W.R. Wang, L.T. Zhang, H. Hu, and E.Z. Yang, "Experimental study on all-optical half-adder based on semiconductor optical amplifier", *Optoelectron. Lett.*, 5(3), 0162–0164 (2009).

[22] Z. Li, Y. Liu, S. Zhang, H. Ju, H. de Waardt, G.D. Khoe, H.J.S. Dorren, and D. Lenstra, "All-optical logic gates using semiconductor optical amplifier assisted by optical filter", *Electron. Lett.* 41, 1397–1399 (2005).

[23] S.H. Kim, J.H. Kim, B.G. Yu, Y.T. Byun, Y.M. Jeon, S. Lee, and D.H. Woo, "All-optical NAND gate using cross-gain modulation in semiconductor optical amplifiers", *Electron. Lett.* 41, 1027–1028 (2005).

[24] X. Zhang, Y. Wang, J. Sun, D. Liu, and D. Huang, "All-optical AND gate at 10 Gbit/s based on cascaded single-port-couple SOAs", *Opt. Express* 12, 361–366 (2004).

[25] Z. Li and G. Li, "Ultrahigh-speed reconfigurable logic gates based on four-wave mixing in a semiconductor optical amplifier", *IEEE Photon. Technol. Lett.* 18, 1341–1343 (2006).

[26] S. Kumar and A.E. Willner, "Simultaneous four-wave mixing and cross-gain modulation for implementing an all-optical XNOR logic gate using a single SOA", *Opt. Express* 14, 5092–5097 (2006).

[27] J.N. Roy and D.K. Gayen, "Integrated all-optical logic and arithmetic operations with the help of a TOAD-based interferometer device-alternative approach", *Appl. Opt.*, 46(22), 5304–5310 (2007).

[28] M. Khorasaninejad and S.S. Saini, "All-optical logic gates using nonlinear effects in silicon-on-insulator waveguides", *Appl. Opt.*, 48(25), F31–F36 (2009).

[29] Y.D. Wu, "All-optical logic gates by using multibranch waveguide structure with localized optical nonlinearity", *IEEE J. Select. Quant. Electron.*, 11(2), 307–312 (2005).

[30] Y. Miyoshi, K. Ikeda, H. Tobioka, T. Inoue, S. Namiki, and K.I. Kitayama, "Ultrafast all-optical logic gate using a nonlinear optical loop mirror based multi-periodic transfer function", *Opt. Express*, 16(4), 2570–2577 (2008).

[31] T. Houbavlis, K. Zoiros, A. Hatziefremidis, H. Avramopoulos, L. Occhi, G. Guekos, S. Hansmann, H. Burkhard, and R. Dall'Ara, "10 Gbit/s all-optical Boolean XOR with SOA fibre Sagnac gate", *Electron. Lett.* 35, 1650–1652 (1999).

[32] J. Wang, Q. Sun, and J. Sun, "All-optical 40 Gbit/s CSRZ-DPSK logic XOR gate and format conversion using four-wave mixing", *Opt. Express*, 17(15), 12555–12563 (2009).

[33] J. Xu, X. Zhang, Y. Zhang, J. Dong, D. Liu, and D. Huang, "Reconfigurable all-optical logic gates for multi-input differential phase-shift keying signals: design and experiments", *J. Lightwave Technol.*, 27(23), 5268–5275 (2009).

[34] Y.D. Wu and T.T. Shih, "New all-optical logic gates based on the local nonlinear Mach-Zehnder interferometer", *Opt. Express*, 16(1), 248–257 (2008).

[35] Y. Zhang, Y. Zhang, and B. Li, "Optical switches and logic gates based on self-collimated beams in two-dimensional photonic crystals", *Opt. Express*, 15(15), 9287–9292 (2007).

[36] G. Berrettini, A. Simi, A. Malacarne, A. Bogoni, and L. Poti, "Ultrafast integrable and reconfigurable XNOR, AND, NOR, and NOT photonic logic gate", *IEEE Photon. Technol. Lett.*, 18(8), 917–919 (2006).

[37] Y.K. Lize, L. Christen, M. Nazarathy, S. Nuccio, X. Wu, A.E. Willner, and R. Kashyap, "Combination of optical and electronic logic gates for error correction in multipath differential demodulation", *Opt. Express*, 15(11), 6831–6839 (2007).

[38] D.M.F. Lai, C.H. Kwok, and K.K.Y. Wong, "All-optical picoseconds logic gates based on a fiber optical parametric amplifier", *Opt. Express*, 16(22), 18362–18370 (2008).

[39] Z. Li, Z. Chen, and B. Li, "Optical pulse controlled all-optical logic gates in SiGe/Si multimode interference", *Opt. Express*, 13(3), 1033–1038 (2005).

[40] Y.A. Zaghloul and A.R.M. Zaghloul, "Complete all-optical processing polarization-based binary logic gates and optical processors", *Opt. Express*, 14(21), 9879–9895 (2006).

[41] L.Y. Han, H.Y. Zhang, and Y.L. Guo, "All-optical NOR gate based on injection-locking effect in a semiconductor laser", *Optoelectron. Lett.*, 4(1), 0034–0037 (2008).

[42] S. Mookherjea and M.A. Schneider, "The nonlinear microring add-drop filter", *Opt. Express* 16, 15130–15136 (2008).

[43] A.H. Gnauck, K.C. Reichmann, J.M. Kahn, S.K. Korotky, J.J. Veselka, and T.L. Koch, "4 Gb/s heterodyne transmission experiments using OOK, FSK, and DPSK modulation", *IEEE Photon. Technol. Lett.* 2, 908–910 (1990).

[44] S.R. Nuccio, O.F. Yilmaz, S. Khaleghi, X. Wu, L. Christen, I. Fazal, and A.E. Willner, "Tunable 503 ns optical delay of 40 Gbit/s RZ-OOK and RZ-DPSK using a wavelength scheme for phase conjugation to reduce residual dispersion and increase delay", *Opt. Letters* 34(12), 1903–1905 (2009).

[45] W. Astar, J.B. Driscoll, X. Liu, J.I. Dadap, W.M.J. Green, Y.A. Vlasov, G. M. Carter, and R. M. Osgood, Jr., "Conversion of 10 Gb/s NRZ-OOK to RZ-OOK utilizing XPM in a Si nanowire", *Opt. Express* 17(15), 12987–12999 (2009).

[46] W. Hong, D. Huang, X. Zhang, and G. Zhu, "Simulation and analysis of OOK-to-BPSK format conversion based on gain-transparent SOA used as optical phase-modulator", *Opt. Express* 15(26), 18357–18369 (2007).

[47] T. Nishitani, T. Konishi, and K. Itoh, "All-optical M-ary ASK signal demultiplexer based on a photonic analog-to-digital conversion", *Opt. Express* 15(25), 17025–17031 (2007).

[48] K. Mishina, S. Kitagawa, and A. Maruta, "All-optical modulation format conversion from on-off-keying to multiple-level phase-shift-keying based on nonlinearity in optical fiber", *Opt. Express* 15(13), 8444–8453 (2007).

[49] I. Kang, "Phase-shift-keying and on-off-keying with improved performances using electroabsorption modulators with interferometric effects", *Opt. Express* 15(4), 1467–1473 (2007).

[50] Y.G. Wen, L.K. Chen, K.P. Ho, F. Tong, and W.S. Chan, "Performance verification of a variable bit-rate limiter for on-off keying (OOK) optical systems", *IEEE J. Lightwave Technol.* 18(6), 779–786 (2000).

[51] T. Mizuochi, K. Ishida, T. Kobayashi, J. Abe, K. Kinjo, K. Motoshima, and K. Kasahara, "A comparative study of DPSK and OOK WDM transmission over transoceanic distances and their performance degradations due to nonlinear phase noise", *IEEE J. Lightwave Technol.* 21(9), 1933–1942 (2003).

[52] C. Xie, L. Möller, H. Haunstein, and S. Hunsche, "Comparison of system tolerance to polarization-mode dispersion between different modulation formats", *IEEE Photon. Technol. Lett.* 15(8), 1168–1170 (2003).

[53] M. Matsumoto, "All-optical signal regeneration using fiber nonlinearity", *Eur. Phys. J. Special Topics* 173, 297–312 (2009).

[54] K. Croussore, C. Kim, and G. Li, "All-optical regeneration of differential phase-shift keying signals based on phase-sensitive amplification", *Opt. Letters* 29(20), 2357–2359 (2004).

[55] A. Akhtar, L. Pavel, and S. Kumar, "Modeling interchannel FWM with walk-off in RZ-DPSK single span links", *IEEE J. Lightwave Technol.* 26(14), 2142–2154 (2008).

[56] C. Xu, X. Liu, and X. Wei, "Differential phase-shift keying for high spectral efficiency optical transmissions", *IEEE J. of Select. Topics in Quantum Electron.* 10(2), 281–293 (2004).

[57] J. Li, L. Li, J. Zhao, and C. Li, "Ultrafast, low power, and highly stable all-optical switch in MZI with two-arm-sharing nonlinear ring resonator", *Opt. Communication* 256, 319–325 (2005).

[58] J.E. Heebner, N.N. Lepeshkin, A. Schweinsberg, G.W. Wicks, and R.W. Boyd, "Enhanced linear and nonlinear optical phase response of AlGaAs microring resonators", *Opt. Letters* 29, 769–771 (2004).

[59] A. Rostami, "Low threshold and tunable all-optical switch using two-photon absorption in array of nonlinear ring resonators coupled to MZI", *J. Microelectronics* 37, 976–981 (2006).

[60] A. Bananej and C. Li, "Parameter controllable all-optical switching in a high-nonlinear microring coupled MZI through a pumped nonlinear coupler". *J. Nonlinear Opt. Phys. and Mater.* 14(1), 85–91 (2005).

[61] S. Mitatha, "Dark soliton behaviors within the nonlinear micro and nanoring resonators and applications". *Progress in Electromagnetics Research, PIER* 99, 383–404 (2009).

[62] D.G. Rabus, M. Hamacher, and H. Heidrich, "Resonance frequency tuning of a double ring resonator in GaInAsP/InP: Experiment and simulation", *Jpn. J. Appl. Phys.* 41, 1186–1189 (2002).

[63] D.G. Rabus, *Integrated Ring Resonators*, Chap. 5.4, pp. 169–173, Heidelberg, Berlin: Springer-Verlag (2007).

[64] OptiFDTD finite difference time domain photonics simulation software, OptiWave systems Inc. © 2008, http://www.optiwave.com/

[65] OptiWave systems Inc., private communication.

# 10
# Drug Delivery

**CHAPTER OUTLINE**
- Introduction
- Optical Vortex Generation
- Drug Trapping and Delivery
- Conclusion
- References

## 10.1 Introduction

Much effort has been made to explain the nature of optical forces and to describe them quantitatively by the establishment of theoretical models. This requires a detailed knowledge of the incident electromagnetic field that impinges on the object and of the properties of the object, which changes the incident field distribution by scattering, absorption or reemission of photons [1]. The theory of optical vortices soliton developed to date assumes a background of constant amplitude and is able to capture a number of interesting features of dynamic vortices. Understanding this dynamics is important for future application of soliton vortices in all steerable optical switching devices based on the concept of dark/bright soliton conversion [2]. Recently, the promising techniques of microscopic volume trapping and transportation within the add/drop multiplexer have been reported in both theory [3] and experiment [4], respectively. To date, the optical tweezer technique has become a powerful tool for manipulation of micrometer-sized particles in three spatial dimensions. Initially, the useful static tweezers are recognized, and the dynamic tweezers are now realized in practical works [5–7]. Schulz et al. [8] have shown that the transfer of trapped atoms between two optical potentials could be performed. In principle, an optical tweezers use forces exerted by intensity gradients in the strongly focused beams of light to trap and move the microscopic volumes of matters. Moreover, the other combination of force is induced by the interaction between photons, which is caused by the photon scattering effects. In practice, the field intensity can be adjusted and tuned to form the suitable trapping potential, in which the desired gradient field and scattering force can be used to form the suitable trapping force.

In biological applications of optical trapping and manipulation it is possible to remotely apply controlled force on living cells, internal parts of cells, and large biological molecules without inflicting detectable optical damage. This has resulted in many unique applications, where one of the most important of them is in the drug delivery and trapping study [9]. Several researchers have developed drug delivery systems that will bring about improvements in pharmacology and therapeutics. The attempts to seek the cures might be achieved through specific surface receptors, where the controlled drug delivery can help parenteral and route drug delivery. For the conventional methods of drug delivery management, usually, drug delivery at constant controlled rate is preferable. However, a better method may be in response to the physiological needs of the body. Moreover, the several advantages over conventional molecules are nanoparticle networks that have been proposed for controlling the release of high molecular weight biomolecules and drug molecules [10–15].

In this chapter, we propose the use of drug delivery system using the optical vortices within a PANDA ring resonator in which dynamic optical vortices are generated using a dark soliton, bright soliton and Gaussian pulse propagating within an add/drop optical multiplexer incorporating two nanoring resonators (PANDA ring resonator). The drug delivery system is formed within a PANDA ring resonator, where the pumped atoms/molecules can be transported into a router network within the waveguide material InGaAsP/InP [16–18]. For the router network, a liquid-core router channel can be used as the fluidic channel, in which the dynamically smooth and easily transportation of the atoms/molecules can be performed [19]. The dynamic behaviors of solitons and Gaussian pulses are analyzed and described [2, 20]. For increasing the channel multiplexing, the dark solitons with slightly different wavelengths are controlled and amplified within the tiny system. The trapping force stability is simulated and seen when the Gaussian pulse is used to control via the add (control) port. In application, the optical vortices (dynamic tweezers) can be used to store (trap) photon, atom, molecule, DNA, ion, or particle, which can perform the dynamic tweezers. By using the hybrid transceiver, where the transmitter and receiver parts can be integrated by a single system. Here, the use of the transceiver to form the hybrid communication of those microscopic volumes of matters in the nanoscale regime can be realized, especially, for drug delivery application.

## 10.2 Optical Vortex Generation

In principle, the optical tweezers concept use force that are exerted by the intensity gradients in the strongly focused beam of light to trap and move the microscopic volume of matter, in which the optical force are customarily and may be described for a trapped mass/volume that give a viscous damping defined by the relationship equate optical force with Stokes force as follows [21, 22]:

$$F = \frac{Qn_m P}{c} = \gamma_0 \dot{x} \qquad (10.1)$$

where

$$\gamma_0 = 6\pi r \rho v \qquad (10.2)$$

where $\dot{x}$ is velocities of volume, $n_m$ is the refractive index of the suspending medium, $c$ is the speed of light, and $P$ is the incident laser power, measured at the specimen. $\gamma_0$ is the Stokes' drag term (viscous damping), $r$ is a particle/volume radius, $v$ is kinematic viscosity and $\rho$ is fluids of density, $Q$ is a dimensionless efficiency. $Q$ represents the fraction of power utilized to exert force. $Q = 1$ for plane waves incident on a perfectly absorbing particle. To achieve stable trapping, the radiation pressure must create a stable, three-dimensional equilibrium. Because biological specimens are usually contained in aqueous medium, the dependence of $F$ on nm can rarely be exploited to achieve higher trapping forces. Increasing the laser power is possible, but only over a limited range due to the possibility of optical damage. $Q$ itself is therefore the main determinant of trapping force. It depends upon the NA, laser wavelength, light polarization state, laser mode structure, relative index of refraction, and geometry of the particle. In the Rayleigh regime, trapping forces decompose naturally into two components. Since, in this limit, the electromagnetic field is uniform across the dielectric particles can be treated as induced point dipoles. The scattering force is given by

$$F_{scatt} = \frac{\langle S \rangle \sigma}{c} \qquad (10.3)$$

where

$$\sigma = \frac{8}{3}\pi(kr)^4 r^2 \left(\frac{m^2 - 1}{m^2 + 2}\right)^2 \qquad (10.4)$$

where is the scattering cross section of a Rayleigh sphere with radius $r$. $S$ is the time-averaged Poynting vector, $n$ is the index of refraction of the particle, $m = n/n_m$ is the relative index, and $k = 2\pi n_m/\lambda$ is the wave number of the light. Scattering force is proportional to the energy flux and points along the direction of propagation of the incident light. The gradient force is the Lorentz force acting on the dipole induced by the light field. It is given by

$$F_{grad} = \frac{\alpha}{2}\nabla\langle E^2 \rangle \qquad (10.5)$$

where

$$\alpha = n_m^2 r^3 \left( \frac{m^2 - 1}{m^2 + 2} \right) \qquad (10.6)$$

is the polarizability of the particle. The gradient force is proportional and parallel to the gradient in energy density (for $m > 1$). The large gradient force is formed by the large depth of the laser beam, in which the stable trapping requires that the gradient force in the $-\hat{z}$ direction, which is against the direction of incident light (dark soliton valley), and it is greater than the scattering force. By increasing the NA (numerical aperture), when the focal spot size is decreased, the gradient strength is increased [23], which can be formed within the tiny system, for instance, nanoscale device (nanoring resonator).

In our proposal, the trapping force is formed by using a dark soliton, in which the valley of the dark soliton is generated and controlled within the PANDA ring resonator by the control port signals. From Figure 1, the output field ($E_{t1}$) at the through port is given by [24]. We are looking for the system that can generate the dynamic tweezers (optical vortices), in which the microscopic volume can be trapped and transmited via the communication link. Firstly, the stationary and strong pulse that can propagate within the dielectric material (waveguide) for a period of time is required. Moreover, the gradient field is an important property required in this case. Therefore, a dark soliton is satisfied and recommended to perform those requirements. Secondly, we are looking for the device that optical tweezers can propagate and form the long distance link, in which the gradient field (force) can be transmitted and received by using the same device. Here, the add/drop multiplexer in the form of a PANDA ring resonator which is well known and is introduced for this proposal, as shown in Figures 10.1 and 10.2. To form the multi function operations, for instance, control, tune, amplify, the additional pulses are bright soliton and Gaussian pulse introduced into the system. The input optical field ($E_{in}$) and the add port optical field ($E_{add}$) of the dark soliton, bright soliton and Gaussian pulses are given by [19], respectively.

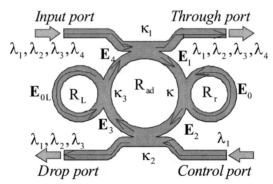

**FIGURE 10.1** A schematic diagram of a proposed PANDA ring resonator.

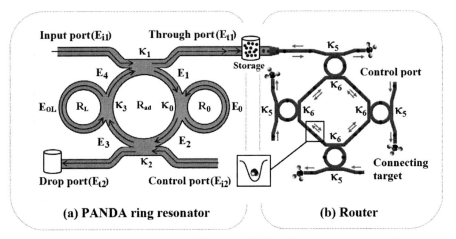

**FIGURE 10.2** A system of drug delivery and distribution using optical vortices.

$$E_{in}(t) = A \tanh\left[\frac{T}{T_0}\right] \exp\left[\left(\frac{Z}{2L_D}\right) - i\omega_0 t\right] \quad (10.7)$$

$$E_{control}(t) = A \operatorname{sech}\left[\frac{T}{T_0}\right] \exp\left[\left(\frac{Z}{2L_D}\right) - i\omega_0 t\right] \quad (10.8)$$

$$E_{control}(t) = E_0 \exp\left[\left(\frac{Z}{2L_D}\right) - i\omega_0 t\right] \quad (10.9)$$

where $A$ and $z$ are the optical field amplitude and propagation distance, respectively. $T$ is a soliton pulse propagation time in a frame moving at the group velocity, $T = t - \beta_1 z$, where $\beta_1$ and $\beta_2$ are the coefficients of the linear and second-order terms of Taylor expansion of the propagation constant. $L_D = T_0^2/|\beta_2|$ is the dispersion length of the soliton pulse. $T_0$ in equation is a soliton pulse propagation time at initial input (or soliton pulse width), where $t$ is the soliton phase shift time, and the frequency shift of the soliton is $\omega_0$. This solution describes a pulse that keeps its temporal width invariance as it propagates, and thus is called a temporal soliton. When a soliton of peak intensity $(|\beta_2/\Gamma T_0^2|)$ is given, then $T_0$ is known. For the soliton pulse in the microring device, a balance should be achieved between the dispersion length $(L_D)$ and the nonlinear length $(L_{NL} = 1/\Gamma\phi_{NL})$. Here $\Gamma = n_2 k_0$, is the length scale over which dispersive or nonlinear effects makes the beam become wider or narrower. For a soliton pulse, there is a balance between dispersion and nonlinear lengths. Hence $L_D = L_{NL}$. For a Gaussian pulse in Eq. (10.9), $E_0$ is the amplitude of optical field.

When light propagates within the nonlinear medium, the refractive index $(n)$ of light within the medium is given by

$$n = n_0 + n_2 I = n_0 + \frac{n_2}{A_{eff}} P \quad (10.10)$$

with $n_0$ and $n_2$ as the linear and nonlinear refractive indexes, respectively. $I$ and $P$ are the optical intensity and the power, respectively. The effective mode core area of the device is given by $A_{\text{eff}}$. For the add/drop optical filter design, the effective mode core areas range from 0.10 to 0.50 μm², in which the parameters were obtained by using the related practical material parameters (InGaAsP/InP) [25]. When a dark soliton pulse is input and propagated within an add/drop optical filter as shown in Figure 10.1, the resonant output is formed. Thus, the normalized output of the light field is defined as the ratio between the output and input fields, [$E_{\text{out}}(t)$ and $E_{\text{in}}(t)$] in each roundtrip. This is given as [26].

$$\left|\frac{E_{\text{out}}(t)}{E_{\text{in}}(t)}\right|^2 = (1-\gamma)\left[1 - \frac{(1-(1-\gamma)x^2)\kappa}{(1-x\sqrt{1-\gamma}\sqrt{1-\kappa})^2 + 4x\sqrt{1-\gamma}\sqrt{1-\kappa}\sin^2\left(\frac{\phi}{2}\right)}\right] \quad (10.11)$$

The close form of Eq. (10.11) indicates that a ring resonator in this particular case is very similar with Fabry–Pérot cavity concept, which has an input and output mirror with a field reflectivity, $(1 - \kappa)$, and a fully reflecting mirror. $\kappa$ is the coupling coefficient, and $x = \exp(-\alpha L/2)$ represents a roundtrip loss coefficient, $\phi_0 = kLn_0$ and $\phi_{\text{NL}} = kLn_2|E_{\text{in}}|^2$ are the linear and nonlinear phase shifts, $k = 2\pi/\lambda$ is the wave propagation number in a vacuum. $L$ and $\alpha$ are the waveguide length and linear absorption coefficient, respectively. In this chapter, the iterative method is introduced to obtain the resonant results and similarly, when the output field is connected and input into the other ring resonators.

In order to retrieve the required signals, we propose to use the add/drop device with the appropriate parameters. This is given in the following details in the optical circuits of ring resonator add/drop filters for the through port and drop port given by Eqs (10.12) and (10.13), respectively [27].

$$\left|\frac{E_t}{E_{\text{in}}}\right|^2 = \frac{\left[(1-\kappa_1) + (1-\kappa_2)e^{-\alpha L} - 2\sqrt{1-\kappa_1}\cdot\sqrt{1-\kappa_2}e^{-\frac{\alpha}{2}L}\cos(k_n L)\right]}{\left[1 + (1-\kappa_1)(1-\kappa_2)e^{-\alpha L} - 2\sqrt{1-\kappa_1}\cdot\sqrt{1-\kappa_2}e^{-\frac{\alpha}{2}L}\cos(k_n L)\right]} \quad (10.12)$$

$$\left|\frac{E_d}{E_{\text{in}}}\right|^2 = \frac{\kappa_1\kappa_2 e^{-\frac{\alpha}{2}L}}{1 + (1-\kappa_1)(1-\kappa_2)e^{-\alpha L} - 2\sqrt{1-\kappa_1}\cdot\sqrt{1-\kappa_2}e^{-\frac{\alpha}{2}L}\cos(k_n L)} \quad (10.13)$$

Here $E_t$ and $E_d$ represent the optical fields of the through port and drop ports, respectively. $\beta = kn_{\text{eff}}$ is the propagation constant, $n_{\text{eff}}$ is the effective refractive index of the waveguide, and the circumference of the ring is $L = 2\pi R$, with $R$ as the radius of the ring. The filtering signal can be managed by using the specific parameters of the add/drop device, and the required signals can be retrieved via the drop port output. $\kappa_1$ and $\kappa_2$ are the coupling coefficients of the add/drop filters, $k_n = 2\pi/\lambda$ is the wave propagation number in a vacuum, and the waveguide loss is $\alpha = 0.5$ dBmm$^{-1}$ for ring resonator waveguide. The fractional coupler intensity loss is $\gamma = 0.1$. In the case of the add/drop device, the nonlinear refractive index does not affect to the system, therefore, it is neglected.

From Eq. (10.12), the output field ($E_{t1}$) at the through port is given by

$$E_{t1} = AE_{i1} - BE_{i2}e^{-\frac{\alpha L}{2} - jk_n\frac{L}{2}} - \left[\frac{CE_{i1}\left(e^{-\frac{\alpha L}{2} - jk_n\frac{L}{2}}\right)^2 + DE_{i2}\left(e^{-\frac{\alpha L}{2} - jk_n\frac{L}{2}}\right)^3}{1 - E\left(e^{-\frac{\alpha L}{2} - jk_n\frac{L}{2}}\right)^2}\right] \quad (10.14)$$

where

$A = \sqrt{(1-\gamma_1)(1-\gamma_2)}$

$B = \sqrt{(1-\gamma_1)(1-\gamma_2)\kappa_1(1-\kappa_2)}E_{0L}$

$C = \kappa_1(1-\gamma_1)\sqrt{(1-\gamma_2)\kappa_2}E_0E_{0L}$

$D = (1-\gamma_1)(1-\gamma_2)\sqrt{\kappa_1(1-\kappa_1)\kappa_2(1-\kappa_2)}E_0E_{0L}^2$ and

$E = \sqrt{(1-\gamma_1)(1-\gamma_2)(1-\kappa_1)(1-\kappa_2)}E_0E_{0L}$

The electric fields $E_0$ and $E_{0L}$ are the field circulated within the nanoring at the right and left side of add/drop optical filter. The power output ($P_{t1}$) at through port is written as

$$P_{t1} = |E_{t1}|^2 \quad (10.15)$$

The output field ($E_{t2}$) at drop port is expressed as

$$E_{t2} = \sqrt{(1-\gamma_2)(1-\kappa_2)}E_{i2} - \left[\frac{\sqrt{(1-\gamma_1)(1-\gamma_2)\kappa_1\kappa_2}E_0E_{i1}e^{-\frac{\alpha L}{2} - jk_n\frac{L}{2}} + XE_0E_{0L}E_{i2}\left(e^{-\frac{\alpha L}{2} - jk_n\frac{L}{2}}\right)^2}{1 - YE_0E_{0L}\left(e^{-\frac{\alpha L}{2} - jk_n\frac{L}{2}}\right)^2}\right] \quad (10.16)$$

where

$$X = (1-\gamma_2)\sqrt{(1-\gamma_1)(1-\kappa_1)\kappa_2(1-\kappa_2)}$$

$$Y = \sqrt{(1-\gamma_1)(1-\gamma_2)(1-\kappa_1)(1-\kappa_2)}$$

The power output ($P_{t2}$) at drop port is

$$P_{t2} = |E_{t2}|^2 \tag{10.17}$$

## 10.3 Drug Trapping and Delivery

By using the proposed design, the optical tweezers via PANDA ring resonator and wavelength router can be generated, trapped, transported and stored as shown in Figures 10.1 and 10.2, which can be used to form the microscopic volume transportation, i.e. drug delivery, via the waveguide [3, 4]. Simulation results of the dynamic optical vortices within the PANDA ring are as shown in Figure 10.3. In this case the bright soliton is input into the control port, and the trapped atoms/molecules are as shown in Figure 10.3(a)–(f). The ring radii are $R_{add}$ = 10 µm, $R_R$ = 4 µm and $R_L$ = 4 µm, in which the evidence of the practical device was reported by the authors in reference [28]. $A_{eff}$ are 0.50, 0.25 and 0.25 µm$^2$. In this case, the dynamic tweezers (gradient fields) can be in the forms of bright soliton, Gaussian pulses and dark soliton, which can be used to trap the required microscopic volume. There are five different center wavelengths of tweezers generated, where the dynamical movements are (a) $|E_1|^2$, (b) $|E_2|^2$, (c) $|E_3|^2$, (d) $|E_4|^2$, (e) through port and (f) drop port signals.

More results of the optical tweezers generated within the PANDA ring are as shown in Figure 10.4, where in this case the bright soliton is used as the control port signal, and the output optical tweezers of the through and drop ports with different coupling constants are shown in Figure 10.4 (e) and (f), respectively, the coupling coefficients are (1) 0.1, (2) 0.35, (3) 0.6 and (4) 0.75. The important aspect of the result is that the tunable tweezers can be obtained by tuning (controlling) the add (control) port input signal, in which the required number of microscopic volume (atom/photon/molecule) can be obtained and seen at the drop/through ports, otherwise, they propagate within a PANDA ring before going through the process of decaying or collapsing into the waveguide.

In application, the trapped microscopic volumes can transport into the wavelength router via the through port, while the retrieved microscopic volumes are received via the drop port (connecting target), which can perform the drug delivery applications. The advantage of this system is that the transmitter and receiver can be fabricated on-chip and alternatively operated by a single device. Result of the dynamic tweezers

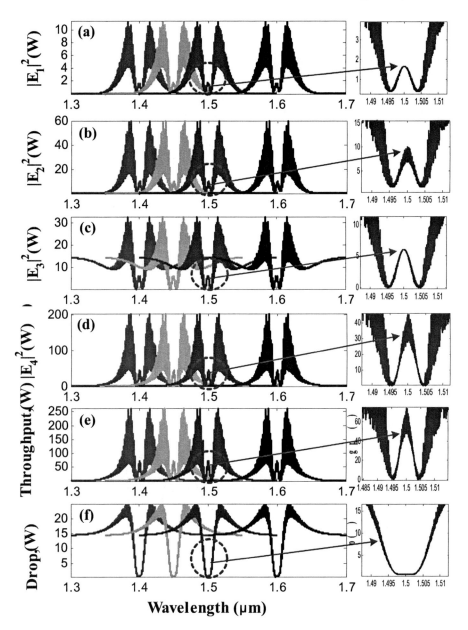

**FIGURE 10.3** Simulation result of four potential wells (optical vortices/tweezers) with four different center wavelengths.

with microscopic volumes is shown in Figure 10.5, where the generated wavelengths are 1.4, 1.45, 1.5, 1.55 and 1.6 µm, in which the manipulation of trapped microscopic volumes within the optical tweezers is as shown in Figure 10.5(d), whereas in this case study, the coupling coefficients are given as $\kappa_0 = 0.1$, $\kappa_1 = 0.35$, $\kappa_2 = 0.1$ and $\kappa_3 = 0.2$, respectively.

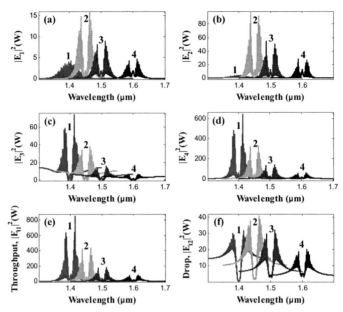

**FIGURE 10.4** Simulation result of the tunable and amplified tweezers by varying the coupling coefficients.

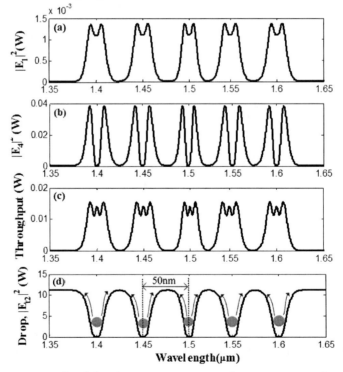

**FIGURE 10.5** Result of the dynamic tweezers with microscopic volumes, where the generated wavelengths are 1.4, 1.45, 1.5, 1.55 and 1.6 μm.

## 10.4 Conclusion

We have shown that the microscopic volumes such as photon, atom, or molecule can be trapped and transported into the optical waveguide by optical tweezers (vortices) by using a PANDA ring resonator and wavelength router. The drug delivery can be transported at long distance. By utilizing the reasonable dark soliton input power, the dynamic tweezers can be controlled and stored within the system. The obtained tweezer spacing with free spectrum range (FSR) of 50 nm is achieved. Moreover, the tweezer amplification is also available by using the nanoring resonators and the modulated signals via the control port as shown in Figure 10.4. In conclusion, we have also shown that the use of a transceiver to form the long distance microscopic volume transportation by using the proposed system, in which the drug delivery to the required (connecting) targets can be performed via the wavelength router.

## REFERENCES

[1] A. Rohrbach and Ernst H.K. Stelzer," Trapping forces, force constants, and potential depths for dielectric spheres in the presence of spherical aberrations", *Appl. Opt.*, vol. 41, pp. 2494–2507 (2002).

[2] K. Uomwech, K. Sarapat, and P.P. Yupapin, "Dynamic modulated Gaussian pulse propagation within the double PANDA ring resonator system", *Microw. and Opt. Technol. Lett.*, vol. 52, pp. 1818–1821 (2010).

[3] B. Piyatamrong, K. Kulsirirat, Techithdeera, S. Mitatha, and P.P. Yupapin, "Dynamic potential well generation and control using double resonators incorporating in an add/drop filter", *Mod. Phys. Lett. B*, vol. 24, pp. 3071–3080 (2010).

[4] H. Cai and A. Poon, "Optical manipulation and transport of microparticle on silicon nitride microring resonator-based add-drop devices", *Opt. Lett.*, vol. 35 (17), pp. 2855–2857 (2010).

[5] A. Ashkin, J.M Dziedzic, J.E Bjorkholm, and S. Chu, "Observation of a single-beam gradient force optical trap for dielectric particles", *Opt. Lett.*, vol. 11, pp. 288–290 (1986).

[6] K. Egashira, A. Terasaki, and T. Kondow, "Photon-trap spectroscopy applied to molecules adsorbed on a solid surface: probing with a standing wave versus a propagating wave", *Appl. Opt.*, 49, 1151–1157 (2010).

[7] A.V. Kachynski, A.N. Kuzmin, H.E. Pudavar, D.S. Kaputa, A.N. Cartwright, and P.N. Prasad, "Measurement of optical trapping forces by use of the two-photon-excited fluorescence of microspheres", *Opt. Lett.*, vol. 28, pp. 2288–2290 (2003).

[8] M. Schulz, H. Crepaz, F. Schmidt-Kaler, J. Eschner, and R. Blatt, "Transfer of trapped atoms between two optical tweezer potentials", *J. Modern Opt.*, 54, 1619–1626 (2007).

[9] A. Ashkin, "Optical trapping and manipulation of neutral particles using lasers," *Proc. Natl. Acad. Sci.*, vol. 94, pp. 4853–4860 (1997).

[10] Lindsey J.E. Anderson, E. Hansen, E.Y. Lukianova-Hleb, H.H Jason, D.O. Lapotko, and J. Lindsey, "Optically guided controlled release from liposomes with tunable plasmonic nanobubbles", *J. Controlled Release*, vol. 144, pp. 151–158 (2010).

[11] H. Shangguan, W.L. Casperson, A. Shearin, K.W. Gregory, and S.A. Prahl, "Drug delivery with microsecond laser pulses into gelatin", *Appl. Opt.*, 35, 3347–3358 (1996).

[12] M. Biondi, F. Ungaro, F. Quaglia, and P.A. Netti, "Controlled drug delivery in tissue engineering", *Adv. Drug Deliv. Rev.*, vol. 60, pp. 229–242 (2008).

[13] N.V. Majeti and Ravi Kumar,"Nano and Microparticles as Controlled Drug Delivery Devices", *J. Pharm. Pharmaceut. Sci.*, vol. 3(2), pp. 234–258 (2000).

[14] M.Z. hang, T. Tarn, and Ning Xi, "Micro-/Nano-devices for Controlled Drug Delivery", in *Proceeding of the International Conference on Robotics 6 Automation*, New Orleans. LA., pp. 2068–2063 (2004).

[15] G. Huanga, J. Gaoa, Z. Hua, John V. St. Johnb, C.P. Bill, and D. Morob, "Controlled drug release from hydrogel nanoparticle networks", *J. Controlled Release*, vol. 94, pp. 303–311 (2004).

[16] J. Hu, S. Lin, L.C. Kimerling, and K. Crozier, "Optical trapping of dielectric nanoparticles in resonant cavities", *Phys. Rev. A*, vol. 82, p. 053819 (2010).

[17] U. Troppenz, M. Hamacher, D.G. Rabus, and H. Heidrich, "All-active InGaAsP/InP ring cavities for widespread functionalities in the wavelength domain", *Proc. 14th Internat. Conf. Indium Phosphide and Related Materials* (IPRM'02), Stockholm, Sweden, pp. 475–478 (2002).

[18] S. Mikroulis, E. Roditi, and D. Syvridis, "Direct modulation properties of 1.55 μm InGaAsP/InP Microring Lasers", *J. Lightwave Technol.*, vol. 26(2), pp. 251–256 (2008).

[19] D.B. Wolfe, R.S. Conroy, P. Garstecki, B.T. Mayers, M.A. Fischbach, K.E. Paul, M. Prentiss, and G.M. Whitesides, "Dynamic control of liquid-core/liquid-cladding optical waveguides", *Proc. Natl. Acad. Sci.* vol. 101(34), pp. 12434–12438 (2004).

[20] T. Phatharaworamet, C. Teeka, R. Jomtarak, S. Mitatha, and P.P. Yupapin, "Random binary code generation using dark-bright soliton conversion control within a PANDA ring resonator", *Lightw. Technol.*, vol. 28(19), pp. 2804–2809 (2010).

[21] K. Svoboda and S.M. Block, "Biological applications of optical forces", *Annu. Rev. Biophys. Biomol. Struct.* 23, 247–283 (1994).

[22] K. Dholakia and M. Gu,"Optical micromanipulation", *Chem. Soc. Rev.*, vol. 37, pp. 42–55 (2008).

[23] M. Tasakorn, C. Teeka, R. Jomtarak, and P.P. Yupapin, "Multitweezers generation control within a nanoring resonator system", *Opt. Eng.*, vol. 49, p. 075002 (2010).

[24] S. Mitatha, N. Pornsuwancharoen, and P.P. Yupapin, "A simultaneous short wave and millimeter wave generation using a soliton pulse within a nano-waveguide", *IEEE J. of Photon. Technol. Lett.*, vol. 21, pp. 932–934 (2009).

[25] Y. Kokubun, Y. Hatakeyama, M. Ogata, S. Suzuki, and N. Zaizen, "Fabrication technologies for vertically coupled microring resonator with multilevel crossing busline and ultracompact-ring radius", *IEEE J. of Sel. Top. Quantum Electron.*, vol. 11, pp. 4–10 (2005).

[26] P.P. Yupapin and W. Suwancharoen, "Chaotic signal generation and cancellation using a microring resonator incorporating an optical add/drop multiplexer", *Opt. Commun.*, vol. 280(2), pp. 343–350 (2007).

[27] P.P. Yupapin, P. Saeung, and C. Li, "Characteristics of complementary ring-resonator add/drop filters modeling by using graphical approach", *Opt. Commun.*, vol. 272, pp. 81–86 (2007).

[28] J. Zhu, S.K. Ozdemir, Y.F. Xiao, L. Li, L. He, D.R. Chen, and L. Yang, "On-chip single nanoparticle detection and sizing by mode splitting in an ultrahigh-Q microresonator", *Nature Photonics*, vol. 4, 46–49 (2010).

# 11
# Hybrid Transistor

**CHAPTER OUTLINE**

- Introduction
- All-Optical Photonic Transistor
- Single-Photon Transistor
- Single-Atom Transistor
- Single-Electron Transistor
- Single-Molecule Transistor
- Photonic Transistor using PANDA Ring
- Conclusion
- References

## 11.1 Introduction

The chapters [1–10] have reported a novel kind of "optical transistor", and obtained substantial publicity for it. However, it made us wonder whether this device really deserves that term. Surprisingly, although many people talk about such things, few seem to care about the semantics. We couldn't find a clear definition. If anybody has seen one, we are keen to learn about it. Obviously, "optical transistor" should mean some kind of photonic analog of an electronic transistor, being able to perform similar functions. Everybody would agree that such a device allows one to control some voltage or current with another voltage or current. But that's not all – there is another essential detail: the controlled voltage or current must be larger than the control voltage or current. In other words, there is amplification. For a standard bipolar transistor, there is e.g. a substantial current gain, defined as the controlled emitter current divided by the required base current. Note that the feature of amplification is essential not only when we use transistors to build amplifiers. It is also required for use in logic circuits. If some logic gate would have to switch a current of 1 µA and need 10 µA gate current to do that, one could obviously not connect many gates in a series. The optical transistor is usually considered important because it would allow the sophisticated optical circuits being constructed. So we see that amplification will be important just as it is for the electronic transistor. It is not sufficient simply to influence one light beam with another beam. In addition, it is required that several photons can be controlled per input photon. Unfortunately, the device described in the above-mentioned previous chapter doesn't have that property. More precisely, it does exhibit some kind of amplification of the controlled beam, but not in the sense discussed above: you need at least one control photon to switch one photon at the output. Note that other kinds of optical transistor [11–15] have been described, which do have the feature of amplification and thus deserve that name. In the concrete case, the label "optical transistor" for the presented device appears to have already caused confusion. Just look at the many reports describing that experiment as a milestone on the way to optical computing, apparently not recognizing that such kind of optical gates could not be combined arbitrarily because they lack amplification. So the label "optical transistor" appears to have raised expectations which cannot be met – not just due to an imperfect implementation, but as a consequence of the method used. Obviously, such confusion should be avoided in science. We have written this comment as an attempt to mitigate that confusion somewhat.

## 11.2 All-Optical Photonic Transistor

Yanik et al. [7] demonstrated an all-optical transistor action in a nonlinear photonic crystal, in which the transmission of a signal can be reversibly switched ON or OFF by a control input with high contrast and low power as shown in Figure 11.1 by creating a nonlinear optical switch with this geometry, introducing Kerr nonlinearity to the rod at the center of the cavity.

Song and Popescu [10] have presented a novel optoelectronic device, called a photonic transistor, and investigated its properties as they pertain to digital electronic applications. The photonic transistor design is based on a heterojunction configuration similar to a semiconductor optical amplifier, and can be used to construct either N-nary digital logic gates or binary Boolean logic gates. The properties of the photonic transistor are investigated using computer simulations, and the numerical results indicate that such a device can be successfully used in digital optoelectronic applications.

One approach to the construction of N-nary photonic logic gates is to use a device that is capable of on/off switching functions for optical signals in a manner similar to the switching operation of a traditional

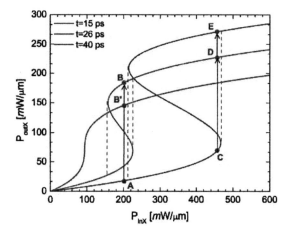

**FIGURE 11.1** Input versus output power for the output power in waveguide. Blue, red, and green curves correspond to the control output power of 0, 151, and 75 mW/µm, respectively, which is appropriate for various times in the switching process [7].

electronic transistor. Based on the operational requirements of the NDP system, the switching capabilities of this device should be controlled by an optical signal covering a domain of wavelengths wide enough to allow for the practical implementation of N-nary logic. With such a device, one can construct the photonic gates that are necessary for the operational and practical implementation of the N-nary digital logical system described earlier. Due to the operational analogy of such a device with its electronic counterpart – the electronic transistor – we have considered it appropriate to call it a *photonic transistor* (PT).

The functional configuration of the photonic transistor is depicted in Figure 11.2. As can be seen from this figure, the photonic transistor consists of a semiconductor core that is similar in structure to a wideband traveling wave SOA structure with two input and one output ports.

The present work, as shown in Figure 11.2, has investigated a new optoelectronic device called an N-nary photonic transistor (PT), as well as its application to the construction of digital O-AND gates.

A class of novel all-optical transistor devices based on a new approach of using optically-controlled gain and absorption to change optical interference resulting in efficient all-optical switching action with very low operating power and switching gain. This class of device can be seen as close as analogue of transistors for photons that can be used to provide a wide variety of all-optical operations including all-optical switching, wavelength conversion, and logical operations. The sizes are small, with area of tens of square micrometers and high operating speed of 10–100 Gb/s (or higher depending on design). This class of photonic transistors is referred to as GAMOI Photonic Transistors [8] (GAMOI stands for Gain and Absorption Manipulation of Optical Interference).

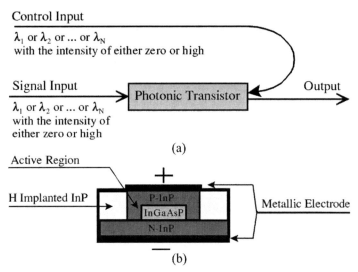

**FIGURE 11.2** (a) The configuration of the photonic transistor. (b) The transistor design for a traditional InP/InGaAsP/InP buried stripe SOA with InP layer ($n_E$ = 3.167) surrounding the InGaAsP active core ($n_A$ = 3.22) [10].

Beside their high speed, all-optical devices could help to substantially reduce the power consumption of current optical network equipments involving optical-electrical-optical (OEO) conversions [16]. The all-optical devices of interest for high-speed networks include all-optical wavelength converters, all-optical switches, and all-optical logic gates. These all-optical devices will also enable ultrafast all-optical signal processing and computing. The reason why all-optical devices have not yet seen large insertion in practice is because they are still marginal in performance for the desired device properties.

The desired properties for an all-optical switching device include [17]:

(a) High speed: High operating speed of 10–100 Gbit/s or faster.
(b) Low switching power: Low switching power of several mW so that on-chip semiconductor lasers can be used.
(c) Low throughput loss: Low throughput loss or signal gain (the output or "switched" signal is of comparable or higher power than input signal), which is important for multiple device cascadability.
(d) Switching gain: Switching gain (the control signal or "pump" is of lower power than the input signal being switched), which is important for efficient wavelength conversions or logic gates.
(e) Small device size: Device size of <500 µm so that integration density can reach 10–100 device/cm$^2$.
(f) Pulse reshaping: Some form of pulse reshaping so that the output pulse will not be worse than the input pulse in terms of pulse shape and frequency chirp.

All optical switching devices with switching gain are often compared to electronic transistors and may be called photonic transistors. Figure 11.3 shows the basic scheme of a photonic transistor. The device operates with two input port and one output port. The two input port include one power supply port and one input signal port, and the output port is where the output signal is drawn. The device can have two different operation modes. In the first operation mode, the power supply beam will be a continuous-wave optical beam, which we will refer to as the "DC" mode of operation for the photonic transistor. The power supply beam will be switched out to the output signal port when the input signal pulse is present. In the second mode of operation, the power supply beam itself is in pulse form. Only when both the power supply beam and the input signal beam are present, the power supply beam will be switched out to the output signal port.

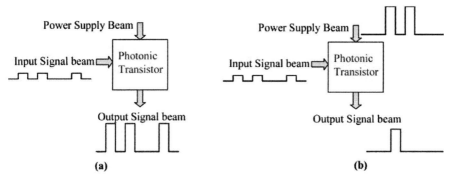

**FIGURE 11.3** (a) Schematics of a photonic transistor in DC mode. (b) Schematics of a photonic transistor in pulse mode [8, 17].

The difference between the photonic transistor and a typical all-optical switching device is that in both operation modes, the optical power of the input signal beam is less than the power supply beam and the output signal beam. So the device will have net signal gain, resulting in the optical circuit having high cascadability.

Specifically, the coupled waveguides based Energy-Up Photonic Transistor (EUPT) device is shown in Figure 11.4(a). It has an unusual geometry in that only one arm of the two coupled waveguides is transparent (passive) while the other arm contains active medium $M$ that is typically absorptive, but is amenable to optical excitation. The coupled waveguides have various input/output ports labeled as PS-IN (power-supply in), SIG-IN/PS-OUT (signal in/power-supply out), and SIG-OUT (signal out). The coupler's length is chosen to be a full coupling length $L_C$ for input wavelength $\gamma_H$ from PS-IN.

Let us first discuss how AMOI works. To illustrate the principle of optical switching via Absorption Manipulation of Optical Interference (AMOI), we simulated the launching of a CW input beam with power $P_{PS-IN}$ into PS-IN. We then vary the absorption coefficient $\alpha$ of medium $M$ and plot the output powers from PS-OUT and SIG-OUT ($P_{PS\text{-}OUT}$ and $P_{SIG\text{-}OUT}$) as a function of $\alpha L_C$ in Figure 11.4(b), for the case where the input intensity $I_{PS\text{-}IN}$ is low compared to the saturation intensity $I_{SAT}$ of the medium (intensity $I=P/A_{WG}$, where $P$ is the beam power and $A_{WG}$ is the waveguide mode area). The normalized outputs depend only on the product $\alpha L_C$ and are independent on the individual values of $\alpha$ and $L_C$. We see that when $\alpha = 0$, we have $P_{PS\text{-}OUT} = P_{PS\text{-}IN}$ and $P_{SIG\text{-}OUT} = 0$ since input beam entering the lower transparent waveguide will be coupled to the top waveguide completely [Figure 11.4(c)]. When $\alpha L_C$ increases from 0 to 5, $P_{SIG\text{-}OUT}$ increases while $P_{PS\text{-}OUT}$ decreases, and the total output decreases as a result of the absorption. When $\alpha L_C$ increases to 50, $P_{SIG\text{-}OUT}$ reaches 80% of the input power [Figure 11.4(d)]. At this point, the system actually sees little loss. This is because at high absorption, the small energy leakage to the top waveguide has no chance to build up constructively, resulting in little energy transfer to the top waveguide and hence little absorption. Our applications will operate around $\alpha L_C \sim 5$. Typical direct-gap semiconductor can have $\alpha \sim 0.5$ μm$^{-1}$, resulting in compact device size of $L_C \sim 10$ μm.

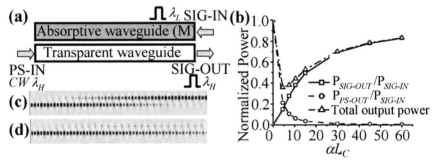

**FIGURE 11.4** (a) EUPT structure; (b) output from EUPT two ports as a function of $\alpha L_C$; (c) electric field showing output to PS-OUT when top waveguide is transparent; (d) output to SIG-OUT when top waveguide is absorptive [8, 17].

In all-optical operation, the absorption coefficient of the medium in Figure 11.15(a) is controlled via a fast control-signal pulse with $P_{SIG\text{-}IN}$ entering the right side of the top waveguide. The medium may be modeled as a multi-level type medium with a bandgap energy $E_G$ for the two lowest levels, and can be provided by a semiconductor [18, 19]. A separate CW beam $P_{PS\text{-}IN}$ at $\gamma_H$, with energy $E_H > E_G$, enters PS-IN. This CW beam functions as a "power supply" beam and is labeled as PS. For the desired operation, we make $P_{PS\text{-}IN}$ to be substantially higher than the medium saturation power $P_{SAT}$ so that it will initially saturate $M$ (i.e. pump $M$ to transparency at $\gamma_H$) and be fully coupled to PS-OUT resulting in $P_{PS\text{-}OUT} = P_{PS\text{-}IN}$. Transparency at $\gamma_H$ means that $M$ is having gain for a lower energy beam. When a short signal pulse $P_{SIG\text{-}IN}$ at $\gamma_L$ with a lower energy $E_L < E_H$ (still above $E_G$) enters SIG-IN/PS-OUT, it will see gain and de-excite $M$ via stimulated emission. This causes $M$ to decrease at $\gamma_H$ and the $P_{PS-IN}$ beam will see losses in the upper waveguide. Due to AMOI, the $P_{PS-IN}$ beam will be partially channeled to exit the lower waveguide at SIG-OUT with a short pulse $P_{SIG-OUT}$ at $\gamma_H$. Thus, a signal pulse at $\gamma_L$ entering SIG-IN/PS-OUT will produce an output pulse from SIG-OUT with $\gamma_H < \gamma_L$, resulting in energy-up conversion. As shown below, the switched out pulse at SIG-OUT can actually have higher power than the input pulse to SIG-IN, resulting in signal amplification. Note in particular that the power supply beam PS-IN to be switched would not self switched through irrespective of how high its power ($P_{PS-IN}$) is. As mentioned above, this is important for the achieving switching gain needed for obtaining signal amplification. As discussed below, the medium of this device has no optical gain at the switched-out wavelength (in fact the medium is absorptive). Hence, the signal amplification is simply due to the switching gain (a weak pulse is used to switch a stronger "power-supply" optical beam).

The simulation assumes the following typical semiconductor parameters: spontaneous decay time $\tau_{SP}$ = 1nsec, intraband relaxation time $\tau'$ = 100 fsec, dipole transverse relaxation rate $\delta\omega = 3.9 \times 10^{13}$ Hz, and a ground-state electron population density that gives an on-resonance absorption coefficient of $\alpha = 0.6$ µm$^{-1}$ and $I_{SAT}$ = 1 kW/cm$^2$, which is within typical experimental values [20].

As an example of the all-optical operation, we assume $L_C$ = 15 µm. The structure is a single-mode semiconductor waveguide with a high refractive index core of $n$ = 3.4 surrounded by $n$ = 1.45 cladding materials (e.g. SiO$_2$) similar to the ones shown in [21]. For the simulation, we assume TM modes, a waveguide width of 0.25 µm, and a waveguide height of 0.35 µm, which give a calculated mode area $A_{WG}$ of 0.043 µm$^2$ (for $\gamma_H$ = 1450 nm) or $P_{SAT}$ = 0.43 µW. We take $\gamma_L$= 1550 nm, $\gamma_H$ = 1450 nm, $P_{PS-IN}$ = 3,000, $P_{SAT}$ = 1.29 mW. At SIG-IN, we send in a 50 psec pulse (20 Gb/s) with a varying pulse power $P_{SIG-IN}$ and plot $P_{SIG-OUT}$ versus $P_{SIG-IN}$ [Figure 11.5(a)] in which we see that the device signal output has a quasi-linear regime with a maximal amplification $G_S > 10$ at $P_{SIG-IN}$ = 0.014 mW, followed by a transition region, and then a saturation region ($P_{SIG-IN}$ = 0.15–1.5 mW) in which the output changes by only 10% with the input changes by >10×.

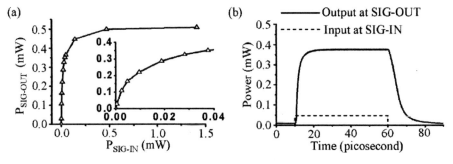

**FIGURE 11.5** (a) $P_{SIG-OUT}$ at different $P_{SIG-IN}$ with constant $P_{PS-IN}$; (b) dynamical simulation with a 50psec input pulse [8, 17].

The temporal shapes for the SIG-IN and SIG-OUT pulses at the quasi-linear regime are shown in Figure 11.5(b). Note that these devices work with a much shorter interacting length. As a result, the carrier induced refractive index variations do not cause any appreciable cross-phase or self-phase modulation. Hence, the pulses switched out remained basically undistorted spectrally.

While EUPT outputs pulse at $\gamma_H$ with input at $\gamma_L$, Energy-Down Photonic Transistor (EDPT) will be able to convert the pulse back to a wavelength $\gamma_L'$ near or at $\gamma_L$. The EDPT device is also in the form of two coupled waveguide arms as shown in Figure 11.6(a). One of the waveguide arms is transparent and the other half transparent and half with gain medium $M'$. For the EDPT, a CW beam $P_{PS\text{-}IN}$ at $\gamma_L$ and energy $E_L$ above the bandgap energy $E_G$ of the active medium enters PS-IN. When the medium gain is below zero, this beam will initially be fully coupled to PS-OUT, resulting in an output power $P_{PS\text{-}OUT} = P_{PS\text{-}IN}$. When a signal pulse $P_{SIG\text{-}IN}$ at shorter wavelength $\gamma_H$ enters SIG-IN, it will excite $M'$. This causes the medium $M'$ to achieve gain at $\gamma_L$ and the $P_{PS\text{-}IN}$ beam will see gain in the upper waveguide. Due to the Gain Manipulation of Optical Interference (GMOI) action discussed below, the $P_{PS\text{-}IN}$ beam will then be channeled to exit the upper waveguide at SIG-OUT with a short output pulse $P_{SIG\text{-}OUT}$ at $\gamma_L$. The result is that a signal pulse at $\gamma_H$ entering SIG-IN will produce an output pulse at SIG-OUT with a higher energy at $\gamma_L$.

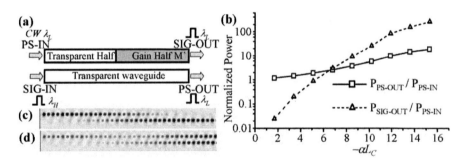

**FIGURE 11.6.** (a) EDPT structure; (b) output from two ports as a function of $\alpha L_C$; (c) electrical field showing when the gain half is transparent; (d) when the gain half has gain = 0.5/μm for the 13 μm long coupler [8, 17].

To illustrate the principle of optical switching via Gain Manipulation of Optical Interference (GMOI), we let the gain coefficient of $M'$ to be $-\alpha$ and plot in Figure 11.6(b) $P_{PS\text{-}OUT}$ and $P_{SIG\text{-}OUT}$ versus $-\alpha L_C$ when $I_{PS\text{-}IN}$ is low compared to $I_{SAT}$. We see that when $\alpha = 0$ so that $-\alpha L_C = 0$, we have $P_{PS\text{-}OUT} = P_{PS\text{-}IN}$ and $P_{SIG\text{-}OUT} = 0$. When $-\alpha L_C$ ($\alpha < 0$ indicates gain) increases as the gain goes higher, $P_{SIG\text{-}OUT}$ will increase faster than $P_{PS\text{-}OUT}$. After $-\alpha L_C$ reaches 7, $P_{SIG\text{-}OUT}$ is actually higher than $P_{PS\text{-}OUT}$, which is a good point to operate. Our dynamical simulation with a 50 psec input pulse at 1450 nm and $P_{SIG-IN}$ = 0.43 mW is shown in Figure 11.7(a). The waveguide and medium parameters are the same as those for EUPT. From Figures 11.5(b) and 11.7(a), one sees that both the EUPT and EDPT can operate at a speed of ~100 GHz (10 psec). Figure 11.7(b) plots the output power versus input pulse power at power-supply $P_{PS-IN}$ = 14 µW. We also see that while EDPT has no signal amplification, high-efficiency conversion (> 75%) can be achieved. Note again that the power supply beam would not be self-switched irrespective of how high its power is.

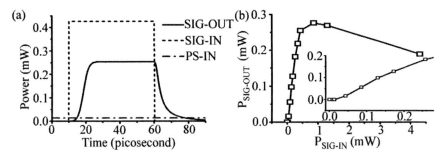

FIGURE 11.7  Dynamical FDTD simulation for the EDPT: (a) Input/output relations for down converters; (b) Output power at different input pulse power [8, 17].

The EUPT and EDPT can be joined in tandem as shown in Figure 11.8(a) to achieve a full photonic transistor (FPT) capable of wide wavelength conversion range, signal amplification, and pulse regeneration with the cascaded input/output relation shown in Figure 11.8(b). The device can operate within the wavelength range from $\gamma_L$ and $\gamma_H$. The EUPT as well as the FPT device behaves similar to an electronic transistor in that it has both a linear and a saturation amplification regime in the response curve. The "power supplies" of EUPT, EDPT, and FPT are CW optical beams, switched to the output via input signals by means of optical absorption or gain. In input-output block diagram form, the Photonic Transistors are like the

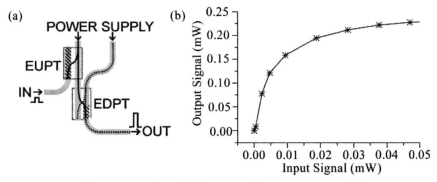

FIGURE 11.8  (a) Layout of a full-function photon transistor; (b) input–output relation [8, 17].

Electronic Transistors but with "power-supply currents" replaced by "CW optical beams". The use of two CW optical beams at different wavelengths provides wavelength conversion and is unique to the optical case. There is no electrical power needed for the Photon Transistors to function. The power-supplies to a photonic transistor circuit are just CW optical beams. Like transistors, logic gates can be formed via cascading EDPTs and EUPTs to realize various all-optical signal-processing functions. In such applications, some of the CW beams will be replaced by other input pulse streams.

## 11.3 Single-Photon Transistor

### 11.3.1 Single-Photon Transistor using Microtoroidal Resonators

A single-photon transistor [2] is a device where the propagation of a single signal photon is under the control of the presence or absence of a single gate photon.

A single-photon transistor can be realized by combining the techniques of state-dependent conditional transmission and single-photon storage. The three-level atom is initialized in state $|g\rangle$ with $\omega_e = \omega_c + h$. Under the action of the control field $\Omega(t)$, the presence or absence of a photon in a "gate" pulse will conditionally flip the internal state of the atom to (remain in) state $|s\rangle$ ($|g\rangle$) during the storage process, and then this conditional flip will affect the propagation of the subsequent "signal" photons arriving at the resonator.

### 11.3.2 Single-Photon Transistor using Nanoscale Surface Plasmons [3, 4]

In analogy with the electronic transistor, a photonic transistor is a device where a small optical 'gate' field is used to control the propagation of another optical 'signal' field via a nonlinear optical interaction. Its limit is the single-photon transistor, where the propagation of the signal field is controlled by the presence or absence of a single photon in the gate field.

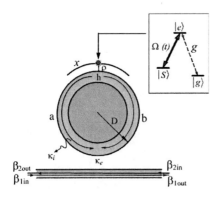

FIGURE 11.9 A schematic diagram of a single-photon transistor composed of a microtoroidal resonator, a tapered fiber, and a three-level atom. State $|s\rangle$ is coupled to $|e\rangle$ by classic control field $\Omega(t)$ and State $|e\rangle$ interacts with state at rate $g$ through the evanescent fields of two internal modes $a$ and $b$, which interacts at rate $h$ by scattering [2].

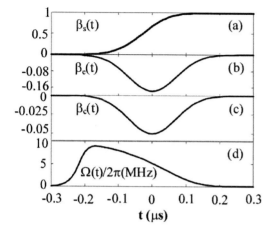

FIGURE 11.10 Numerical simulation of a coherent storage of a single photon in the system comprised a three-level atom, a toroid, and a tapered fiber. (a) Amplitude of the atomic state $c_s$. (b) Amplitude of the atomic state $\beta_e$. (c) Amplitude of the cavity state $\beta_c$. (d) The control field $\Omega(t)$ [2].

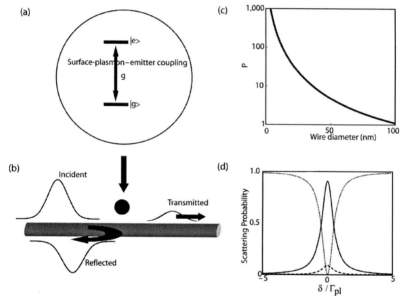

FIGURE 11.11 Interaction of single surface plasmons with a single emitter. (a) Two-level emitter interacting with the nanowire. States $|g\rangle$ and $|e\rangle$ are coupled via the surface-plasmon modes with a strength $g$. (b) Schematic diagram of a single incident photon scattered off a near-resonant emitter. The interaction leads to reflected and transmitted fields whose amplitudes can be exactly calculated. (c) The maximum Purcell factor of an emitter positioned near a silver nanowire ($\varepsilon \approx -50 + 0.6i$) and surrounded by a uniform dielectric ($\varepsilon = 2$), as a function of wire diameter. The plot is calculated using the method in refs [15, 28] and the silver properties used correspond to a free-space wavelength of $\lambda_0 = 1$ μm. (d) Probabilities of reflection (solid line), transmission (dotted line) and loss (dashed line) for a single photon incident on a single emitter, as a function of detuning. The Purcell factor for this system is taken to be $P = 20$ [4].

**270** Nanoscale Nonlinear PANDA Ring Resonator

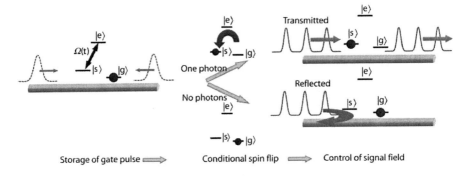

**FIGURE 11.12** A schematic diagram of transistor operation involving a three-level emitter. In the storage step, a gate pulse consisting of zero or one photon is split equally in counter-propagating directions and coherently stored using an impedance-matched control field $\Omega(t)$. The storage results in a spin flip conditioned on the photon number. A subsequent incident signal field is either transmitted or reflected depending on the photon number of the gate pulse, owing to the sensitivity of the propagation to the internal state of the emitter [4].

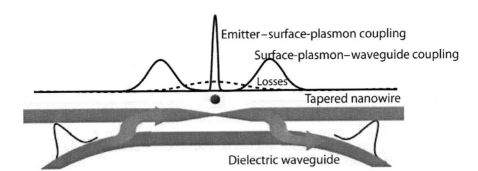

**FIGURE 11.13** A schematic diagram of in- and out-coupling of surface plasmons on a tapered nanowire to an evanescently coupled low-loss dielectric waveguide. Here, a single photon originally in the waveguide is transferred to the nanowire, where it interacts with the emitter before being transferred back into the waveguide. The coupling between the nanowire and waveguide is efficient only when they are phase-matched (in the regions indicated by the blue peaks). The phase-matching condition is poor in the regions of the wire taper and in the bending region of the waveguide away from the nanowire. Dissipative losses (- - - curve) are concentrated to a small region near the nanowire taper, owing to a large concentration of fields here [4].

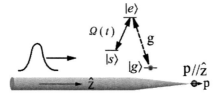

**FIGURE 11.14** Schematic description of coherent storage of a single photon (SP) in the system consists of an emitter and a nanotip. The emitter is initially in the ground state $|g\rangle$ and the dipole moment $p$ of the emitter is parallel to the axis of the nanotip. Under the action of the control field $\Omega(t)$ dependent on the wave packet of the incoming photon, the capture of the incoming single photon may be realized while a state flip from $|g\rangle$ to $|s\rangle$ is induced [3].

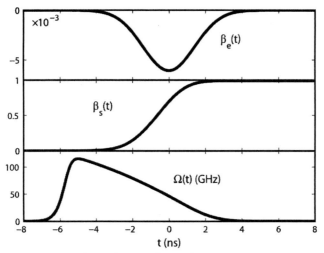

**FIGURE 11.15** Numerical simulation of a coherent storage of a single photon in the system of an emitter and a nanotip. (a) Amplitudes of the state $c_{e1}$. (b) Amplitude of the state $\beta_{s1}$. (c) The control field $\Omega(t)$ [3].

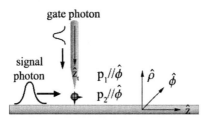

**FIGURE 11.16** A schematic diagram of a single photon transistor. The nanotip is perpendicular to the nanowire. The emitter has four energy levels, with dipole moments $p_1 \parallel \hat{\phi}$ and $p_2 \parallel \hat{\phi}$. The single 'gate' photon propagating along the nanotip is coherently absorbed under the action of the control field $\Omega(t)$, which results a state flip from $|g\rangle$ to $|s\rangle$. This conditional state flip can control the propagation of the 'signal' photon traveling along the nanowire [3].

Combining the techniques of state-dependent conditional reflection and single-photon storage, a single-photon transistor can be realized. First, the emitter is initialized in state $|g\rangle$. Under the action of the control field $\Omega(t)$, the presence or absence of a photon in a 'gate' pulse with frequency $\omega_1$ traveling along the nanotip flips the internal state of the emitter to state $|s\rangle$ or remains in state $|g\rangle$ during the storage process. Then, this conditional flip can control the propagation of subsequent 'signal' photons with frequency $\omega_2$ propagating along the nanowire. Thus, the interaction of subsequent signal pulse and the emitter depends on the internal state of emitter after the storage. If the emitter is in the state $|g\rangle$, the signal field is near, completely reflected by the emitter. Otherwise, the emitter is in the state $|s\rangle$, then the field is near-completely transmitted because $|s\rangle$ does not interact with the surface plasmon. The storage and conditional spin flip makes the emitter either highly reflecting or completely transparent depending on the gate field containing none or one single-photon. Thus, the presence or absence of a single incident photon in a 'gate' field is sufficient to control the propagation of the subsequent 'signal' field, and the system therefore can serve as an efficient single-photon switcher or transistor.

We have presented a scheme for a single-photon transistor, where the 'gate' field propagates along a nanotip and the 'signal' field travels along a nanowire perpendicular to the nanotip. A single 'gate' photon can control the propagation of a single 'signal' photon through changing the internal state of an emitter assisted by classic control field. This transistor may find many important applications in areas such as efficient single-photon detection and quantum information science. Based on this scheme, the controlled-phase gate for photons can be made; furthermore, a CNOT gate which is a key part of an optical quantum computer is available. This system may also be a promising candidate for realizing electromagnetically induced transparency-based nonlinear schemes.

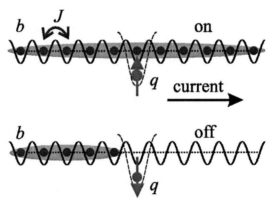

**FIGURE 11.17** A single-atom-transistor in a 1D optical lattice. A single spin-1/2 impurity atom $q$ separately trapped at a particular lattice site is transparent to probe atoms $b$ in one state ("on"), but in the other acts as a single atom mirror ("off") [6].

## 11.4 Single-Atom Transistor

Recently, a single-atom transistor (SAT) setup [5] has been proposed and numericals [6] provide new opportunities to experimentally examine the coupling of a spin-1/2 system with bosonic and fermionic modes.

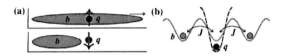

**FIGURE 11.18** (a) A spin-1/2 impurity used as a switch: in one spin state it is transparent to the probe atoms, but in the other it acts as a single atom mirror. (b) Implementation of the SAT as a separately trapped impurity $q$ with probe atoms $b$ in an optical lattice [5].

We consider a spin-1/2 atomic impurity which is used to switch the transport of either a 1D Bose–Einstein condensate or a 1D degenerate Fermi gas initially situated to one side of the impurity. In one spin state the impurity is transparent to the probe atoms, while in the other it acts as a single atom mirror, prohibiting transport via a quantum interference mechanism reminiscent of electromagnetically induced transparency (EIT) [Figure 11.1(a)]. Observation of the atomic current passing the impurity can then be used as a quantum nondemolition (QND) measurement of its internal state, which can be seen to encode a qubit, $|\psi_q\rangle = \alpha|\uparrow\rangle + \beta|\downarrow\rangle$. If a macroscopic number of atoms pass the impurity, then the system will be in a macroscopic superposition, $|\psi(t)\rangle = \alpha|\uparrow\rangle|\phi_\uparrow(t)\rangle + \beta|\downarrow\rangle|\phi_\downarrow(t)\rangle$ which can form the basis for a single-shot readout of the qubit spin. Here, $|\phi_\sigma(t)\rangle$ denotes the state of the probe atoms after evolution to time $t$, given that the qubit is in state $\sigma$ [Figure 11.1(a)]. In view of the analogy between state amplification via this type of blocking mechanism and readout with single electron transistors used in solid state systems, refer to this setup as a single atom transistor (SAT).

We propose the implementation of a SAT using cold atoms in 1D optical lattices. We consider probe atoms $b$ to be loaded in the lattice to the left of a site containing the impurity atom, which is trapped by a separate (e.g., spin-dependent) potential [Figure 11.1(b)]. The passage of $b$ atoms past the impurity $q$ is then governed by the spin-dependent effective collisional interaction $\hat{H}_{\text{int}} = \sum_\sigma U_{\text{eff},\sigma} \hat{b}_0^\dagger \hat{b}_0 \hat{q}_\sigma^\dagger \hat{q}_\sigma$.

The quantum interference mechanism needed to engineer $U_{\text{eff}}$ can be produced using an optical or magnetic Feshbach resonance. For the optical case a Raman laser drives a transition on the impurity site 0 from the atomic state $\hat{b}_0^\dagger \hat{q}_\sigma^\dagger |\text{vac}\rangle$ via an off-resonant excited molecular state to a bound molecular state back in the lowest electronic manifold $\hat{m}_\sigma^\dagger |\text{vac}\rangle$ [Figure 11.2(a)]. We denote the effective two-photon Rabi frequency and detuning by $\Omega_\sigma$ and $\Delta_\sigma$, respectively. For the magnetic case, the Hamiltonian will have the same form, but with $\Omega_\sigma$ the coupling between open and closed channels and

$\Delta_\sigma$ the magnetic field detuning. The Hamiltonian for our system is then given ($h \equiv 1$) by $\hat{H} = \hat{H}_b + \hat{H}_0$, with

$$\hat{H}_b = -J\sum_{\langle ij \rangle} \hat{b}_i^\dagger \hat{b}_j + \frac{1}{2}U_{bb}\sum_j \hat{b}_j^\dagger \hat{b}_j \left(\hat{b}_j^\dagger \hat{b}_j - 1\right)$$

$$\hat{H}_0 = \sum_\sigma \left[ \Omega_\sigma \left( (\hat{m}_\sigma^\dagger \hat{q}_\sigma \hat{b}_0 + H.c.) + \Delta_\sigma \hat{m}_\sigma^\dagger \hat{m}_\sigma \right] + \sum_\sigma \left[ U_{qb,\sigma} \hat{b}_0^\dagger \hat{q}_\sigma^\dagger \hat{q}_\sigma \hat{b}_0 + U_{bm,\sigma} \hat{b}_0^\dagger \hat{m}_\sigma^\dagger \hat{m}_\sigma \hat{b}_0 \right]$$

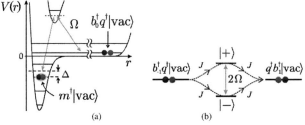

**FIGURE 11.19** (a) An optical Feshbach resonance for a single spin channel ($\Omega = \Omega_\sigma$, $\Delta = \Delta_\sigma$.). One probe atom and the impurity atom, in an atomic state $\hat{b}_0^\dagger \hat{q}_\sigma^\dagger |\text{vac}\rangle$ which is quantized by the trapping potential of the lattice site, are coupled by an optical Feshbach setup to a bound molecular state $m_\sigma^\dagger |\text{vac}\rangle$ of the Born–Oppenheimer potential $V(r)$. The coupling has the effective two-photon Rabi frequency $\Omega$ and detuning $\Delta$. (b) The sequence as (left) a probe atom approaches the impurity site and is located on site $i = -1$, (center) the probe atom is on the impurity site $i = 0$, and (right) the probe atom has tunneled past the impurity and is located on site $i = 1$. Quantum interference occurs in this process because the two dressed states on the impurity site, $|\pm\rangle = \left(\hat{b}_0^\dagger \hat{q}_\sigma^\dagger |\text{vac}\rangle \pm m_\sigma^\dagger |\text{vac}\rangle\right)/\sqrt{2}$, give rise to two separate paths with equal and opposite amplitudes [5, 6].

For a more detailed analysis, we solve the Lippmann–Schwinger equation exactly for scattering from the impurity of an atom $b$ with incident momentum $k > 0$ in the lowest Bloch band. The resulting forwards and backwards scattering amplitudes, $f^{(\pm)}(k)$ respectively, are

$$f^{(\pm)}(k)\left[1 + \left(\frac{iaU_{\text{eff}}(k)}{v(k)}\right)^{\pm}\right]^{-1}$$

where the energy dependent interaction $U_{\text{eff}} = U_{qb} + \Omega^2/[\varepsilon(k) - \Delta]$ and the phase velocity $v(k) = \partial \varepsilon/\partial k = 2Ja \sin ka$. The corresponding transmission probabilities, $T(k) = |f^{(+)}(k)|^2$, are plotted in Figure 11.20(a) as a function for various $\Omega$ and $\Delta$. For $\Omega \sim J$, these are Fano profiles with complete reflection at $\varepsilon(k) = \Delta$ and complete transmission at $\varepsilon(k) = \Delta - \Omega^2/U_{qb}$. The SAT thus acts as an energy filter, which is widely tunable via the laser strength and detuning used in the optical Feshbach setup. For $\Omega > 4J$, $T$ is approximately independent of $k$, and we recover the previous result, i.e., that transport can be completely blocked or permitted by appropriate selection of $\Delta$. Note that this mechanism survives when higher energy Bloch bands are included, and is resistant to loss processes, which are discussed below.

In Figure 11.20(b) we plot the number of particles on the right of the impurity $N_R(t)$ for fermions and for bosons with $U_{bb}/J \to \infty$, starting from a Mott insulator (MI) state with $n = 1$, for $\Delta = 0$, $\Omega/J = 0, 1, 2$. For $\Omega = 0$ the results for bosons and fermions are identical, while for $\Omega/J = 1, 2$ we observe an initial period for the bosons in which the current is similar to that for the fermionic systems, after which the bosons settle into a steady state with a significantly smaller current. The initial transient period for the bosons incorporates the settling to steady state of firstly the molecule dynamics, and secondly the momentum distribution on the right of the impurity. These transients are suppressed if $\Omega$ is ramped slowly to its final value from a large value $\Omega > 4J$.

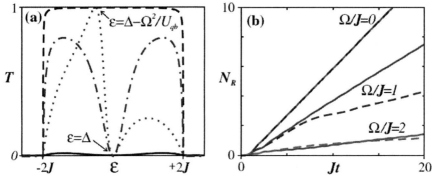

**FIGURE 11.20** (a) SAT transmission coefficients $T \equiv |f^{(+)}|^2$ for a particle $b$ as a function of its energy $\varepsilon(k)$ for $\Omega/J = 4$, $\Delta = 0$, $U_{qb}/J = 0$, (solid line), $\Omega/J = 8$, $\Delta/J = 4$, $U_{qb}/J = 2$ (dashed line), $\Omega/J = 1$, $\Delta = 0$, $U_{qb}/J = 2$ (dotted line), and $\Omega/J = 1$, $\Delta = 0$, $U_{qb}/J = 0$ (dash-dotted line). (b) The number of particles to the right of the impurity, $N_R(t)$, from exact numerical calculations for bosons in the limit $U_{qb}/J \to \infty$ (dashed lines) and fermions (solid lines) in a 1D Mott insulator state with $n = 1$, for $\Delta = 0$, $\Omega/J = 0, 1, 2$ [5].

## 11.5 Single-Electron Transistor

Rodrigues and Armour [22] have analyzed the quantum dynamics of a nanomechanical resonator coupled to a normal-state single-electron transistor (SET). Starting from a microscopic description of the system, we derive a master equation for the SET island charge and resonator which is valid in the limit of weak electromechanical coupling. Using this master equation we show that, apart from brief transients, the resonator always behaves like a damped harmonic oscillator with a shifted frequency and relaxes into a thermal-like steady state. Although the behavior remains qualitatively the same, we find that the magnitude of the resonator damping rate and frequency shift depend very sensitively on the relative magnitudes of the resonator period and the electron tunneling time. Maximum damping occurs when the electrical and mechanical timescales are the same, but the frequency shift is greatest when the resonator moves much more slowly than the island charge. We then derive reduced master equations which describe just the resonator dynamics. By making slightly different approximations, we obtain two different reduced master equations for the resonator. Apart from minor differences, the two reduced master equations give rise to a consistent picture of the resonator dynamics which matches that obtained from the master equation including the SET island charge.

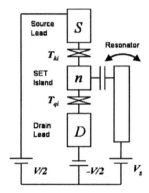

FIGURE 11.21 A schematic circuit diagram for the SET resonator system [22].

The nanomechanical resonator–SET system we consider is shown schematically in Figure 11.21. The SET consists of a metallic island, with total capacitance $C_\Sigma$, connected via tunnel junctions to two leads, with a bias voltage $V$ applied across it. The nanomechanical resonator is adjacent to the SET island and is coated with a metal layer so that it forms a mechanically compliant voltage gate to which a voltage $V_g$ as shown in Figure 11.22. Motion of the resonator modulates the gate capacitance and hence the charging energy of the SET island, while changes in the charge on the SET island modulate the equilibrium position of the resonator. The resonator has a mass $m$ and is treated as a single-mode harmonic oscillator with frequency $\omega_0$.

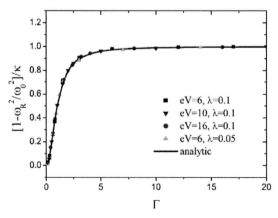

FIGURE 11.22 Plot of the relative frequency shift (squared) as a function of $\Gamma$. The junctions are chosen to have equal resistance $g_D = g_S = g$, and we choose $n'_g = 1/2$ (the degeneracy point). The voltages given for the various data sets are measured in units of $\hbar\omega_0$ [22].

The quantum master equations we derive for the SET–resonator system will provide useful tools for further investigations into the quantum dynamics of this system. In particular, we can use them as a starting point for investigations of how an individual SET and resonator (rather than an ensemble of such systems) evolves. However, it will also be interesting to try to extend the ensemble-averaged master equations derived here to obtain descriptions of the low-bias and strong-coupling regimes where the close connection between quantum and classical treatments found here is unlikely to persist.

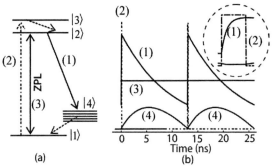

**FIGURE 11.23** (a) Energy level scheme of a molecule with ground state ($\langle 1|$), and ground ($\langle 2|$) and first excited ($\langle 3|$) vibrational states of the first electronic excited state. Manifold $\langle 4|$ shows the vibronic levels of the electronic ground state, which decay rapidly to $\langle 1|$. (1) arrow, excitation by the gate beam; (3) double-headed arrow, coherent interaction of the CW source beam with the zero-phonon line (ZPL); (2) arrow, Stokes-shifted fluorescence; dashed arrows, non-radiative internal conversion. (b) Time-domain des-cription of laser excitations and corresponding response of the molecule simulated by the Bloch equations with periodic boundary conditions. (1) spikes and (2) curve represent the pump laser pulses and the population of the excited state $\langle 2|$, respectively. (4) curve shows the time trajectory of Im($\rho_{21}$). Straight (3) line indicates the constant probe laser intensity that is on at all times. Inset, magnified view of curves during a laser pulse [1].

## 11.6 Single-Molecule Transistor

Hwang et al. [1] have presented the transistor is one of the most influential inventions of modern times and is ubiquitous in present-day technologies. In the continuing development of increasingly powerful computers as well as alternative technologies based on the prospects of quantum information processing, switching and amplification functionalities are being sought in ultra small objects, such as nanotubes, molecules or atoms [3, 5, 23–29]. Among the possible choices of signal carriers, photons are particularly attractive because of their robustness against decoherence, but their control at the nanometer scale poses a significant challenge as conventional nonlinear materials become ineffective. To remedy this shortcoming, resonances in optical emitters can be exploited, and atomic ensembles have been successfully used to mediate weak light beams [29]. However, single-emitter manipulation of photonic signals has remained elusive and has only been studied in high-finesse microcavities [30–31] or waveguides [3, 32]. Here we demonstrate that a single dye molecule can operate as an optical transistor and coherently attenuate or amplify a tightly focused laser beam, depending on the power of a second 'gating' beam that controls the degree of population inversion. Such a quantum optical transistor has also the potential for manipulating non-classical light fields down to the single-photon level as shown in Figure 11.23. We discuss some of the hurdles along the road towards practical implementations, and their possible solutions.

The data points in Figure 11.24 summarize the transistor characteristic curve by plotting the visibility (the ratio of the peak or dip magnitude to the off-resonance signal) for the transmission of the source as a function of the gate power.

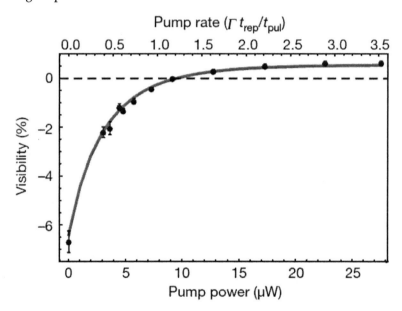

**FIGURE 11.24** Attenuation and amplification of a laser beam by a single molecule. Filled circles, visibility of the transmission spectra with respect to the pump power. Solid curves in all panels represent the results Error bars determined by propagating errors based on the 95% confidence intervals of the amplitude and background in Lorentzian fits the experimental spectra [1].

The operation realm of single-emitter quantum optical transistors can be extended in several ways. We have shown that such a device can control a source beam that has a weaker flux than its saturated emission rate. To push the power range of the source to higher values, one could enhance the excited-state spontaneous emission rate by coupling to nano-antennas [33, 34] or simple microcavity geometries [35]. This would also diminish the impact of Stokes-shifted fluorescence, which reduces the coherent scattering cross-section of dye molecules. At the other extreme, because the geometric coupling of a light beam to a single emitter can reach unity [36, 37], it is expected that single-emitter transistors can also manipulate non-classical source beams with only a few photons. Indeed, preliminary experiments have demonstrated 10% attenuation of a light beam with as few as about 1,000 photons per second [38]. In addition, the gate power can be reduced to only a few tens of photons per pulse if the molecular population is inverted in a resonant pumping scheme through the narrow ZPL [39], instead of the broad $\langle 1| \leftrightarrow \langle 3|$ transition. Such a coherent pumping could also prepare superpositions of states $\langle 1|$ and $\langle 2|$, which would be then stamped onto photons scattered in the forward and backward directions. Given that a molecule occupies only a few nanometres in the solid state, it should be possible to package a huge number of single-emitter transistors in a nanoscopic area and exploit the inhomogeneous distribution of frequencies to operate many signals simultaneously.

## 11.7 Photonic Transistor using PANDA Ring

Recently, Yupapin et al. [40–46] have presented an all-optical device, which consists of an add/drop optical filter and is known as a PANDA ring resonator using optical dark-bright soliton control within a semiconductor add/drop multiplexer which has promising applications. It has been investigated clearly by the authors in references [47, 48]. One of the advantages is that the dark soliton peak signal is always at a low level, which is useful for secured signal communication in the transmission link. The other is formed when the high optical field is configured as an optical tweezer or potential well [49, 50].

In this section, we present a novel system of the photonic transistor using PANDA ring resonator by using a dark-bright soliton pulses propagating within an add/drop optical multiplexer (PANDA ring). The multiplexing signals with slightly different wavelengths of the dark solitons are controlled and amplified within the system. The dynamic behaviors of dark bright soliton interaction are analyzed and described. Finally, the use of optical switching to form a hybrid photonic transistor using the Gaussian controlled at the add port is discussed in detail.

We are looking for a stationary dark soliton pulse, which is introduced into the add/drop optical filter system as shown in Figure 11.25. The input optical field ($E_{in}$) and the add port optical field ($E_{add}$) of the dark, bright soliton or Gaussian pulses are given by [51].

$$E_{in}(t) = A \tanh\left[\frac{T}{T_o}\right]\exp\left[\left(\frac{x}{2L_D}\right) - i\omega_0 t\right] \quad (11.1a)$$

$$E_{add}(t) = A \operatorname{sech}\left[\frac{T}{T_o}\right]\exp\left[\left(\frac{x}{2L_D}\right) - i\omega_0 t\right] \qquad (11.1b)$$

$$E_{add}(t) = E_0 \exp\left[\left(\frac{x}{2L_D}\right) - i\omega_0 t\right] \qquad (11.1c)$$

where $A$ and $x$ are the optical field amplitude and propagation distance, respectively. $T$ is a soliton pulse propagation time in a frame moving at the group velocity, $T = t - \beta_1^* x$, where $\beta_1$ and $\beta_2$ are the coefficients of the linear and second-order terms of Taylor expansion of the propagation constant. $L_D = T_0^2/|\beta_2|$ is the dispersion length of the soliton pulse. $T_0$ in equation is a soliton pulse propagation time at initial input (or soliton pulse width), where $t$ is the soliton phase shift time, and the frequency shift of the soliton is $\omega_0$. This solution describes a pulse that keeps its temporal width invariance as it propagates, and is thus called a temporal soliton. When a soliton of peak intensity ($|\beta_2/\Gamma T_0^2|$) is given, then $T_0$ is known. For the soliton pulse in the microring device, a balance should be achieved between the dispersion length ($L_D$) and the nonlinear length ($L_{NL} = 1/\Gamma\phi_{NL}$). Here $\Gamma = n_2^* k_0$, is the length scale over which dispersive or nonlinear effects makes the beam become wider or narrower. For a soliton pulse, there is a balance between dispersion and nonlinear lengths. Hence $L_D = L_{NL}$. For Gaussian pulse in Eq. (11.1c), $E_0$ is the amplitude of optical field.

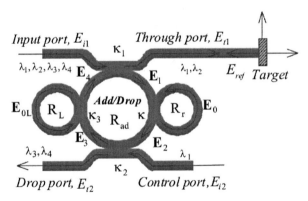

**FIGURE 11.25** A schematic diagram of light signal multiplexer controlled by light.

When light propagates within the nonlinear medium, the refractive index ($n$) of light within the medium is given by

$$n = n_0 + n_2 I = n_0 + \frac{n_2}{A_{eff}} P \qquad (11.2)$$

with $n_0$ and $n_2$ as the linear and nonlinear refractive indexes, respectively. $I$ and $P$ are the optical intensity and the power, respectively. The effective mode core area of the device is given by $A_{eff}$. For the add/drop optical filter design, the effective mode core areas range from 0.50 to 0.10 μm², in which the parameters were obtained by using the related practical material parameters [(InGaAsP/InP) [52]. When a dark soliton pulse is input and propagated within an add/drop optical filter as shown in Figure 11.1, the resonant output is formed.

The resonator output field, $E_{t1}$ and $E_1$ consists of the transmitted and circulated components within the add/drop optical filter system, which can perform the driven force to photon/molecule/atom.

For the first coupler of the add/drop optical filter system, the transmitted and circulated components can be written as

$$E_{t1} = \sqrt{1-\gamma_1}\left[\sqrt{1-\kappa_1}E_{t1} + j\sqrt{\kappa_1}E_4\right] \quad (11.3)$$

$$E_1 = \sqrt{1-\gamma_1}\left[\sqrt{1-\kappa_1}E_4 + j\sqrt{\kappa_1}E_{t1}\right] \quad (11.4)$$

$$E_2 = E_0 E_1 e^{-\frac{\alpha L}{2 2}jk_n\frac{L}{2}} \quad (11.5)$$

where $\kappa_1$ is the intensity coupling coefficient, $\gamma_1$ is the fractional coupler intensity loss, $\alpha$ is the attenuation coefficient, $k_n = 2\pi/\lambda$ is the wave propagation number, $\lambda$ is the input wavelength light field and $L = 2\pi R_{ad}$, $R_{ad}$ is the radius of add/drop device.

For the second coupler of the add/drop system,

$$] E_{t2} = \sqrt{1-\gamma_2}\left[\sqrt{1-\kappa_2}E_{i2} + j\sqrt{\kappa_2}E_2\right] \quad (11.6)$$

$$E_3 = \sqrt{1-\gamma_2}\left[\sqrt{1-\kappa_2}E_2 + j\sqrt{\kappa_2}E_{i2}\right] \quad (11.7)$$

$$E_4 = E_{0L} E_3 e^{-\frac{\alpha L}{2 2}jk_n\frac{L}{2}} \quad (11.8)$$

where $\kappa_2$ is the intensity coupling coefficient, $\gamma_2$ is the fractional coupler intensity loss. The circulated light fields, $E_0$ and $E_{0L}$ are the light field circulated components of the nanoring radii, $R_r$ and $R_L$ which coupled into the right and left sides of the add/drop optical filter system, respectively. The light field transmitted and circulated components in the right nanoring, $R_r$, are given by

$$E_2 = \sqrt{1-\gamma_2}\left[\sqrt{1-\kappa_0}E_1 + j\sqrt{\kappa_0}E_{r2}\right] \quad (11.9)$$

$$E_{r1} = \sqrt{1-\gamma}\left[\sqrt{1-\kappa_0}E_{r2} + j\sqrt{\kappa_0}E_1\right] \quad (11.10)$$

$$E_{r2} = E_{r1} E_3 e^{-\frac{\alpha}{2}L_1 - jk_n L_1} \quad (11.11)$$

where $\kappa_0$ is the intensity coupling coefficient, $\gamma$ is the fractional coupler intensity loss, $\alpha$ is the attenuation coefficient, $k_n = 2\pi/\lambda$ is the wave propagation number, $\lambda$ is the input wavelength light field and $L_1 = 2\pi R_r$, $R_r$ is the radius of right nanoring.

The output optical field of the through port ($E_{t1}$) is expressed by

$$E_{t1} = AE_{i1} - BE_{t2}e^{-\frac{\alpha L}{2 2}jk_n\frac{L}{2}}\left[\frac{CE_{i1}e^{-\frac{\alpha}{2}L-jk_nL} + DE_{i2}e^{-\frac{3\alpha L}{2 2}jk_n\frac{3L}{2}}}{1 - Fe^{-\frac{\alpha}{2}L-jk_nL}}\right] \quad (11.12)$$

where

$A = x_1 x_2$

$B = x_1 x_2 y_2 \sqrt{\kappa_1} E_{0L}$

$C = x_1^2 x_2 \kappa_1 \sqrt{\kappa_2} E_0 E_{0L}$

$D = (x_2 x_2)^2 y_1 y_2 \sqrt{k_1 k_2} \sqrt{\kappa_1 \kappa_2} E_0 E_{oL}^2$

$F = x_1 x_2 y_1 y_2 E_0 E_{0L}$

The power output of the through port ($P_{t1}$) is written by

$$P_{t1} = (E_{t1}) \cdot (E_{t1})^* = |E_{t1}|^2 \quad (11.13)$$

Similarly, the output optical field of the drop port ($E_{t2}$) is given by

$$E_{t2} = x_2 y_2 E_{i2} - \left[\frac{x_1 x_2 \sqrt{\kappa_1 \kappa_2} E_0 E_{i1}e^{-\frac{\alpha l}{2 2}jk_n\frac{L}{2}} + x_1 x_2^2 y_1 y_2 \sqrt{\kappa_2} E_0 E_{0L} E_{i2}e^{-\frac{\alpha}{2}t-jk_nL}}{1 - x_1 x_2 y_1 y_2 E_0 E_{0L}e^{-\frac{\alpha}{2}i-jk_nL}}\right] \quad (11.14)$$

The power output of the drop port ($P_{t2}$) is expressed by

$$P_{t2} = (E_{t2}) \cdot (E_{t2})^* = |E_{t2}|^2 \quad (11.15)$$

In order to retrieve the required amplification and optical switching signals of photonic transistor, we propose to use the add/drop device with the appropriate parameters. This is given in the following details. The optical circuits of ring resonator add/drop filters for the through port and drop port can be given by Eqs (11.13) and (11.15), respectively. The chaotic noise cancellation can be managed by using the specific parameters of the add/drop device, and the required signals can be retrieved by the specific users. $\kappa_1$ and $\kappa_2$ are the coupling coefficients of the add/drop filters, $k_n = 2\pi/\lambda$ is the wave propagation number for in a vacuum, and the waveguide (ring resonator) loss is $\alpha = 5 \times 10^{-5}$ dBmm$^{-1}$. The fractional coupler intensity loss is $\gamma = 0.01$. In the case of the add/drop device, the nonlinear refractive index is neglected.

Simulation results of the dynamic optical amplification and switching signals within the light signal multiplexer are as shown in Figure 11.26. In this case the dark soliton is input into the add port, and the dynamic optical photonic transistor are seen as shown in Figure 11.26(a)–(f).

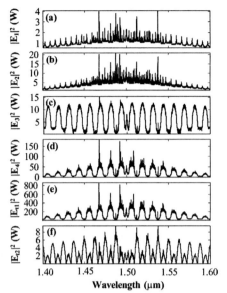

**FIGURE 11.26** Simulation result of the optical amplification and switching signals with three different center wavelengths, where (a) $|E_1|^2$, (b) $|E_2|^2$, (c) $|E_3|^2$, (d) $|E_4|^2$ and (e) are the through port and (f) drop port signals.

In Figure 11.27, the optical switching power signal multiplexer can be used to form a hybrid photonic transistor, whereas photons/molecules/atoms can be fed into the light signal multiplexer by dark soliton (optical switching power). The output photon/atoms/molecules from the through port pass through the target and are reflected back to the light signal

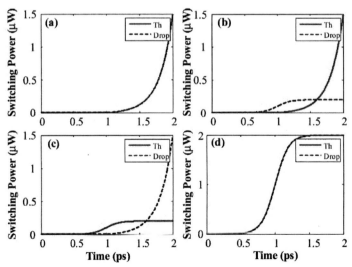

**FIGURE 11.27** Switching power of photonic transistor using PANDA ring where input and control signals are (a) dark-dark soliton, (b) dark-bright soliton, (c) bright-dark soliton and (d) bright-bright soliton.

multiplexer. The induced change of the collision (coupling effects) can be controlled by the add port input signal. Finally the interference signal is seen at the drop and through ports. The measurement can be made by balancing and adjusting the controlled parameters via the add port. The interesting characteristics is as shown in Figure 11.28, where the plots of output and input signal relationships have shown that the linear gains can be obtained by using the coupling coefficient variations, in which the linear gain can be obtained by using the desired coupling coefficient values. The input power is varied from 1 to 15 mW, where the other parameters are given in the Figure 11.29.

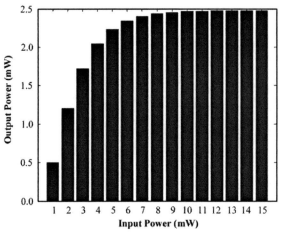

**FIGURE 11.28** Switching power of photonic transistor using PANDA ring where controlled the Gaussian input at the add port and we fixed $\kappa = \kappa_1 = \kappa_2 = \kappa_3 = 0.5$, $R_{ad} = 15$ μm, $R_{r,L} = 3$ μm.

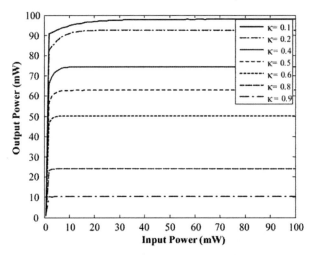

**FIGURE 11.29** Switching power of photonic transistor using PANDA ring where controlled the Gaussian input at the add port and we fixed $\kappa = \kappa_2 = \kappa_3 = 0.5$, $R_{ad} = 15$ μm, $R_{r,L} = 3$ μm. By varies $\kappa_1 = 0.1$–$0.9$.

## 11.8 Conclusion

In this chapter, we have shown that the hybrid photonic transistor can be formed by using the dark-bright solitons conversion. This can be formed by using PANDA ring resonator incorporating into the proposed light signal multiplexer. In this case, the dynamic behavior can be controlled and used to form the hybrid devices. In application, such behavior can be used to confine the suitable size of light pulse, atom or molecule, which is then employed in the same way as the optical tweezers. But in this case the terms dynamic probing is becoming the realistic, where the trapped pulses or molecules within the period of time (memory) is plausible. In applications, the use of the proposed concept and system can be used to form a new nanoscale interpretation of hybrid interferometer, whereas the trapped photon/molecules/atoms can be used to process the nanoscale signal processing interpretation based on the photonic transistor, where the balancing parameters can be found (measured). Thus, by using the dynamic tweezers, the hybrid photonic transistor using photons/atoms/molecules trapping and transportation within the system can be realized.

## REFERENCES

[1] J. Hwang, M. Pototschnig, R. Lettow, G. Zumofen, A. Renn, S. Gotzinger, and V. Sandoghdar, "A single-molecule optical transistor", *Nature*, vol. 460, pp. 76–80 (2009).

[2] F.Y. Hong and S.J. Xiong, "Single-photon transistor using microtroidal resonators", *Phys. Rev. A*, vol. 72, 01381 (2008).

[3] F.Y. Hong and S.J. Xiong, "Single-photon transistors based on the interaction of an emitter and surface plasmons", *Nanoscale Res. Lett.*, vol. 3, pp. 361–364 (2008).

[4] D.E. Chang, A. S. Sorensen, E.A. Demler, and M.D. Lukin, "A single-photon transistor using nanoscale surface plasmons", *Nat. Phys.*, vol. 3, pp. 807–812 (2007).

[5] A. Micheli, A.J. Daley, D. Jaksch, and P. Zoller, "Single atom transistor in a 1D optical lattice", *Phys. Rev. Lett.*, vol. 93, no. 14, 140408 (2004).

[6] A.J. Daley, S.R. Clark, D. Jaksch, and P. Zoller, "Numerical analysis of coherent many-body currents in a single atom transistor", *Phys. Rev. A*, vol. 72, 043618 (2005).

[7] M.F. Yanik, S. Fan, M. Soljacic, and J.D. Joannopoulos, "All-optical transistor action with bistable switching in a photonic crystal cross-waveguide geometry", *Opt. Lett*, vol. 28, no. 24, pp. 2506–2508 (2003).

[8] Y. Huang and S.T. Ho, "High-speed low-power photonic transistor devices based on optically-controlled gain or absorption to affect optical interference", *Opt. Express*, vol. 16, no. 21, pp. 16806–16824 (2008).

[9] S. Medhekar and Ram Krishna Sarkar, "All-optical passive transistor", *Opt. Lett.*, vol. 30, pp. 887–889 (2005).

[10] S. Song and E.M. Popescu, "N-nary optical semiconductor photonic transistor", *Opt. Eng.*, vol. 46, no. 11, 115201, Nov. (2007).

[11] H. Nakanotani, M. Saito, H. Nakamura, and C. Adachi, "Tuning of threshold voltage by interfacial carrier doping in organic single crystal ambipolar light-emitting transistors and their bright electroluminescence", *Appl. Phys. Lett.*, vol. 95, 103307 (2009).

[12] M. Yamagishi, Y. Tominari, T. Uemura, and J. Takeya, "Air-stable $n$-channel single-crystal transistors with negligible threshold gate voltage", *Appl. Phys. Lett.*, vol. 94, 053305 (2009).

[13] T. Takenobu, K. Watanabe, Y. Yomogida, H. Shimotani, and Y. Iwasa, "Effect of post-annealing on the performance of pentacene single-crystal ambipolar transistors", *Appl. Phys. Lett.*, 93, 073301 (2008).

[14] S.Z. Bisri, T. Takenobu, T. Takahashi, and Y. Iwasa, "Electron transport in rubrene single-crystal transistors", *Appl. Phys. Lett.*, vol. 96, 183304 (2010).

[15] K. Sawabe, T. Takenobu, S.Z. Bisri, T. Yamao, S. Hotta, and Y. Iwasa, "High current densities in a highly photoluminescent organic single-crystal light-emitting transistor", *Appl. Phys. Lett.*, vol. 97, 043307 (2010).

[16] G.I. Papadimitriou, C. Papazoglou, and A.S. Pomportsis, "Optical switching: switch fabrics, techniques, and architectures", *J. Lightw. Techn.*, vol. 21, no. 2, pp. 384–403, Feb. (2003).

[17] Y. Huang, "Simulation and Experimental Realization of Novel High Efficiency All-Optical and Electrically Pumped Nanophotonic Devices", PhD dissertation, Northwestern University, Evanston, IL, USA (2007).

[18] Y. Huang and S.T. Ho, "Computational model of solid-state, molecular, or atomic media for FDTD simulation based on a multi-level multi-electron system governed by Pauli exclusion and Fermi-Dirac thermalization with application to semiconductor photonics", *Opt. Express*, vol. 14, 3569 (2006).

[19] Y. Huang and S. T. Ho, "A numerically efficient semiconductor model with Fermi-Dirac thermalization dynamics (Band-Filling) for FDTD simulation of optoelectronic and photonic devices", QELS (2005).

[20] L.A. Coldren and S. W. Corzine, *Diode Lasers and Photonic Integrated Circuits*, John Wiley & Sons. New York (1995).

[21] J.P. Zhang, D.Y. Chu, S.L. Wu, W.G. Bi, R.C. Tiberio, C.W. Tu, and S.T. Ho, "Photonic-Wire Laser", *Phys. Rev. Lett.*, vol. 75, pp. 2678–2681 (1995).

[22] D.A. Rodrigues and A.D. Armour, "Quantum master equation descriptions of a nanomechanical resonator coupled to a single-electron transistor", *New J. Phys.*, vol. 7, p. 251 (2005).

[23] S.J. Tans, A.R.M. Verschueren, and C. Dekker, "Room-temperature transistor based on a single carbon nanotube", *Nature*, 393, 49–52 (1998).

[24] J. Park, "Coulomb blockade and the Kondo effect in single-atom transistors", *Nature* 417, 722–725 (2002).

[25] W. Liang, M.P. Shores, M. Bockrath, J.R. Long, and H. Park, "Kondo resonance in a single-molecule transistor", *Nature* 417, 725–729 (2002).

[26] L. Davidovich, A. Maali, M. Brune, J.M. Raimond, and S. Haroche, "Quantum switches and nonlocal microwave fields", *Phys. Rev. Lett.* 71, 2360–2363 (1993).

[27] S.E. Harris and Y. Yamamoto, "Photon switching by quantum interference", *Phys. Rev. Lett.* 81, 3611–3614 (1998).

[28] A.M.C. Dawes, L. Illing, S.M. Clark, and D.J. Gauthier, "All-optical switching in rubidium vapor", *Science* 308, 672–674 (2005).

[29] J. Vaishnav, J. Ruseckas, C.W. Clark, and G. Juzeliunas, "Spin field effect transistors with ultracold atoms", *Phys. Rev. Lett.* 101, 265302 (2008).

[30] A. Imamoglu, H. Schmidt, G. Woods, and M. Deutsch, "Strongly interacting photons in a nonlinear cavity", *Phys. Rev. Lett.*, vol. 79, pp. 1467–1470 (1997).

[31] L.-M. Duan and H.J. Kimble, "Scalable photonic quantum computation through cavity-assisted interactions", *Phys. Rev. Lett.* vol. 92, 127902 (2004).

[32] J.-T. Shen and S. Fan, "Strongly correlated two-photon transport in a one-dimensional waveguide coupled to a two-level system", *Phys. Rev. Lett.*, vol. 98, 153003 (2007).

[33] S. Kühn, U. Hakanson, L. Rogobete, and V. Sandoghdar, "Enhancement of single molecule fluorescence using a gold nanoparticle as an optical nano-antenna", *Phys. Rev. Lett.*, vol. 97, 017402 (2006).

[34] L. Rogobete, F. Kaminski, M. Agio, and V. Sandoghdar, "Design of nanoantennae for the enhancement of spontaneous emission", *Opt. Lett.*, vol. 32, pp. 1623–1625 (2007).

[35] A. Chizhik, et al. "Tuning the fluorescence emission spectra of a single molecule with a variable optical subwavelength metal microcavity", *Phys. Rev. Lett.*, vol. 102, 073002 (2009).

[36] G. Zumofen, N.M. Mojarad, V. Sandoghdar, and M. Agio, "Perfect reflection of light by an oscillating dipole", *Phys. Rev. Lett.*, vol. 101, 180404 (2008).

[37] M. Stobinska, G. Alber, and G. Leuchs, "Perfect excitation of a matter qubit by a single photon in free space", *Europhys. Lett.*, vol. 86, 14007 (2009).

[38] G. Wrigge, I. Gerhardt, J. Hwang, G. Zumofen, and V. Sandoghdar, "Efficient coupling of photons to a single molecule and the

observation of its resonance fluorescence", *Nature Phys.*, vol. 4, pp. 60–66 (2008).

[39] I. Gerhardt, et al. "Coherent state preparation and observation of Rabi oscillations in a single molecule", *Phys. Rev. A*, vol. 79, 011402(R) (2009).

[40] K. Uomwech, K. Sarapat, and P.P. Yupapin, "Dynamic modulated Gaussian pulse propagation within the double PANDA ring resonator system", *Microw. Opt. Techn. Lett.*, vol. 52, no. 8, pp. 1818–1821 (Aug. 2010).

[41] T. Phatharaworamet, C. Teeka, R. Jomtarak, S. Mitatha, and P.P. Yupapin, "Random Binary Code Generation Using Dark-Bright Soliton Conversion Control Within a PANDA Ring Resonator", *J. Lightw. Techn.*, vol. 28, no. 19, pp. 2804–2809 (2010).

[42] N. Suwanpayak, M.A. Jalil, C. Teeka, J. Ali, and P.P. Yupapin, "Optical vortices generated by a PANDA ring resonator for drug trapping and delivery applications", *Biomed. Opt. Express*, vol. 2, no. 1, pp. 159–168 (January 2011).

[43] P.P. Yupapin, N. Suwanpayak, B. Jukgoljun, and C. Teeka, "Hybrid Transceiver using a PANDA Ring Resonator for Nano Communication", *Global Journal of Physics Express*, vol. 1(1) (2010). (invited paper)

[44] M. Tasakorn, C. Teeka, R. Jomtarak, and P. P. Yupapin, "Multitweezers generation control within a nanoring resonator system", *Opt. Eng.*, vol. 49, no. 7, 075002 (July 2010).

[45] B. Jukgoljun, N. Suwanpayak, C. Teeka, and P.P. Yupapin, "Hybrid Transceiver and Repeater using a PANDA Ring Resonator for Nano Communication", *Opt. Eng.*, vol. 49, no. 12, 125003 (2010).

[46] P. Youplao, T. Phattaraworamet, S. Mitatha, C. Teeka, and P.P. Yupapin, "Novel optical trapping tool generation and storage controlled by light", *Journal of Nonlinear Optical Physics & Materials*, vol. 19, no. 2, pp. 371–378 (June 2010).

[47] K. Sarapat, N. Sangwara, K. Srinuanjan, P.P. Yupapin, and N. Pornsuwancharoen, "Novel dark-bright optical solitons conversion system and power amplification", *Opt. Eng.*, 48, 045004-1-5 (2009).

[48] S. Mitatha, N. Chaiyasoonthorn, and P.P. Yupapin: "Dark-bright optical solitons conversion via an optical add/drop filter", *Microw. Opt. Technol. Lett.*, 51, 2104–2107 (2009).

[49] T. Threepak, X. Luangvilay, S. Mitatha and P.P. Yupapin, "Novel quantum-molecular transporter and networking via a wavelength router", *Microw. Opt. Technol. Lett.*, 52(6), 1353–1357 (2010).

[50] K. Kulsirirat, W. Techithdeera and P.P. Yupapin, "Dynamic potential well generation and control using double resonators incorporating in an add/drop Filter", *Mod. Phys. Lett. B*, 24(32), 3071–3080 (2010).

[51] S. Mitatha, N. Pornsuwancharoen, and P.P. Yupapin, "A simultaneous short-wave and millimeter-wave generation using a soliton pulse within a nano-waveguide", *IEEE Photon. Technol. Lett.*, 21, 932–934 (2009).

[52] Y. Kokubun, Y. Hatakeyama, M. Ogata, S. Suzuki, and N. Zaizen, "Fabrication technologies for vertically coupled microring resonator with multilevel crossing busline and ultracompact-ring radius", *IEEE J. Sel. Top. Quantum Electron.*, 11, 4–10 (2005).

# 12

# Electron-Hole Pair Manipulation

**CHAPTER OUTLINE**

- Introduction
- Single Electron-Hole Pair Generation
- Multi Electron-Hole Pair Generation
- Conclusion
- References

## 12.1 Introduction

Recently, the electron-hole pair generated in 1.06 μm separate-absorber-avalanche (multiplier) InP-based devices [1], SiGe/Si planar waveguides [2] fabricated with a Ge concentration ranging from 2% to 6% and different thicknesses ranging from 200 nm to 2 μm, generating electron-hole pairs with a 100 fs laser pulse emitted at 810 nm, and monitoring the free-carrier absorption transient with a continuous wave (CW) probe beam at 1.55 μm, bipolar transistors [3], CMOS process [4], InAs-GaSb superlattice (SL) photodiodes [5], resonant microcavity [6], a cavity-QED (using a single InAs quantum dot, and a high-Q whispering gallery mode) [7]. Yupapin et al. [8] have shown that the large bandwidth light pulses can be generated and compressed optically within a microring resonator, and the use of such a proposed device in various applications has also been reported [9–10]. They have also reported the interesting results of light pulse, i.e. a soliton pulse propagating within a nonlinear microring device, where the transfer function of the output at resonant condition is derived and used.

They found that the broad spectrum of light pulse can be transformed to the discrete pulses. The use of soliton, i.e. bright soliton in long distance communication link has been implemented for nearly two decades. However, the interesting work of using bright soliton in communications still remain, whereas the use of a soliton pulse within a microring resonator for communication security has been studied [11].

The purpose of this chapter is to investigate the single and multi electron-hole pair manipulation using dark-bright solitons conversion controlled in nanoring resonator coupled to one arm of Mach–Zehnder Interferometer (MZI) and PANDA ring resonator, respectively. Both electron-hole pair are placed by dark-bright solitons conversion proposed.

## 12.2 Single Electron-Hole Pair Generation

Dark and bright solitons are the short optical pulses that can propagate into the optical medium for long period of time due to their nonlinear properties, for instance, self phase and cross phase modulations. The lack of phase with ($\pi/2$) between dark and bright solitons can be used to form the orthogonal photon components (entangled photon). Here, the proposed model for single electron-hole pair generation system using dark-bright soliton conversion is as shown in Figure 12.1. Single electron-hole pair can be generated by dark-bright optical soliton conversion within the system, in which the zero logical state $|0\rangle$ (or dark soliton input pulse: $|h\rangle$) and the logical one $|1\rangle$(bright soliton input pulse: $|e\rangle$) are encoded into MZI. The random states between dark and bright solitons can be established within the photonic circuit, in which the output of the orthogonal states can be randomly detected via the MZI output ports.

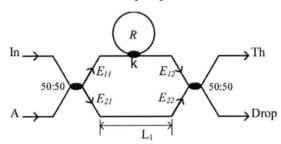

**FIGURE 12.1** Scheme of microring resonator coupled to MZI for single electron-hole pair generation.

When the dark and bright soliton pulses are generated and input into the system, the optical fields propagate through the photonic circuit are expressed by Eqs (12.1)–(12.3).

$$E_{11} = \tfrac{1}{2} In + j\tfrac{1}{2} A$$
$$E_{21} = \tfrac{1}{2} A + j\tfrac{1}{2} In$$
(12.1)

$$E_{12} = E_{11}\left\{(1-\gamma)^{1/2}\left[\frac{(1-\kappa)^{1/2} - (1-\gamma)^{1/2} e^{-\frac{\alpha}{2}L - jk_n L}}{1 - (1-\gamma)^{1/2}(1-\kappa)^{1/2} e^{-\frac{\alpha}{2}L - jk_n L}}\right]\right\} \quad (12.2)$$

$$E_{22} = E_{21} e^{-\frac{\alpha}{2}L_1}$$

$$\text{Th} = \frac{1}{2}E_{12} + j\frac{1}{2}E_{22}$$
$$\text{Drop} = \frac{1}{2}E_{22} + j\frac{1}{2}E_{12} \quad (12.3)$$

where $In$ is an input field, $A$ is an added field (control), $E_{11}$ and $E_{21}$ are fields spitted into two arm of MZI passes through the first coupler as shown in Figure 12.1, $E_{12}$ and $E_{22}$ are fields before merge into the second coupler, $\gamma$ is the coupler loss, $\kappa$ is coupling coefficient, $\alpha$ is the attenuation/loss coefficient of microring resonator, (definition of $L$) $L = 2\pi R$, where $R$ is radius of microring resonator, $k_n = 2\pi/\lambda$ is the wave propagation number in a circuit. Th is an output field and Drop is the output field signals of single electron-hole pair. $L_1$ is a lower MZI arm, with length 8 µm.

We are looking for a stationary dark soliton pulse, which is introduced into the MZI as shown in Figure 12.1. The input optical field ($E_{in}$) of the dark soliton pulse input and the add optical field ($E_{add}$) of the bright soliton pulse at $A$ port are given by

$$E_{in}(t) = A\tanh\left[\frac{T}{T_0}\right]\exp\left[\left(\frac{z}{2L_D}\right) - i\omega_0 t\right]$$
$$E_{add}(t) = A\,\text{sech}\left[\frac{T}{T_0}\right]\exp\left[\left(\frac{z}{2L_D}\right) - i\omega_0 t\right] \quad (12.4)$$

where $A$ and $z$ are the optical field amplitude and propagation distance, respectively. $T$ is a soliton pulse propagation time in a frame moving at the group velocity, $T = t - \beta_1^* z$, where $\beta_1$ and $\beta_2$ are the coefficients of the linear and second-order terms of Taylor expansion of the propagation constant. $L_D = T_0^2/|\beta_2|$ is the dispersion length of the soliton pulse. $T_0$ in equation is a soliton pulse propagation time at initial input (or soliton pulse width), where $t$ is the soliton phase shift time, and the frequency shift of the soliton is $\omega_0$. This solution describes a pulse that keeps its temporal width invariance as it propagates, and thus it is called a temporal soliton. When a soliton peak intensity ($|\beta_2/\Gamma T_0^2|$) is given, then $T_0$ is known.

When $In$ field of dark soliton $|h\rangle$ or bright soliton $|e\rangle$ are input into the $In$ and $A$ ports, the input fields are combined and split into two parts with the same amount via a 3 dB coupler (50:50), which they are $E_{11}$ and $E_{12}$. The field $E_{12}$ is the output field of $E_{11}$ which generates in microring resonator and controlled by phase shifted device. Finally, the fields are combined again at the second 3 dB coupler, in which the random combination of the orthogonal pulses (electron-hole pair) can be obtained via the MZI ports as shown in Figures 12.2–12.5.

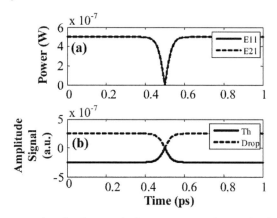

FIGURE 12.2  Single electron-hole pair manipulation for h-h input.

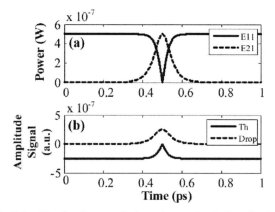

FIGURE 12.3  Single electron-hole pair manipulation for h-e input.

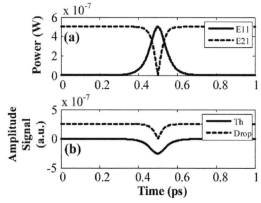

FIGURE 12.4  Single electron-hole pair manipulation for e-h input.

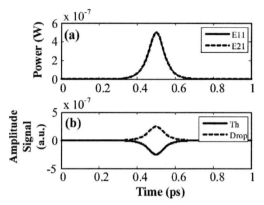

**FIGURE 12.5** Single electron-hole pair manipulation for e-e input.

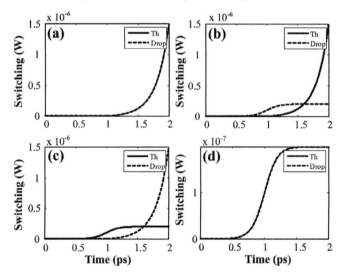

**FIGURE 12.6** Switching time of electron-hole pair manipulation using solitons input are (a) h-h, (b) h-e, (c) e-h, and (d) e-e, respectively.

Figure 12.2 shows the single electron-hole pair manipulation generated by using dark-dark solitons (h-h) input in In and A ports which can measure h-h conversed as shown in Figure 12.2(a) and we found that the single electron-hole pair manipulation are seen in Th and Drop port as shown in Figure 12.2(b), respectively. When using dark-bright solitons (h-e) input in In and A ports we found that the electron manipulated only in Th and Drop port as shown in Figure 12.3(b), respectively. Bright-dark solitons (e-h) input in In and A ports, we found that the hole only manipulated in Th and Drop port as shown in Figure 12.4(b), respectively. Similarly, when using bright-bright solitons (e-e) input in In and A ports which can measure 'e-e' conversed as shown in Figure 12.5(a) and we found that the single electron-hole pair manipulation in Drop and Th port are as shown in Figure 12.5(b), respectively. The switching time for the single electron-hole pair manipulation generation when using h-e and e-h input has shown in Figure 12.6(b) and (c). The switching time is different seen (difference delay), whereas, when using h-h and e-e input as shown in Figure 12.6(a) and (d), the switching time are same.

## 12.3 Multi Electron-Hole Pair Generation

Figure 12.7 consists of add/drop optical multiplexing used for generated random binary coded light pulse and other is add/drop optical filter device for decoded binary code signal. The resonator output field, $E_{t1}$ and $E_1$ consists of the transmitted and circulated components within the add/drop optical multiplexing system, which can perform the driven force to photon/molecule/atom.

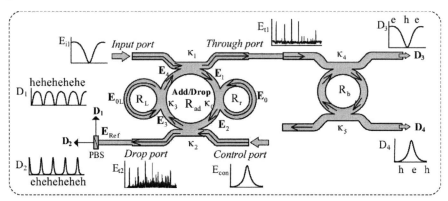

**FIGURE 12.7** Scheme of multi electron-hole pair generation. $D_n$: photodetectors, $\kappa_n$: coupling coefficients, $E_{Ref}$: referencing fields, $E_n$: electric fields, PBS: Polarized Beam Splitter.

The circulated roundtrip light fields of the right nanoring radii, $R_r$, are given in Eqs (12.5) and (12.6), respectively.

$$E_{r1} = \frac{j\sqrt{1-\gamma}\sqrt{\kappa_0}E_1}{1-\sqrt{1-\gamma}\sqrt{1-\kappa_0}e^{-\frac{\alpha}{2}L_1-jk_nL_1}} \qquad (12.5)$$

$$E_{r2} = \frac{j\sqrt{1-\gamma}\sqrt{\kappa_0}E_1 e^{-\frac{\alpha}{2}L_1-jk_nL_1}}{1-\sqrt{1-\gamma}\sqrt{1-\kappa_0}e^{-\frac{\alpha}{2}L_1-jk_nL_1}} \qquad (12.6)$$

Thus, the output circulated light field, $E_0$, for the right nanoring is given by

$$E_0 = E_1 \left\{ \frac{\sqrt{(1-\gamma)(1-\kappa_0)} - (1-\gamma)e^{-\frac{\alpha}{2}L_1-jk_nL_1}}{1-\sqrt{(1-\gamma)(1-\kappa_0)}e^{-\frac{\alpha}{2}L_1-jk_nL_1}} \right\} \qquad (12.7)$$

Similarly, the output circulated light field, $E_{0L}$, for the left nanoring at the left side of the add/drop optical multiplexing system is given by

$$E_{0L} = E_3 \left\{ \frac{\sqrt{(1-\gamma_3)(1-\kappa_3)} - (1-\gamma_3)e^{-\frac{\alpha}{2}L_2 - jk_n L_2}}{1 - \sqrt{(1-\gamma_3)(1-\kappa_3)}e^{-\frac{\alpha}{2}L_2 - jk_n L_2}} \right\} \quad (12.8)$$

where $\kappa_3$ is the intensity coupling coefficient, $\gamma_3$ is the fractional coupler intensity loss, $\alpha$ is the attenuation coefficient, $k_n = 2\pi/\lambda$ is the wave propagation number, $\lambda$ is the input wavelength light field and $L_2 = 2\pi R_L$, $R_L$ is the radius of left nanoring.

From Eqs (12.5)–(12.8), the circulated light fields, $E_1$, $E_3$ and $E_4$ are defined as $x_1 = (1 - \gamma_1)^{1/2}$, $x_2 = (1 - \gamma_2)^{1/2}$, $y_1 = (1 - \kappa_1)^{1/2}$, and $y_2 = (1 - \kappa_2)^{1/2}$.

$$E_1 = \frac{jx_1\sqrt{\kappa_1}E_{i1} + jx_1x_2y_1\sqrt{\kappa_2}E_{0L}E_{i2}e^{-\frac{\alpha L}{2 2} - jk_n \frac{L}{2}}}{1 - x_1x_2y_1y_2E_0E_{0L}e^{-\frac{\alpha}{2}L - jk_n L}} \quad (12.9)$$

$$E_3 = x_2y_2E_0E_1e^{-\frac{\alpha L}{2 2} - jk_n \frac{L}{2}} + jx_2\sqrt{\kappa_2}E_{i2} \quad (12.10)$$

$$E_4 = x_2y_2E_0E_{0L}E_1e^{-\frac{\alpha}{2}L - jk_n L} + jx_2\sqrt{\kappa_2}E_{0L}E_{i2}e^{-\frac{\alpha L}{2 2} - jk_n \frac{L}{2}} \quad (12.11)$$

Thus, from Eqs (12.4)–(12.11), the output optical field of the through port ($E_{t1}$) is expressed by

$$E_{t1} = x_1y_1E_{i1} + \begin{pmatrix} jx_1x_2y_2\sqrt{\kappa_1}E_0E_{0L}E_1 \\ -x_1x_2\sqrt{\kappa_1\kappa_2}E_{0L}E_{i2} \end{pmatrix} e^{-\frac{\alpha L}{2 2} - jk_n \frac{L}{2}} \quad (12.12)$$

The power output of the through port ($P_{t1}$) is written by

$$P_{t1} = (E_{t1}) \cdot (E_{t1})^* = |E_{t1}|^2 \quad (12.13)$$

Similarly, the output optical field of the drop port ($E_{t2}$) is given by

$$E_{t2} = x_2y_2E_{i2} + jx_2\sqrt{\kappa_2}E_0E_1e^{-\frac{\alpha L}{2 2} - jk_n \frac{L}{2}} \quad (12.14)$$

The power output of the drop port ($P_{t2}$) is expressed by

$$P_{t2} = (E_{t2}) \cdot (E_{t2})^* = |E_{t2}|^2 \quad (12.15)$$

In simulation, the parameter used for the first add/drop filter (optical multiplexer, PANDA ring resonator) are fixed to be $\kappa_0 = 0.15$, $\kappa_1 = 0.35$, $\kappa_2 = 0.7$, and $\kappa_3 = 0.2$ respectively. The ring radii are $R_{ad} = 30$ μm, and $R_r = R_l = 1.5$ μm. $A_{eff}$ are 0.50, 0.25 and 0.25 μm$^2$. The referencing multi electron-hole pair generation is generation as shown in Figure 12.8(a) and (b), multi electron-hole pair generation detected by $D_1$ is 'ehehehehehehehehehehehehehehehehehehehehehehehehehe' and 'hehehehehehehehehehehehehehehehehehehehehehehehehe' manipulation. The multi electron-hole pair generation detected by $D_2$ is 'ehehehehehehehehehehehehehehehehehehehehehehehehehh' manipulation.

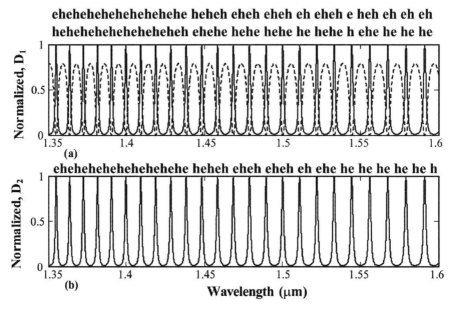

**FIGURE 12.8** Multi electron-hole pair generation for referencing output signals, when $E_{Ref}$ passes through PBS at the drop port, where (a) detected by $D_1$ and (b) detected by $D_2$.

In this chapter, the obtained pulse switching time is 12.8 ns. In operation, the multi electron-hole pair generation can be retrieved in the form of dark-bright soliton conversion via the add/drop optical filter output port, as shown in Figure 12.9, where the manipulation of dark soliton conversion is 'ehe', and the bright soliton is 'heh', which they are multi electron-hole pair generation relating to the referencing in Figure 12.8. In Figure 12.7, to perform the multi single photon switching, the add/drop filter is replaced by a beam splitter, and the attenuator is required and placed before the beam splitter. The bright soliton power is attenuated to reach the single photon power, while the dark soliton power is split by (50:50) ratio and detected by both the photo detectors. The random single photon output can be seen by a photo detector $D_3$ or $D_4$, which is formed a state "e". This results in no photon is detected by $D_4$ or $D_3$, which forms a state "h", i.e. no photon (h). This means that the single photon switching is randomly formed. Moreover, the obtained switching photons can be referred to as referencing signals as shown in Figure 12.8, where the output photo states can be confirmed.

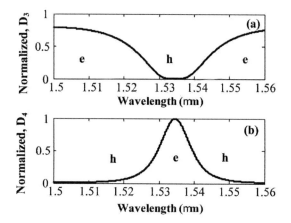

**FIGURE 12.9** Multi electron-hole pair generation, where (a) detected by $D_3$ and (b) detected by $D_4$.

## 12.4 Conclusion

In order to obtain the required multi electron-hole pair generation by using dark-bright soliton collision, we used the modified add/drop filter, which it is known as PANDA ring resonator. By using the dark-bright soliton conversion control, the obtained outputs of the dynamic states can be used to form the multi electron-hole pair, which is available for communication security application, setting the dark and bight soliton pulses first to be 'h' and 'e', respectively. By using the dark-bright soliton pulse trains, the dark and bright solitons can be separated by using PBS, which can be used to set the referencing multi electron-hole pair generation, where finally, the required multi electron-hole pair generation can be relatively obtained from the referencing multi electron-hole pair generation via the add/drop filter output ports. In this chapter, we have derived and presented a new concept of single electron-hole pair generation. The scheme for single electron-hole pair manipulation can simply be made ultrafast by using dark-bright solitons conversion controlled.

## REFERENCES

[1] J.P. Donnelly et al., "Design considerations for 1.06 μm InGaAsP-InP Geiger-mode avalanche photodiodes", *IEEE J. Quantum Electronics*, vol. 42 no. 8, pp. 797–809 (Aug. 2006).

[2] A. Trita, I. Cristiani, V. Degiorgio, D. Chrastina, and H. von Kanel, "Free carrier lifetime measurements in SiGe/Si planar waveguides", *Lasers and Electro-Optics*, 2007 and the International Quantum Electronics Conference. CLEOE-IQEC 2007. European Conference on 17–22, pp. 1–2 (June 2007).

[3] J.H. Klootwijk, J.W. Slotboom, M.S. Peter, V. Zieren, and D.B. de Mooy, "Photo carrier generation in bipolar transistors", *IEEE Transactions on Electron Devices*, vol. 49, no. 9, pp. 1628–1631 (Sept. 2002).

[4] Y.J. Kook, J.H. Cheon, J.H Lee Park, and Y.J. Hong-Shick Min, "A novel bipolar imaging device – BASIC (BAse stored imager in CMOS Process) ", *IEEE Transactions on Electron Devices*, vol. 50, no. 11, pp. 2189–2195 (Nov. 2003).

[5] S.M. Li, J.V. Shun, and L. Chuang, "Quantum efficiency analysis of InAs–GaSb type-II superlattice photodiode", *IEEE J. Quantum Electronics*, vol. 45, no. 6, pp. 737–743 (June 2009).

[6] D.H. Marti, M.-A. Dupertuis, and B. Deveaud, "Feasibility study for degenerate two-photon gain in a semiconductor microcavity", *IEEE J. Quantum Electronics*, vol. 39, no. 9, pp. 1066-1073 (Sept. 2003).

[7] P. Michler, A. Kiraz, C. Becher, Z.H. Lidong, E. Schoenfeld, W.V. Petroff, and P.M. Imamoglu, "A cavity-QED using a single InAs quantum dot and a high-Q whispering gallery mode", Quantum Electronics and Laser Science Conference (2001).

[8] P.P. Yupapin, N. Pornsuwanchroen, and S. Chaiyasoonthorn, "Attosecond pulse generation using nonlinear microring resonators", *Microw. Opt. Technol. Lett.*, vol. 50, pp. 3108–3111 (2008).

[9] C. Fietz and G. Shvets, "Nonlinear polarization conversion using micro ring resonators", *Opt. Lett.*, vol. 32, pp. 1683–1685 (2007).

[10] P.P. Yupapin and W. Suwancharoen, "Chaotic signal generation and cancellation using a micro ring resonator incorporating an optical add/drop multiplexer", *Opt. Commun.* vol. 280, pp. 343–350 (2007).

[11] D.N. Christodoulides, T.H. Coskun, M. Mitchell, Z. Chen, and M. Segev, "Theory of incoherent dark solitons", *Phys. Rev. Lett.*, vol. 80, pp. 5113–5115 (1998).

# Index

Add/Drop Filter(ADF), 5, 24, 27, 43
All optical logic gate, 217
All-optical transistor, 260
Arbitrary Waveform Generation (AWG), 71

Bandwidth(BW), 2
Birefringence, 202
Bistability, 32
Bright soliton, 84

Continuous-wave (CW), 36, 60, 61
Controlled-NOT (CNOT), 214
Coupled Resonator Optical Waveguide (CROW), 67
Cross phase modulation(XPM), 74

Dark soliton, 84, 91
Dark-bright soliton conversion, 93
Dispersion Compensating Fibre (DCF), 126
Dispersion management, 99
Distributed
    quantum sensors, 189
    spatial sensors, 185
DNA, 122
Drug delivery, 252
Drug trapping, 252

Dense Wavelength-division Multiplexing (DWDM), 13, 100
Dynamic
    optical tweezers, 138
    potentials, 93

Electron-hole pair, 292, 299
Entangled photon, 203
Evanescent waves, 2

Fabry–pérot (FP), 2, 3, 40, 68
Fiber Bragg Grating (FBG), 72
Finesse, 4
Finite Difference Time Domain (FDTD), 192
Free Carrier Absorption(FCA), 36
Free Carrier Dispersion(FCD), 36
Free Spectral Range (FSR), 2, 8, 13, 102
FWHM, 9, 12, 29, 30, 102

Gaussian pulse, 134

Hybrid
    interferometer, 142
    repeater, 157
    transceiver, 153, 6:154
    transistor, 259

Knill-laflamme-milburn (KLM), 214

Living micro-organisms, 122
Lorentz force, 149

Mach–Zehnder interferometer, 71
Mason's rule, 6, 15
Microring (MR), 3
Microring resonator(MRR), 3, 14, 16, 43, 85
Microring sensing transducer, 196
Modulation depth, 89
Modulator, 1
Molecular transporter, 169

NA (Numerical Aperture), 149
Nano-waveguide, 184
Network sensors, 192
Nonlinear Coupled-mode (NLCM), 125
Nonlinear
    kerr type, 83
    length, 55, 85
    switching, 1

On Off Keying (OOK), 228
Optical buffer, 2
Optical Channel Dropping Filter (OCDF), 95
Optical Circulator (OC), 126
Optical force, 123
Optical Spectrum Analyzer (OSA), 127
Optical tweezers, 93
Optical vortices, 148
OptiFDTD, 58

PANDA, 53, 54, 58, 61, 67, 68, 69, 70, 73, 74, 77, 80
Phase shifter, 200
Photo Detector (PD), 72
Photonic Bandgap (PBG), 2
Photonic Crystal (PC), 2
Photonic Integrated Iircuits (PICs), 71
Poynting vector, 123
Propagation constant, 55
Pulse Repetition Rate Multiplication (PRRM), 70

Quantum Electrodynamics (QED), 2, 93
Q-factor, 4, 5, 8, 11, 12, 43
Quantum
    SWAP gate, 217
    tweezers, 162
    well, 32
Qubits, 217

Raman scattering, 91
Ring Resonator(RR), 4, 5, 7, 8
Round-trip, 96
Round-trip gain ($a$), 32

Scattering force, 149
Self-calibration, 199
Self-phase modulation (SPM), 85, 125
Signal flow graph (SFG), 5, 7, 8, 14, 17, 44
Silicon-on-insulator (SOI), 36
Single
    atom transistor, 273
    electron transistor, 275
    photon sources, 214
Soliton, 75, 76, 84, 85, 87, 90
    -soliton interactions, 99
Static tweezers, 122
Stokes parameter, 124

Temporal soliton, 163
Thermo-optic effect (TOE), 36, 39, 41, 42
Total internal reflection (TIR), 2
Two-photon absorption (TPA), 35, 38, 39

Vernier effect, 13

Wavelength Division-multiplexes (WDM), 2, 4
Whispering gallery mode, 3

XOR/XNOR logic gate, 221

Young's Modulus, 198